The Architecture of Ideas

The Life and Work of
Ranulph Glanville
Cybernetician

Edited by Bill Seaman

ia
IMPRINT ACADEMIC

Copyright © Imprint Academic, 2022

The moral rights of the authors have been asserted
No part of any contribution may be reproduced in any form
without permission, except for the quotation of brief passages
in criticism and discussion.

Published in the UK by Imprint Academic
PO Box 200, Exeter EX5 5YX, UK

Distributed in the USA by
Ingram Book Company
One Ingram Blvd., La Vergne, TN 37086, USA

ISBN 9781788360784 (cloth)

A CIP catalogue record for this book is available from the
British Library and US Library of Congress

www.imprint-academic.com

The Architecture of Ideas

The Life and Work of Ranulph Glanville — Cybernetician

Edited by Bill Seaman
Including **Composing Composing,** and **A Long Conversation** Between Bill Seaman and Ranulph Glanville
With Related Texts By Aartje Hulstein, Albert Müller, Søren Brier, Bernard Scott, Ted Krueger, and Others

Table of Contents

Acknowledgments v

Introduction — Seaman 1

Living In, With and Through Cybernetics — Aartje Hulstein 9

In Memoriam — Albert Müller 13

A Long Conversation — Glanville | Seaman 15
• Phone Ringing   15 • A Multi-voiced Hypertext Resource Document Concerning Central Cybernetics Terms and Their Relations 15 • Replacing Observing with Composing 19 • Object Theory 22 • Spencer-Brown and an Infinite Regress of Distinctions 28 • Finnish Language | Finnish Architecture 32 • Working With Pask 33 • Spencer-Brown 35 • Creativity 37 • Ashby — The Law of Requisite Variety and the Black Box 38 • Interview Next to the Water (and Ranulph's Favorite Mobile Coffee Shop) 39 • The Thesis Has Two Different Names — Ranulph's Writing and Curiosity   40 • "The Most Curious Person I Ever Met" 41 • Some Talk About Music and Sound Art 42 • Circling Back to Objects 50 • Analogue vs Digital in Recording and in Bodies 54 • The First PhD Thesis 55 • The Second PhD 58 • A Conversation with Ourselves Through Pencil and Paper 72 • Student Projects in Music 73 • The Environment Modified by Sound 79 • edition echoraum 83 • Portsmouth 85 • Professor of Odd Jobs 86 • The Glanville Name 91 • A Contrary Friendship 92 •   Architecture Student at the AA, Assorted Architecture Offices, Then Abroad to Finland, and the Introduction to Cybernetic Concepts, and on to the PhD — Object Theory 93 • Finding Analogies   94 • The Architecture of Delight 99 • Continental Philosophy 103 • Getting Tired 106 • Listening and Impossible Problems   106 • Architectural Moments 112 • To Not Be In Control   119 • Some Key Thoughts About Cybernetic Concepts 122 • Mercurial 125 • Spirituality 129 • AA not AA 129 • The Longer Route 131 • Family 143 • The Most Cheerful Author Ever 144 • Mental Maps of Living 148 • Famous Dictums 149 • Emptiness 149 • Delight in Exploring the Senses and Their Temporary Removal 150 • Postscript: Room For Delight 150 • References for The Long Conversation 155

Ranulph Glanville: The Cybernetician of Ignorance (Expanded) — Søren Brier 159

Ranulph Glanville's *Objekte* — Bernard Scott 171

On The Contrary — Ted Krueger 188

Table of Contents

Monument — The Writings of Glanville — Albert Müller   195

The Mentoring Process: 203

Difficult Student
Ted Krueger   204

We Had Never Met Before
Craig Bremner   206

Listening
Ben Sweeting   208

Generosity, Bombast, Love, and the Third Eye
Caroline Vains   211

My Experience of Ranulph
Michael Trudgeon   214

Tracks Left by a Walk Walked
Thomas Fischer   215

My PhD with Ranulph
Tim Jachna   218

Composing Composing — Seaman | Glanville Quotes   221
Introduction 221 • Composing Composing Composing 221 • Parts and Wholes 222 • The One and the Many 223 • The Five Friends 224 • The Black Box 225 • The Black B∞x ↔ The Black Box 227 • Thinking Companions 229 • Five Themes 230 • Transcendence and Release Through Idea Development 231 • The Principle of Mutual Reciprocity 232 • Subject and Metasubject 233 • Assembling Concepts 233 • A Way of Thinking About Mind and Matter 234 • Cybernetics and the Machine 235 • Cybernetics — A Brief Pragmatic Definition 236 • Second-Order Cybernetics — A Philosophical Positioning 237 • Characteristics 239 • Circularity 240 • Second-Order Cybernetics and Design 240 • Designing Design Variety 241 • Anti-Cybernetician | Cybernetician 243 • Unmanageability 243 • Ranulph's Music and Unmanageability 244 • Unmanageability in Teaching/Learning 244 • The Unruly Definition of Design 245 • Novelty 246 • To Do Design Research is to Design Design 246 • Cybernetics ↔ Design 247 • Definition of Design Writ Large 247 • Design Process 247 • Wandering 248 • Thinking as a Design Activity 249 • Cybernetics Applied to Art and Design 249 • Homo Designans 250 • Pattern Relations 250 • Piaget's – Constant Objects

251 • Piaget and Constructivist Design Production 252• Patterns of Experience and Piaget's Constant Objects 252 • Teaching as Derived From A Process of Knowing About Knowing 253 • Cybernetically Informed Learning Tenets 253 • Teaching and Learning — Some Imperatives 254 • Teaching and Creativity 254 • Beauty 255 • Cybernetics Is My Art 256 • Different Observers/Composers 256 • Computation 257 • Computational Intelligence 258 • Creative Computation 258 • The Computer as Environment for Making 259 • Conversation 260 • Pask's Conversation Theory 261 • Design Conversation — A Conversation With the Self 262 • Conversation With the Self — Accretive Processes — "Try Again, Fail Again, Fail Better" 262 • Different Modes of Conversation in Pask's Work 263 • Switching Roles 264 • Object Theory 265 • Varieties of Variety 266 • Variety Constantly Increasing... 266 • Art and Design Variety 266 • Variety and Novelty 267 • Error and Variety 267 • Variety and Unmanageability 268 • Control 268 • Creativity 269 • Emergence 270 • Recursion 272 • Intelligence and the Turing Test 272 • Looping Back to Conversation Theory 273 • Delight 274 • Wonder 275 • Summary 276 • References for Composing Composing 277

Appendices	283
1. Curriculum Vitae	283
2. Papers, Publications and Writings	292
3. Research Experience	315
4. Software Developed	318
5. Major Conferences Attended (& Major Sponsors)	319
6. Main Teaching Experience	326
7. Lecture Venues	327
8. Visiting Critic/Professor (Shorter Stays, Irregular)	331
9. Supervisory Experience	335

Dedicated to Ranulph Glanville and Albert Müller

Acknowledgments

I want to acknowledge a series of people who helped in the composing of this book. Ranulph himself put huge energy into meeting with me, and initially helped me to define an outline of topics over the phone, to cover in our long conversation. Especially important was the input of the now deceased Albert Müller. I am thankful for his generous manner of contributing, as well as the support of the Cybernetics Archive at the University of Vienna which he led. Albert was incredibly helpful in contextualizing events, finding texts, fully articulating references, making oblique connections, and giving me a warm welcome on each of my visits to the archive in Vienna. After Ranulph's death, Albert was central in helping me find the most obscure of references and often gave feedback to me as things progressed. I am also thankful for his text in this book, Monument — The Writings of Glanville, which was one of the last texts he wrote. I am thankful to the Aartje Hulstein who also passed during the process of writing this book, for providing me with final readings and corrections as well as contributing lovely texts, images and feedback; Ted Krueger for support, conversation, and contrarianism; special thanks goes to Ben Migirditch for his transcription work, and the American Society for Cybernetics for their participation; in particular I would like to deeply thank Jeanette Bopry for her patience, proofreading, photo-editing and corrections; Marianne Ertl for her generous support at the Cybernetics Archive in Vienna and in helping with informational and photo-related processes; Paul Pangaro for contributing the images of Gordon Pask; The Emergence Lab which I co-direct with the highly supportive John Supko; and The Department of Art, Art History & Visual Studies, as well as the Computational Media, Arts and Cultures PhD program—at Duke University; I am highly appreciative of AAHVS and Duke University given their longstanding support of my ongoing research funding. I am deeply thankful to Søren Brier and Bernard Scott for contributing their thoughtful writings. I am appreciative of comments related to Ranulph's mentoring from Craig Bremner, Ben Sweeting, Caroline Vains, Michael Trudgeon, Thomas Fischer, and Tim Jachna. I am thankful for Volkmar Klien's friendship and for helping to provide a beautiful space for me to stay while doing research in Vienna. I am happy for the initial support of Werner Korn, and for his personal commitment to the creative work of Ranulph. I am very excited to have this book be published by Imprint Academic. I thank both Graham Horswell, Managing Editor, and Keith Sutherland, Publisher, for helping this book see the light of day! I am thankful to Michael Oatman and Bruce Bergmann for friendship and ongoing discussion and support. I also wish to thank my family—Maura Walsh-Seaman, Frieda Mae Walsh-Seaman and Fenn Walsh-Seaman who have put up with me continuously working away in my little studio for so many years, and Luca DeMarco who comes in, refocuses my attention, and pulls me out of this continuum as needed…

Bill Seaman, January 10, 2022

Ranulph Resting. Saint Hubert's Chapel, Idesworth. Photo © Bill Seaman.

Introduction

Bill Seaman

Ranulph and I met while I was attending CAIIA (The Center for Advanced Inquiry in the Interactive Arts) for my PhD around 1997. He often functioned as a visiting critic during symposia and conferences. He was very insightful about his feedback, a strong contributor to discussions and he always had numerous references to share that were drawn from many different fields. I found his kind of multi-perspective approach to knowledge production as central to my own way of coming to know the world. He was a generous critic in the sense that he would spend a great deal of energy discussing the matter at hand, pushing back when needed to clarify specific points. As time went on I ended up joining the American Society for Cybernetics where he was President. He attracted a number of his students to join the society. They were to breathe a breath of fresh air into the organization, whose illustrious heroes covered multiple past generations. Names like Norbert Wiener, John von Neumann, Warren McCulloch, Walter Pitts, W. Ross Ashby, Stafford Beer, Gordon Pask, Claude Shannon, Warren Weaver, Heinz von Foerster, Gregory Bateson, Margaret Mead, Humberto Maturana, Francisco Varela, and Ernst von Glasersfeld, to name just some of the participants in this impressive field of fields. They all played a part in the continuing development of the identity of cybernetics.

Ranulph helped generate through conversation, publications, works of art, design, music, modes of teaching, and, in particular, his manner of addressing problem solving and creativity, a cybernetic way of coming to understand the world. He was to some extent a *contrarian*—someone who wanted to question the status quo in an ongoing manner, and he would do so utilizing the multiple strategies and methods discussed throughout this document. Ted Krueger will illuminate this notion of the contrarian in his text "On the Contrary." Ranulph had a vital and mischievous sense of humor which had him singing songs, delivering impersonations, providing limericks, playing with language, reciting lyrics, and making up elaborate stories throughout our talks. I have left a number of these moments in our "Long Conversation." He was also very fond of playful yet pointed titles, often exploring paired reciprocal relations. This was part of his warmth and erudite wit as well as another of his ways of continuously living in and through cybernetics.

He was extremely well read across a series of fields and this will be discussed by Søren Brier in his text "The Cybernetician of Ignorance (Expanded)"—a playful title in its own right. Along with this playfulness Ranulph wanted to be very clear about language and our sharing of ideas through conversation. In particular he wanted to point to the nature of language and meaning production, as well as its limits and perhaps hidden qualities—in particular in relation to constructivist and second-order cybernetic notions.

He also wanted to define a new kind of logic to help give second-order cybernetics a means for a form of grounded proceeding. This was embodied in his ideas surrounding Object theory. Bernard Scott will unpack this concept in his text "Ranulph Glanville's *Objekte*." Ranulph and I return to the concept multiple times in our conversational meanderings. I also talked to Ranulph at length in our long conversation about many of his diverse views, drawing from knowledge related to many different fields. He was quite interested in *composing* these views into meaningful, knowledge producing processes which were open to error, iteration, and the ongoing exploration of creativity and clear thought.

Aartje Hulstein, his brilliant and loving wife has provided a very personal reflection on the breadth of his life and work in "Living In, With and Through Cybernetics." Ranulph called her his greatest critic. His friend Albert Müller, a talented and erudite historian who ran the archive in Vienna, provided the text "Monument: The Writings of Glanville." Ranulph, Aartje, and Albert all died along the path leading up to this document. A fact that I still cannot quite grasp.

I have asked a series of chosen former students of his—people that Ranulph suggested—to write about his mentorship. He drew on a deep intuitive sense of listening and forming meaningful pointed conversations with them. Interestingly, his approaches were a bit different for each of the people he mentored.

He wondered why I wanted to write this book. For me it is obvious that there needed to be one very focused work that could bring many of his major ideas into proximity. In a certain way, during his lifetime his ideas quietly lived a bit behind the scenes, sometimes overshadowed by other cyberneticians like Gordon Pask under whom he studied and came to work with, and Heinz von Foerster with whom he came in contact. He didn't have a giant ego and he also somewhat quietly pushed back against self-promotion. He was, in essence, trying to continue to articulate the fields of both cybernetics and in particular second-order cybernetics as an ongoing project—to bring new individuals into the fold, and to generate contemporary conversations to bring new life into the ongoing becoming of cybernetic thought. This was not to define a specific set of definitions for the field, but to work toward defining a dynamic network of complementary perspectives that one could draw upon as needed. He describes this particular project in detail at the beginning of our long conversation and I hope that it will get fully played out by others who are inspired by his thoughts. No small task.

I discussed my own ideas surrounding *open-order cybernetics*[1] with him, which sought to take into account new forms of *machinic* intelligence—new varieties of observer, new machinic methodologies for poly-sensing, and new language to discuss these fields, which would grow over time in an open manner. He pushed back and told me these ideas always collapsed back into second-order cybernetics, which he helped to create as well as clarify in terms of its limits and undertakings through his Object theory and his many writings.

Our dialogue took place over a number of meetings, starting with a phone call to set out the overarching set of points that we might cover—a technically clumsy beginning. It must be noted that I then visited Ranulph and we had a number of discussions in many of his favorite locations. Over the course of this long conversation we sat by the water drinking coffee; ate at multiple restaurants; drove into the county; visited a favorite chapel; entered some fantastic, majestic woods; sat in his special room at home; and also visited some upstairs spaces including his office and a lovely room where his tube amp and record player lived. I believe he was very fond of the infinity inherent to analogue forms. Aartje, his charming partner, also joined us and contributed bits and pieces on occasion, always adding to the conversation at succinct moments.

I believe Ranulph was quite fond of the mixture of long drives with focused conversation. It must be noted that we allowed the conversation to cycle and move quite naturally. We thus circled back to a number of topics. Interestingly, this kind of dynamic feedback loop, where each new pass took the conversation in a slightly different direction, added new layers of clarity. This was very much a cybernetic form. Our conversation about cybernetics became a cybernetic conversation. I believe this notion is very much in line with his own manner of thinking and exploring, and in his living in cybernetics and through his ongoing creative production exploring and re-articulating an *architecture of ideas* composed of a number of different perspectives or *views* as he called them.

---

1. Gaugusch & Seaman (2004).

I am very proud and honored to have had this chance to bring this set of conversations, ideas, thoughts and reflections together in this book. For me it was very much a learning process. The text includes a major bibliography provided by Ranulph as well as additional bibliographic information provided by multiple of the authors. This book also includes information about some of his artistic and design production, so that the reader (and I) can, in time, delve much more deeply into Ranulph's way of thinking and proceeding, enabling us to explore ideas that are just beginning to be touched upon here.

Much Love To You Ranulph…
Bill Seaman
December 30, 2019

Reference

Gaugusch, A. & Seaman, B. (2004). (Re)sensing the observer—Offering open order cybernetics. *Technoetic Arts: Journal of Speculative Research, 2*(1), 17–31.

Aartje in Ranulph's Studio. Courtesy Aartje Hulstein and The Glanville Archives, University of Vienna.

Aartje Hulstein. Courtesy Aartje Hulstein
and The Glanville Archives, University of Vienna.

Albert Müller. Courtesy Aartje Hulstein
and The Glanville Archives, University of Vienna.

Albert Müller in Ranulph's Studio. Courtesy Aartje Hulstein
and The Glanville Archives, University of Vienna.

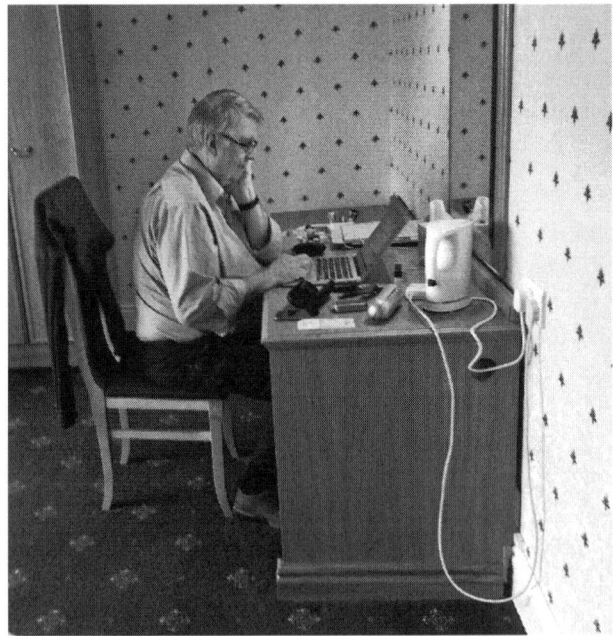

Ranulph, Traveler. Courtesy Aartje Hulstein
and The Glanville Archives, University of Vienna.

Introduction 7

Glanville Speaking. Courtesy Aartje Hulstein
and The Glanville Archives, University of Vienna.

Ranulph, Breakfast Work. Courtesy Aartje Hulstein
and The Glanville Archives, University of Vienna.

Ranulph, The Air Traveler. Courtesy Aartje Hulstein
and The Glanville Archives, University of Vienna.

Werner Korn with Albert Müller. Courtesy Aartje Hulstein
and The Glanville Archives, University of Vienna.

Living In, With, and Through Cybernetics

Aartje Hulstein

Ranulph came to Arnhem on the 6th January 1995, to see if, when we met without my brother Gert and wife Liewke around, we would be able to compose a life together.

We both carried a past with us, Ranulph's included his son Severi and Tuuliki, Severi's mother, who both lived in Finland. And his aging mum Cecil and stepfather Rodie.

He also was a cybernetician, taught architecture, had done very early electronic music, was interested in modern classical music, but also an enormous range of other music, from early church music right up to Sir Peter Maxwell Davies and Sir Harry Birtwhistle, Messiaen and Stockhausen. Not to forget David Bowie, Beefheart and Ian Dury. I got a whole day of music on my first visit to Portsmouth. Other people will share this memory with me, going from Perotin to Feldman, to Stockhausen to Laurie Anderson.

Ranulph went to a Frobelschool, then a prepschool, Papplewick and later went to Bryanston, then the Architectural Association in London in the wild sixties. He went to see Samuel Beckett in Paris when he was 16, and met him in his Flat: One of Ranulph's favourite stories was about saying goodbye to Beckett, he even did a performance in Berlin about that some years ago. Somewhere there are some letters Sam wrote to Mr. Glanville.

He met Gordon Pask in his last year and, when Gordon summarised his work in two minutes, after Ranulph talked for some hours, he knew he wanted something of the discipline Gordon used, cybernetics. He was then offered a scholarship to do a PhD with Gordon at Brunel University. This became the base of Objects. He was examined by Heinz von Foerster, the start of a long relationship.

Then he did a second PhD with Laurie Thomas about architectural space. At the same time, he was teaching at the AA, who always asked their most talented and most difficult students back to teach. A lot of what he did then with the students was later used as research material for his PhD.

In 2007 he was awarded a higher doctorate, a DSc by Brunel University. Ranulph went to teach in Portsmouth. Geoffrey Broadbent, who employed him there, said he needed an irritant to keep the department moving.

At some point Ranulph's drinking became a problem and he went to one of the Priory hospitals, and this summer would have achieved 25 years of sobriety, one day at a time. We have some very special friends from this time. Ranulph gave himself in the clinic the name of Commander of the Universe and was part of the Potty Club, they still meet every year.

When I came into Ranulph's life he felt he was being side tracked in Portsmouth University because of his very modern way of teaching, teaching from behind as he called it. Always expecting students to do better than they thought they could do. Not giving directions, but creating opportunities. I, at the time, was living life from an alternative angle, anthroposophy and homeopathy was what I was interested in, experiencing movement, acting, organic farming and rhythmical massage. I had got used to taking risks and starting from nothing. So, when Ranulph was offered early retirement by the Uni we decided he should take that and I would be a physio in England as a back-up to live on together with his pension.

Travelling has always been a big part of Ranulph's life—Europe, America and later Australia, Hong Kong, Brazil and China. He had connections all of over the world and often used those to make connections between people and then got annoyed when they forgot he had made the link in the first place.

A physio job at a school for disabled students found me and became very much part of our life. Ranulph was very touched by the students, often I would talk about what happened in treatment and frustrations about equipment.

Looking at cathedrals with Ranulph and having explained how forces helped to keep walls up and support the vaults helped me to finally understand physics and how to apply it to bodies and wheelchairs.

Often we talked about quality, what is it, can you measure it? I came with Ranulph to London when students showed their portfolios, travelled with him to conferences and asked endless questions if I did not understand.

Ranulph pointed me in the direction of child development and Piaget, constructing our world, creating it. This influenced the way I worked with the students at Treloar a lot. If movement is at the start of all learning, what does it mean to move even more differently when you have cerebral palsy? How hard is it when everything takes so much longer?

These were questions we talked about when driving, going to places, visiting Cecil and Rodie. Every aspect of life was looked at and talked about and concepts developed in the conversation. We both respected the others point of view and at least tried to understand it as best we could, although I don't think Ranulph ever took my gnomes and elves seriously.

Ranulph would spend long times away from home, working in Melbourne with Leon van Schaik and in Hong Kong with John Frazer. Sometimes I joined for a bit, or just stayed home and worked at Treloar. For

Ranulph it was important to have somebody at the centre, where he could come back to and where he was accepted as he was, however frustrating that sometimes was for each of us.

My colleagues often said I had the best of two worlds, a very engaging and interesting husband, but one who went away a lot. Ranulph was a very caring person, he felt responsible for Cecil and Rodie and was worried and fascinated by what happened to them growing older.

When Cecil died Rodie moved to Southsea and Ranulph was a very regular visitor who organised a lot of fun trips with Rodie and sometimes the three of us. When Ranulph was away Rodie's nieces would step up their visiting and I did a bit more.

We did a lot of trips, right up till the end. Ranulph always managed to find interesting people and join trips, holidays, family visits, teaching, conferences and meeting old friends, together. Visits to Richard Jung in Kutna Hora come to mind.

Richard told me there were more types of intelligence and mine was in my hands. In reflection I could then, often with Ranulph's help, make sense of what I was doing.

Ranulph was also the first non-American president of the American Society for Cybernetics. His involvement with the ASC started when I was diagnosed with breast cancer and we decided that should not stop either of us living the life we wanted. I continued work and did my most creative work in the following years.

Ranulph started to think about new types of conferences, where people would not talk about what was already done, but have discussions, exercises to open up creativity, conversations to find the new questions, which would then lead to new research. Ranulph did not like coffee breaks, as he thought they killed the flow of conversation and afterwards he needed to herd cats to get people back in their group. Then the students, they became part of our life, they visited, we visited them, talked about cybernetics, but also learned to live by what we talked. Ted Krueger gave Ranulph the venue for the first new type conference, in Troy, New York. I will never forget the music, the concert hall, the joy of playing with experiences, looking for cybernetic concepts, or using them. Living in cybernetics became part of how we tried to live. When you take seriously that you create or compose the world you live in, you decide where boundaries are, and what you believe or not. You are responsible for yourself.

5 years ago Ranulph was diagnosed with Prostate cancer, he took very advanced treatment, did not really change his way of living and it all stayed calm till around Christmas 2013 when he came back with yet another

cold that would not get better and in the end was diagnosed as a separate primary cancer and then another one was discovered in his oesophagus. He never wanted to know what would happen to him, he would find out when it happened. So we started to build a room to die in and to live in. It is now a wonderful monument to the conversations between Ranulph, the builder and me.

Ranulph wanted to live life to the full and he did. He developed a wonderful friendship with his oncologist and together treatment and possibilities were discussed. He only could not travel back to America for his last conference as President. He observed that from home, having treatment and being very proud of what the executive board did without him being there.

Ranulph and I were presented in Washington with a special award, for dedication to each other and the ASC. We were given it by Tim Jachna and Tom Fischer in Oslo where Ranulph gave his last lecture in late October. After that we visited the Witches Monument in Vardo, Ranulph was very moved by that, there was wind and fire, snow, earth, water, movement and light at the very top of Norway.

Ranulph's last job was his dream job. The Royal College of Art provided him with wonderful colleagues, interesting students and even a computer and his first job appraisal. He leaves behind a connection between Treloar, my old school, and the RCA and hopefully those projects will continue in the way he dreamed of or perhaps even better.

I hope you got some of the joy, excitement and constant learning from each other that for me was and is the essence of living with Ranulph, in cybernetics, living what you talk and think about, in circularity and deep honesty. When Ranulph died Liewke, who suggested the match in the beginning, and I were sitting at his bedside, he died peacefully, it was a very special moment in my life, like the beginning.

In Memoriam:
Ranulph Glanville (1946-2014)

Albert Müller

When I had to complete a hospitalization due to an accident in Vienna, I was visited by Ranulph Glanville, who was just a guest here. As a visitor's gift he brought me a pair of pajamas. As a first class traveler, coming from Australia, these pajamas were obtained. This pair of Qantas pajamas was for someone like me, of course, much too large. I am clearly of smaller stature than Ranulph. Nevertheless, this gift has so stunned me that I kept it until today. They have been preserved, without actually being able to wear them.

Much of Ranulph was equally amazing, his versatility as a person and as a scientist: he was an architect, art historian, cybernetician, semiotician, writer, musician, composer, photographer, designer, epistemologist, programmer, computer expert, and so much more.

Near Windsor, Ranulph enjoyed the education that a certain someone took on as a member of the British middle class. After school he participated in choral singing (his stepfather was a significant choral conductor and music educator there). He learned several languages over Latin and mathematics. He became a student of architecture in London and became a member of AA, the Architectural Association, about the same time as Zaha Hadid and Peter Cook. As a student, he organized, in London, music concerts with bands that were world famous four weeks after their first appearance. He was among the first to be intensely concerned with computer-aided design. During his studies, he met one of the most important English cyberneticians, Gordon Pask, who directed Ranulph's thought and work in new and different paths. His interest in epistemology began in the foreground to this contact. With his dissertation, *The Object of Objects, the Point of Points or Something About Things* he was able to create a new system of logical analysis, at the same time compatible with the tradition of British cybernetics, with Wittgenstein, and the teaching of George Spencer-Brown. His auditors in the promotion process were Gordon Pask and Heinz von Foerster. With both he maintained lifelong allegiances, and with these two he was a co-creator of second-order cybernetics.

Ranulph has more degrees, including a DSc for his scientific life's work. He had several professorships in Europe, Asia and Australia. He oversaw worldwide dissertations as a supervisor. At the same time, he was far away from being a one-sided academic specialist. It was unlikely for him to be with other people with whom he could not converse with in an intellectual sense. Such topics a conversation might cover included Captain Beefheart, Frank Zappa, Bach, Haydn, Mozart, Wagner, Schoenberg, Berg and Webern, Josef Matthias Hauer, Stockhausen, Xenakis, Sir Peter Maxwell Davies, and [John] Cage. Immediately following conversations related to these individuals one could talk about Swift, Joyce, Beckett and a few more unusual Irish writers. And still much more.

Ranulph was a Pontifex Maximus, a bridge builder, in his academic work, as in his private life. His contrasts are not to be ignored or passed over. He built bridges on the basis of his admirable knowledge and his special personal abilities, his integrity, and his generosity. As president of the American Society for Cybernetics he did just about everything in order to open up this venerable organization, whose 50-year anniversary was celebrated in 2014, to young people and new audiences.

Ranulph wrote countless scientific articles, on the occasion of meetings and conferences and journals in various fields; in the journal *Cybernetics & Human Knowing*, he oversaw a regular column. He was editor of numerous anthologies. In echoraum editions which he helped to publish, there appears an edition of his writings. Merve Verlag, in Berlin, in 1988, presented a selection of his work in the German language—spurred on by Niklas Luhmann and translated by Dirk Baecker.

In Vienna, Ranulph had a series of decade-long lasting relationships. Since the 1970s. He participated in the European Meetings on Cybernetics and Systems conferences since 2003; gave presentations to the Heinz von Foerster Congresses; was an honorary member of the Heinz von Foerster Society, and prepared the transfer of scientific papers of Gordon Pask and Richard Jung at the Institute, prevailing over much of the Contemporary History at the University of Vienna. Aware of his serious illness, about which he spoke openly, he saw to it that his own scientific estate would come to the archive in Vienna.

Nevertheless, the news of his death came as a surprise.

We have lost a great scientist and a dear friend.

A Long Conversation — Glanville | Seaman

Participants:
Ranulph Glanville
Bill Seaman
Aartje Hulstein

(Phone Ringing)

Ranulph: OK, did it work?
Bill: No, but I am trying it just in a manual way now.
Ranulph: Bill, Bill, Bill, Have you got an ordinary tape recorder there?
Bill: Yes, I am just recording it with an ordinary one now.
Ranulph: So it will hear your voice and hear me off the loud speaker.
Bill: Exactly.
Ranulph: OK.
Bill: The old-fashioned way.
Ranulph: Sometimes the simplest things are best.

A Multi-Voiced Hypertext Resource Document Concerning Central Cybernetic Terms and Their Relations

Ranulph: I would like just to talk about that *Insight Engine*™ (program by Seaman), if in fact that is what it is. I sent you some stuff, I don't know if you could make head or tail of it. The central idea is this: Many people say we should have a Cybernetics 101 program and I think this is ludicrous. I think any program that you write or I write or anyone else writes will satisfy 5% of the readers and put off 95%. I don't think what we need is a Cybernetics 101… I think what we need is a multi-voiced hypertext resource document, by multi-voiced I mean that you sound as Bill, not a Bill forced into someone else's framework, or forced into a marriage with someone else's thinking. Not that sort of thing …but you come across as Bill and people can say I really like that voice, I'll follow it.
Bill: How is that different from a Wordpress blog let's say?
Ranulph: Well, blogs just tend to be linear scripts don't they? This is quite different. This is a structure which is full of nodes that have content and these are initially defined beforehand. There is a development stage of this which comes later. But at least to start off with you might choose 20 or 30 key cybernetic concepts and you would get a group of experts, that is willing participants, to agree on these. You don't have to agree on every one

and you don't have to be interested in every one. But you have to be interested in a good chunk of them. And then everyone would write their own short version, that is 50 word, summary of what something means. It might say *variety* and you would write a 50 word summary and then a 500 word article about what variety means. And this should have attached to it some key references, so it would have URLs, linked. Now you do this for…well I did this with Bernard Scott. We did it for 16 terms I think.
Bill: What terms?
Ranulph: *Variety, consistency, state, feedback, control, communication* —very standard cybernetic words.
Bill: It sounds a bit like the *Cybernetics of Cybernetics*[1] book, it has those blurbs by people. Definitions.
Ranulph: Yes, they are not meant to be definitions—that is of course the point. I would characterize it, knowing that other people were going to characterize it, and accepting the difference. There is no attempt to resolve differences.
Bill: So it becomes a kind of dialogical process automatically. You read one thing, then you read the other thing…
Ranulph: You choose the ones that make sense to you. This is partly a matter of the voice of the person, its partly how they speak. You know I tend to speak in very abstract terms that probably won't appeal to most people. Others speak very much in terms of examples. That may help for some. For some it is social, for some it is very engineering oriented, for some it is creative…artistic, and so on. Those voices come out, and those different interests come out. If I were talking about variety I would probably talk about it very much in terms of abstract principles. If you were talking about *variety* you would probably talk about it in terms of giving examples, from the sorts of systems that you build. And the richness of them.
Bill: Exactly.
Ranulph: Now you take these terms, there is always an extra term called A or X depending on how I am feeling that day. And that is every term that you need but you haven't defined. It is the portmanteau term, it means everything else.
Bill: Have you seen *Insight Engine* yet. We can strategize whether to do it on top of that or not. The nice thing about doing it on top is that we could then search other papers that relate to that term.
Ranulph: That of course is useful, but of course there are several things that do that… Lissack's stuff does that, the bibliographic software Sente does it. But let me just finish, because this is just the beginning of the thing.

---

1. Von Foerster (1974, 1995).

Bill: Alright.

Ranulph: Then I also build connections between the terms. Now I should do this in a Pask style, but I may not. Because it is too onerous. Anyhow, if I say that variety connects to control, I have to say which way. Is it variety to control or control to variety or both to the other? Second thing, I have to explain what the connection is. I can't say this goes to that…I have to explain what process is carried out that converts variety into control, for instance. Variety is the measure of controllability of a system. I may say variety is the number of states a system can actually take, or discuss the comparison between the number it actually takes and the conceivably possible number and I can say the *law of requisite variety* says that any controlling system must have more variety than the system that it is controlling, otherwise there are states in the controlled system that are not modeled and are therefore excluded from its behavior by the controller. But I can also say that there is a way of looking at this business of control which turns this around… I can do the second-order cybernetic thing and say that *A* controls *B*, and *B* controls *A*, the variety must be the same. I can then go on to this thing—when there isn't enough variety, and I can say when I don't have enough variety to control things, which is 99.9% of the time, the universe is full of infinitely more variety than I have. I am essentially out of *control* in the universe. What we spend our time trying to do is to control the universe so we find clever ways of reducing the variety—there is a wonderful paper about how Newton did this. And we do all sorts of things. We force the universe to be only as we want it to be, and we exclude everything else. Well we know this… If we don't do that, and rather say there is infinitely more variety and everything that I don't already know that comes to me, is a gift and is part of my creativity, you've got a completely different way of looking at this. It turns being out of control into something creative and positive. We can do that. Control and variety fit together in the following way: Variety is the measure of controllability. The way I move from variety to control is to say if I know the variety, I know the variety I have to have for something to be controllable. That is the thing that turns variety into control in my way of putting it. I would do this between any two terms in this thing or maybe 3 or 4 terms working together…that I could see connected. I might have to add in a new term as well. Effectively these descriptions of what happens in the "arrow" might be new terms. We're not going to call them that for the moment. There are 20 terms and you describe 15 of them and I describe 15 of them, and they are not the same 15 but we have 12 in common. For 12 of these terms there will be Bill's description, Bill's characterizations. Short and long. And there will be Bill's URLs. We might put all of our URLs together. And then there will be mine of these as well. Your text on variety might not go to control at all. It might go just to creativity. But mine would go to control. Now what we have is not only the different individual voices, how you write and how I write, we also have the different choice of your URLs of source material, and we have the different way we explain what these terms are, and then we have the different things that we see connected to them. And so for instance, say that I had made no connection between variety and creativity, but you had, say someone was following me…they might at that point say: "That looks really interesting." How does that work? And pop up onto your level. For any term that you go to you get all of the possible descriptions, and you can look at any of them. You are not stuck with person *A.* You just stay with whomever. You don't have to follow roots. You can hop around between topics. You can go and read about this, then that, then the other and say, "I wonder if there is a connection between these two"? "What would it be like," and guess what it is, and look and see if someone has described it in that way. This becomes…it is not a book, it is not a program, there is no authority in it. It becomes something that provides a resource that other people can take. The idea is that for instance that teachers might lift whole bits from this, and stick them into the programs they are doing at the moment, because this stuff was "really helpful and relevant" just now. It is very much what we were trying to do in the last bit of the conference (Living in Cybernetics). Only everyone forgot what they were trying to do and produce projects, which really were not helpful. We ended up actually, sort of, without anything. Which was a pity. There should be places where people can add comments to these, but to start off with, it is just this—we invite 10 people, say, to do this. And then we start looking at it. We let other people suggest new topics, and everyone is told about this and can add their bit. We allow other people to become authors but we probably have to be a little bit constrained in this, otherwise you might turn up with a thousand authors, and a lot of the stuff might be complete crap. It may be that you have to say to people: "Well, maybe your voice isn't there but maybe there is enough stuff which is close enough to your voice for you to be happy with it." Otherwise the whole thing becomes unmanageable.

    Now, what your thing can do [*The Insight Engine*] perhaps, is to build connections between the way each author sees their things, and then possibly across authors. I don't see the difficulty of explaining the links. If you are going to do this, if you are going to have 20 terms, you are going to get 10 or 12 people who are seriously going to do it. I don't think they will have a problem. I think the problem is when some people think, they think all of these things should be done automatically, rather than be generated to express important personal differences. It matters a great deal that you connect *A* to *B* and *C* whereas I connect to *C* and *D*. That is part of the portrait of who we are. And it is part of a portrait of how people understand cybernetics.

Replacing Observing with Composing

Ranulph: By the way. I have just come up with a new word to replace *observing* in second-order cybernetics. After 45 years of thinking about this I have come up with the word. The word is *composing*. And I have come up with a new definition for second-order cybernetics. It is the art of balancing complements. As in complementarity.

Bill: In the paper I did on Pask he was describing his intelligent system and he was talking quite a bit about these complementary ideas and how they resolve and you could have a system of many of these, and that would be close to thinking.

Ranulph: I think the thing that Gordon completely misses…for me all thinking is *black box* thinking anyway…is the moving across levels. I think that what makes humans intelligent is the ability to be an observer, observing their own observing. Ross Ashby used the term *investigator*. Ashby didn't talk about the observer, he talked about the investigator. Ashby in 1954 said something pretty close to that. When Ashby talks about the black box[2] he is light years ahead of what anyone says. I keep finding out that when I have found something new, and radical and fantastic, then bloody Ashby probably said it already.

Bill: Also in the *The Embodied Mind*[3]—Varela, Thompson and Rosch talk about *mindful awareness*, which is a very similar kind of idea. That you become mindful of what you are doing as you do it.

Ranulph: There is quite a bit of Francisco's stuff that mirrors mine, and usually came later! I'm not saying he took it from me. I am saying there is maybe a Zeitgeist. But I very often got there first, as indeed I got off to a definition of radical constructivism independently around the same time as Ernst von Glasersfeld.[4]

Bill: Did you write a paper about that?

Ranulph: It is in the opening section of my Ph.D. It is just one statement. It says "that just because you think something is the case doesn't mean to say

---

2. Ashby (1964).
3. Varela, Thompson, & Rosch (1993).
4. Glasersfeld (1974).

it is the case, and it doesn't mean it isn't either." It is that which is the key of radical constructivism, which people just don't get. It is the thing that radical constructivism is not denying a hard physical reality, it is saying this is undecidable. It is certainly possible, and is also possibly impossible. The key to radical constructivism is to be able to maintain a position that doesn't deny these, or if it does, it allows you to take a position opposite to the one you took. Varela's paper, "Not One. Not Two,"[5] which I have just been rereading, which is from 1978, every move that he makes in that paper is made in my PhD from 1974/75.[6] I think that he wrote that before we met at the Binghamton conference. And I am sure he didn't read my PhD although, I am also sure that bits of it crept out, for instance Heinz started writing about *Objects* after my PhD [Von Foerster was on Ranulph's PhD Review Committee], which is all about Objects [see Bernard Scott's chapter elsewhere in this book]. There is a characteristic of my Objects, one characteristic which is really what Heinz concentrates on, and he gives absolutely no acknowledgement whatsoever. That was what he did—he was very poor at acknowledging others. He had a very very small set of people he accepted. Anyway...

Bill: This is interesting also about your ideas concerning language use, and how that comes into his writings, maybe later also, after you had written many papers.

Ranulph: Yeah, the reflexive titles [like *Understanding Understanding*].[7]

Bill: Yeah, I love those—as you can imagine. In my classes I do mind mapping where salient terms are brought out. And then we do a second stage of the mind mapping, where people have to do a bridging term or idea between any two of those terms. In a certain way it is very close to what you are describing.

Ranulph: It is, you are doing it in words instead of pages. It's the same sort of thing. If you could compress the whole of a page into a word. It would be just fine.

Bill: The way the *Insight Engine* works now is that you do bring up two different titles, then it goes into the system looking for other papers and media objects that are relevant to those two papers [and the language behind them, e.g., the paper itself or the abstract of a media object]. But you don't add your own notes between them, you don't add your own bridging idea. It wouldn't be that hard to have a slot...for every time you call up two things. Right now you see a "swirl" of all of the papers and so on, but you could

---

5. Varela (1976).
6. See Müller (2015).
7. See Von Foerster (2003a). See also Krueger, in this volume.

have a different color swirl for quotes or paragraphs or these key terms. I think you could add that in fairly easily.

Ranulph: That might be good because it would help people with the authoring and it would sort of oblige them to do it at the time. What I managed to do, when I was President at the ASC [American Society for Cybernetics] our net assets went from $38,000 to about $80,000. I stuck money in there every year. I feel that it is not unreasonable that I should have access to spend some of this. One of the things I got the executives to agree to is giving me access to money for programming the thing I am talking about. There is a possibility that the people at Bolton might do it. But there is also a possibility that we might do it this way if it seems interesting and valuable. What is important is that whoever does it, apart from everything that they want to do with it, does the thing that I want to do with it [laughter]. We can talk about this more. I think I have probably explained it well enough.

Bill: I think it is very clear. It is just working out how it would get implemented...the interface, and things like that. It could be its own new menu or a different place in the whole system and it could still use the back end of the system that lets you do the search.

Ranulph: It might be the system presenting itself through a different interface and a different style. It is likely to be much more static than most of your things are. It is essentially about chunks of word. It might be videos of people talking that would be just as good. There might be little experiments and demonstrations. Lou [Kauffman] doing his rope tricks and what have you.

   I did do a trial run of this...Bernard Scott did a trial run. What was interesting was how...well, I think Bernard didn't really understand what I was getting at. He produced something which is essentially rather taciturn, and closed up and not expansive towards readers. But anyway, we've got it. We have some material to sort of have a look at.

Bill: Yeah, he contributed many of his papers to the Insight Engine project.

Ranulph: I haven't. I'm very lazy. I don't find I have time. Academia[8] keeps telling me to do new papers and what have you and I go yeah, great. Who is following me. Well I don't care. [laughter] Then people write and ask for papers and I never respond. You become life-long bosom buddies and you have to communicate with them every day...and that sort of thing. No thank you. I don't have time for this. Anyway.

---

8. Ranulph potentially meant academia.edu and similar platforms.

Object Theory

Ranulph: OK—well we got through that bit, now let's shift on.
Bill: Would you like to talk about Objects? Where that came from and the major paper or book that people could refer to, to follow up?
Ranulph: Yeah, well…
Bill: It seems like one of your most important conceptual approaches.
Ranulph: You were there when I spoke about it at Bolton. I am going to rewrite it… What I think is the great weakness of second-order cybernetics is not as Pickering says, that it is entirely about the self, this is nonsense. Just look at papers like "Through The Eyes of the Other," or "The Self and the Other: The Purpose of Distinction."[9]
Bill: Are these your papers?
Ranulph: The second one is by myself, "Through the Eyes of the Other" is Heinz Von Foerster's. It is in a book compiled and edited by Fred Steier called "Research and Reflexivity."[10] Published by Sage in 1991. The other is very present in second-order cybernetics. It is the presence of the other that leads to this really acute problem. Pickering[11] is very lazy. He reads things as he wants. He lets words mean what he wants them to, not what others have meant. He keeps on talking about ontology when he is clearly talking about epistemology. And he keeps talking about *performative* when he just means acting. He wants to get in there and set the agenda with the vocabulary. He distorts things so that nobody can understand anything anymore. If I talk about ontology my architecture students think I mean robots that move. That is a performative ontology. This is an ecology of performative ontologies, and I look at them and say crap. [laughter] Nothing of the sort.

The problem is this: Using the old word *observe*, the argument in second-order cybernetics has two parts. And the first part is the only bit that people remember. That is, the observer is always present. Well, there is an earlier part which is consistency. Margaret Mead's criterion for the cybernetics of cybernetics is consistency.[12] That we behave in a way that is consistent with what we are describing. Heinz changed it all into observing because he is a physicist, I think. It became the cybernetics of observing systems as opposed to observed systems. The observer is always present. The observer is the observing system.

---

9. Glanville (2012). *Black B∞x, Vol. 1*.
10. Steier (1991).
11. Pickering (2010).
12. See Mead (1968).

Bill: And this is a break with the canon of historical science?

Ranulph: Well, I think you have to look at why that is there. Science wants to be universal and true. Or that is what it has wanted to be until very recently. And being universal and true means that any result will be repeated, exactly the same, anywhere, when performed by anyone, provided the conditions are properly taken into account. Repeatability destroys the individuality of the observer. To get repeatability you have to kill the observer (so to speak). Now it is fine to do this for certain things. It is very helpful. But it is not for everything. And it is not the only way of doing these things. That is the first part of second-order cybernetics, the observer cannot be ignored. The second part is that each observer is different. And therefore, each observation that is made by an observer who is unavoidable, and omnipresent, is always different, because you are not me. This is a recognition of the other. Now here is the thing. I have an experience and I describe it. I give it some words, and I point to it…it is some sort of physical thing. A chair, whatever. Every part of that apart from uttering the sound of *chair* happens within my internal world. It is entirely mine and you cannot have any of that. It is entirely private, it is personal and it is unique. The sound that I think I make saying *chair* and the fact that I think it's a sound, is also something that is unique to me. Now, you, hearing me look at something and say, yeah chair. How do you and I know that we share anything in this? Because everything in you forming this view, you interpreting what you hear me say, is your interpretation. To argue, while we both said *chair*, is absolutely unsustainable, because there is no way we can compare them. I cannot compare what happens in your private world to what happens in my private world. Now we have a real problem, which is if we wish to believe that we develop ideas together, that there is a society of ideas, we benefit from other and all of these sorts of things, then we have to be able to imagine a structure that lets us pretend that we are talking about and observing the same thing. It allows us to say, when you say, I heard you say chair, and when I say I said chair, it allows us to take those two statements and say, that we will pretend there is a structure that supports commonality in them. We are not saying it is there, we are saying this is a necessary requirement if we are to be able to share anything. It is somehow the assumption of a structure that is going to support sameness, or is going to allow us to act as if we could believe that we had seen the same thing. That we meant the same by seeing, we meant the same by words, and so on. Its recursive, it is infinite, and it is in a sense an entirely pragmatic fix. But without it second-order cybernetics gets nowhere, and not even conversation will get us out of this. Though conversation is quite clever, in that conversation never claims that we understand each other, merely that we build things which when presented to the other, appear to run along a parallel stream.

Bill: What would you say that the value of second-order cybernetics is, if in historical science you remove the ambiguity because you make something that is repeatable and clear, and we can both say this is exactly the same thing: Why is second-order cybernetics important?

Ranulph: Because for 400 years we lived in what is a ridiculous lie. [Seaman laughter] We have lived in a world where we deny the very thing that makes it possible for us to do any of this. If we couldn't observe, each of us, then, there would be nothing, and we wouldn't have made science. I think we can say that it is inconceivable that you can have observations without an observer. Heinz [Von Foerster] and Ernst [Glasersfeld] each said something like that, each attributing it to the other, at the same time. Objectivity is an observer's delusion. What I am saying is that I produced a structure that can be used to allow us to believe we are talking about the same things in second-order cybernetic terms. It therefore makes everything else that is done in second-order cybernetics, possible.

Without this, then you don't really have second-order cybernetics. It is these things called Objects [capital O]… I derived them from an extremely simple move. It is a move that is so simple that it is has been overlooked as far as I can make out since the beginning of time. I have asked philosophers about this and so on. Has anyone ever done anything like this. They usually end up saying, "I can't think of anything." Of course, that isn't to say nobody has.

Bill: Do you call this Object philosophy?

Ranulph: No I call it the theory of Objects.[13] It is a philosophical theory but it is very strangely stated. It is stated in a very formal way. I start with a statement like this… If I want to say "I observe this," I am talking in terms of the universe of observing—the universe of observation. That, is my universe of discourse. The universe of discourse has an entry criterion. You have to have a certain quality to be a member of a universe of discourse. You have to be part of the discourse, that *it* is a *universe of*. In this case you have to be part of a universe of observing—of observation, or of the observed (or whatever). Now the question is, what is the minimum circumstance, the minimum condition under which something can join the universe of the discourse of observing, of observations? The answer is, is that it observes itself. If it needs something else to observe it, then how did that get into the universe? If you say it is *A* observing *B* and *B* observing *A*,

---

13. Glanville (2015).

then I would say, that is a unity. Now, an Object is, that it observes itself. In other words, an Object is its own subject and its own object, using grammatical terms. Now you have the thing that Varela calls *star cybernetics*.[14]

Bill: I would call that subject ↔ object unity.

Ranulph: Yes, exactly. The word *object* is a peculiar word. I chose it because it totally contradicts itself…you have something which observes itself. Now you ask, How can one thing be two things? And my answer to this was to say well…time. It switches roles between being an observer and being observed. Or self-observer, and self-observed.

Bill: It sounds like *attention*…from which kind of attention you are looking—

Ranulph: No, here is what happens. You have got two roles and one of those roles is always empty. Either I am observing or being observed, but I am not both at the same time. That means that when I am observing, my being observed role is vacant. You could look in there, and see me in my being observed state. You would look in there and for that moment you would also observe me, but you would observe me from outside. Then when I switch back, you would go and observe yourself. When you are in your observed state, you are also able to observe something else. Now I can observe another Object, and I can observe many other Objects, and if the time of observing them is different, then I can talk about how the time overlaps, how it synchronizes. They might share a bit in the middle. When you draw sets that share things then you call the bit in the middle the intersection. If you put it in logical terms it is an *and*. Now by use of this time element, I have managed to create a logic. If I can make a logic, then I can do representation. Representation consists of me saying this = that. The simple act of representation is to say, the image on my screen = Bill. This is saying there is an identity between the two. It is also saying there is a difference, but that is something else.

Bill: Would you say this is a particular kind of relationality?

Ranulph: Sorry, which?

Bill: When you make this representation. When you make this substitution. This equality.

Ranulph: It has a particular purpose, yes, in terms of what it requires. All of the others require only me as an observer. A representation assumes there is someone else as observer as well. What you do is you take the two and put them together, and say they are the same. That is the act of representation. But if you are doing it with green and tree, you could put those together and

---

14. Varela (1976).

then you would have another Object which is called *green tree*. Now you begin producing or bringing together other Objects. This is like what Maturana calls *coordination*. Coordination of coordination. What you are getting with Objects is the extraordinary property that the way that I have derived, defined and characterized my Objects, makes them capable of computation and relation. It enables me to say that although it is what I am computing with, I am doing the computing. What I am computing with is my observation of the Object and the times of those. I can say that it is a property of the Object, because that is the whole point of having these things. By pretending that they are self-reproducing, they generate themselves, they become inhabitants of this universe of discourse, and I am out of my problem. I am out of my problem because there is always a slot you can fit into to have a look through, and when you do that, it is your observation of it which is different to my observation of it, but we assume that the *it* is the same.

Bill: When you say you make a representation of something is that the same as making a model of something?

Ranulph: I think a model is a representation which has a purpose. It is usually intended to simplify, to enable you to focus on certain things and act on them. A lot of this thinking came from being concerned with models.

Bill: It is interesting because you are also saying it can have an infinite complexity.

Ranulph: What could have an infinite complexity?

Bill: The relationship between the representation and the things they are equal to.

Ranulph: There is no complexity. Complexity has nothing to do with this. Complexity is a Johnny-come-lately which comes in billions of years later. Complexity is a construct to earn money. This has nothing to do with complexity. There is no notion of this being complex, just you looking or me or whoever it is, looking at something. It is not you as being an incredibly complex human being, it is just Bill Seaman. There is no notion of complexity in here. There is no notion of complexity in the thing being observed. If there is complexity it comes about with me attributing to this thing, many many many properties. And almost certainly contrarian…contradictory properties, which we do all of the time. Which is fine. I am all in favor of it. This is not about complexity. The notion of complexity has no place in this. Neither does chaos, neither do any of those money earning terms. [Ranulph laughs].

Bill: What does this give you? Why is this valuable do you think?

Ranulph: What it gives you is what I was arguing at the beginning—it gives you common ground. It gives you the ability to assume that there is common ground and therefore that you are referring to the same thing, and therefore that you can develop together understandings and knowledge, and all these sorts of things because you can believe that the referent of them all is the same. The system is designed in order that you can do this. Without that it doesn't matter how many observers you do or don't have. Its all a waste of time, because I have no idea when I read something that Heinz has written, that we are talking about anything that is shared at all. Under those circumstances, then you can start playing the Pickering game, you know that it is all solipsistic and what have you and its not. Not that solipsism is bad.
Bill: What is the role of inference to this process?
Ranulph: This has nothing to do with how you value what you have done. It is working at the lowest imaginable level. There is nothing sophisticated in here. There is nothing that would distinguish between inference and implication and all of these sorts of things. Guessing, futurology, all of those come later as you build up concepts. What Piaget talks about is that we have experiences, that we isolate those experiences and begin to find commonalities between them, and then we then exteriorize those commonalities in what we call *constant objects*.
Bill: Would you say Piaget had a kind of precursor to your Object theory?
Ranulph: Yes,
Bill: Or stimulated it.
Ranulph: Heinz was the external examiner and he said this is the first calculus for Piaget's objects.
Bill: Where Piaget was talking primarily about language, this becomes mathematical.
Ranulph: Piaget was interested in concepts. He was interested in concepts and the complexity in concepts. What sort of world, what sorts of bigger concepts were predicated by earlier ones and the stages these formed into. That is why he talks about developmental psychology. You start off with one type of cognition and after you have built up a certain amount, you switch to a second stage, and 3a and 3b and so on. I forget what they were because it never terribly interested me. I think there are endless squabbles now. Was Piaget correct to distinguish this way or should he have done it that way? My little boy did this, and this sort of thing. Piaget was interested in how we come to form concepts that repeat themselves. And in order to do this he talked about populating a world with the objects that embody them. What we do when we have created a constancy between certain concepts and

percepts, we place the constant thing outside of us. We say it exists independently of us. We objectify and place into the outer world. The paper I am writing at the moment has something about that too.

Spencer-Brown and an Infinite Regress of Distinctions

Ranulph: For many years I have been squabbling with Lou Kauffman, with most of the people who do Spencer-Brown.[15] One of my objections is that Spencer-Brown talks about things coming into existence because we draw distinctions. Then he says the distinction has a mark and a value. And I say, "What distinguishes the mark and value'?" And people go it doesn't really matter—it does! It is another distinction. If you regard the distinction as having an inside and hence an outside, then I argue that you have to constantly re-draw that distinction. There is an infinitude of distinctions being drawn inside and outside the original one, and I think it is that which, when you are talking about things in this way, it allows you to talk about distinguishing concepts. You have re-distinguished and re-distinguished until you decide that it is the same. Which is a way of saying I am just going on doing the same thing forever. The difference makes almost no difference. I would say this gives you an inside and an outside. It allows you to talk about the view from outside and the view from inside. That is a very important thing to be able to do. Since that distinction is vital to science. Even though scientists would not talk about the inside view. It is very important to them that there can be a view from outside, which can tell you what is happening inside. For me the magic of the black box is that it doesn't tell you what is happening inside. But you can still make a viable model. You don't have to look inside you don't even have to believe there is an inside. It is all a construct. Now, if you say there is no difference between the mark and the value. Then you are talking about a boundary that is not a circle but is a Moebius strip. The value is on the mark, the mark is the value. And that contains no space. It is just the mark and that is all that is there. You don't have an external observer.

Bill: Would this again be the idea of a subject $\leftrightarrow$ object unity?

Ranulph: What I would like is to be able to look in both ways. What I would say that if you imagine making a circle or a Moebius strip out of a very thin bit of string, and you are looking at it from above. You would have no idea if it was a circle or if it was a Moebius strip anyway. You can take it either way. Or you can have both. You can switch between them as they

---

15. Spencer-Brown (1969/1979).

bring you convenience. One brings you science and the other brings you human experience. Now you have both.

Ashby's Elementary NTM (Non Trivial Machine).
Photo courtesy Jamie Hutchinson, University of Illinois.

Bill: There is a similar reference from Charles Sanders Peirce, when you try to define the meaning of something there is an infinite set of diaphanous veils but you can never get at the thing in itself. [Peirce's Quotation]:

> But an endless series of representations, each representing the one behind it, may be conceived to have an absolute object at its limit. The meaning of a representation can be nothing but a representation. In fact it is nothing but the representation itself conceived as stripped of irrelevant clothing. But this clothing never can be completely stripped off; it is only changed for some more diaphanous. So there is an infinite regression here. Finally, the interpretant is nothing but another representation to which the torch of truth is handed along; and as representation, it has its interpretant again. Lo, another infinite series.[16]

---

16. CP 1:339

Ranulph: The onion. It is recursive, the more people go on with this stuff and Spencer-Brown, the more recursive they become. Lou is talking about nothing but recursion really. Francisco Varela and I did a paper together.[17] One paper. Which is exactly about this and it talks about the need to constantly re-draw the distinction, and being recursive. We argue that at infinity, if you are drawing the distinction inward, that is in logical intention, or outwards, that is logical extension, it makes no difference because at infinity they come together. There is re-entry.
Bill: I like this so much better than Spencer-Brown, much more real… I mean applicable.
Ranulph: It has taken me a long time to make the last few steps. That is what I am making in this paper at the moment.
Bill: I made a silly list of different kinds of Architectures. Would you talk about how this Object theory might be related to architecture? [Seaman's initial list…]:

Architecture:

- of thought
- of knowing
- of physical properties – physics
- of proportion – art and music
- of sharing concepts – approach to second-order cybernetics
- of music
- of space
- of non-standard approaches to knowing
- of psychology/behavior – new Psychology [against behaviorism] (Cybernetics was also about behaviors)
- of language
- of creativity
- of buildings

Ranulph: It has got nothing to do with architecture whatsoever. My interest in architecture, I was never interested as a student…my interest came through teaching it. And through discovering that architecture provides you with a very good model of the world. It covers almost everything and anything. You can teach anything you are interested in. And it has something to do with architecture. I have no interest in architectural theory whatsoever. Except occasional little things like Panofsky's Gothic

---

17. Glanville & Varela (2012).

Architecture and scholasticism. And I write about distinction drawing and thick walls and zero space in Mayan architecture. And the way that Finnish farm houses are organized. I like architecture for its sensory qualities. I like a lot of the type of thinking that architects do when they are designing and working on things. Generally, I like architects but I don't like the vain ones. This is a completely different world. If it has any connection at all it is so deep and so far…the theory of Objects touches everything because it permits everything. It is like a universal key that opens every door in the city. It doesn't mean to say that opening doors, people are interested in the universal key. Or that the architect of the universal key is interested opening the door (and what is behind them). What I am interested in is the question of, how on earth could we believe we are talking about the same thing, given the set-up of second-order cybernetics. I don't believe that any one in second-order cybernetics has dealt with that question…has even recognized how powerful it is behind everything, and how undermining it is. I think I am the only person who has attended to this. I think that is why all of that work is so extraordinarily important. And why it is monstrous that it is ignored.

Bill: I think that the bridge or association that I am making has to do with this idea of a conversation where you are working with a client. And in that conversation you are trying to articulate this shared understanding of space somehow. The client / architect relation may be the pragmatic playing out of the Object theory.

Ranulph: No no, no no. Object theory exists so that you can have conversation theory. But as it happens it was Gordon (Pask) who was interested in talking with the client. Never me. The conversation I am interested in is the creative conversation all people hold. It is not the only way they get creativity but it is a key way. Which is to converse with yourself. And the way it works is like this: You have at least two personae. One of them draws a line. You come back a bit later and say: "Good heavens I never realized I had done this. Everyone who is creative has done that. In a very simple, old fashioned way, you start sketching, and the pencil takes you places. One of the things that does worry me now-a-days is whether in computer-aided designing for architects, there is that element. And I think there isn't and that is one of the reasons we get very bad buildings. And designers don't know anything about design because they don't understand that the whole nature of design is circular (read cybernetic), recursive, and about making mistakes. It is not about having a clever idea and drawing that up, and that's it. It is about having a clever idea and having a cleverer one…

Bill: We talked a little bit about you wanting to learn a new language and then you went to Finland…and you went about studying the structural aspects of language and applying that, or finding a relationality to the structural aspects of architecture. Did I say that correctly?
Ranulph: Finnish architecture, and particularly the vernacular architecture.
Bill: What can you say about that, or what was your interest?
Ranulph: Oh boy, Bill, that, you know—
Bill: Or we can come back to it if you want.
Ranulph: Put very very simply, Finnish is an agglutinative language which has vowel harmony and consonant gradations. And these qualities mean that almost all words exist in a form that is intended to add things to. But that the adding to has to be particularly suited to (for instance) the vowel class of the word you are dealing with. It means that, whereas in other cultures they just add on rooms or they have comprehensive plans, in Finnish they are always creating the possibility of an extension, and of a proper joint. Things aren't just banged together. But it is much more complicated than that… There is always a central space around which expansion can take place.

Finnish Language | Finnish Architecture

Ranulph Next to Finnish Architecture. Courtesy Aartje Hulstein and The Glanville Archives, University of Vienna.

Finnish Architecture. Courtesy Aartje Hulstein
and The Glanville Archives, University of Vienna.

Bill: This "funny" list that I made maybe has to do with architecture as metaphor. And I think you felt uncomfortable with the list or you don't think of yourself that way. But when I imagine you in the complexity of all of the different things that you do, that list that I made was me trying to take stock of how these interests might play (in relation to each other)…for a moment.
Ranulph: We can start with that next time.
Bill: OK. Thank you for your time.

Working with Pask

Bill: Pask threw in everything…
Ranulph: The typical thing from Gordon, "Have you thought of this." [Reply from Gordon]: "Oh yes we did this last night." He never thought of it. He had a tendency to over inflate and over-include. He tried to put everything in there. Gordon's way of covering everything was to list it. My way of covering everything is to distill it into an abstract that can then be applied. One of the reasons we worked so well together as professor and student was that my way of thinking was so different to his. What used to

happen, and it was partly because of the hours he kept… You know that he worked all night. So he would go to bed setting a problem, a set of things for his colleagues to do, and then they would brief him at the end of the day. Then he would go away and do what he wanted with them. Well the result was that all of his colleagues sort of lost their identity, their contribution, it all got Gordon written all over it.

Left: Young Gordon Pask. Copyleft. Right: Gordon Pask with Computer. Courtesy the Gordon Pask Archive at the Institute for Contemporary History at the University of Vienna.

Bill: That was kind of the science model at the time for many PhD students.
Ranulph: Well yes and no. This was a freelance research organization.
Bill: A little different than academia.
Ranulph: Many people ended up being unable to face Gordon because of it. I don't think it was malicious on Gordon's side. I think he was just getting on.
Bill: It was just his nature?
Ranulph: He couldn't do that with my stuff. It was just too abstract for him. Gordon was really, in my mind, an engineer. He was always better when he was making things. And he was really good when it was valves or what you call tubes. Transistors were already too abstract. Too intangible. And there was always something about Gordon that had to do with the very tangible, a very real world problem. With me it was… I am interested in just the beauty of the thought. I don't care if it works in the "real world" or not. That is not my stuff. I did this abstraction, Gordon did this inclusion. I would distill things down into a tear drop and Gordon would put the encyclopaedia in

there. I think there was always this difference. I recognize his extraordinary genius and his many wonderful qualities as a person as well as an academic. But I really was very different. The people that worked in his lab were mostly either Gordonically-inclined or Gordonized. I did do some work for him. Which has me ending up with top clearance with the US Air force. Because I reinvented NATO's defense system by accident...by designing an experiment. Then I left it on a bus in Finland and found it 2 years later on the back shelf of a bus in Finland.
Bill: You left it and found it in the same bus?
Ranulph: It must have been the same bus...
Bill: That is bizarre.
Ranulph: Isn't it. That is weird.

Spencer-Brown

Bill: You earlier talked to me about Spencer-Brown and the difference between how you and he talked about distinction. Can you talk about this. You have a beautiful way of showing this infinite regress.
Ranulph: Well, Spencer-Brown does this very strange thing. He says, "Draw a distinction." He talks about becoming and he says that you draw a distinction and things come into existence. Yet, he distinguishes himself between the mark of the distinction and the value of the distinction. And I just say, if you have to draw a distinction to distinguish things, what distinguishes between the mark of the distinction and the value of the distinction? The answer it is that it is another mark. Another distinction. For me this is bizarre. He really doesn't like to admit to this...he won't agree [Editor's note: Spencer-Brown was still alive at the time of this conversation.] This is what he's done. But I think it is. I think there are other things too, for instance, he requires that there is already paper on which you draw the distinction. He is not making the paper, so how is the paper there. It is things like that that concern me about Spencer-Brown. If you distinguish between the mark and the distinction, then there is an infinite regress of marks, being drawn, distinguishing the mark from the distinction. And they are drawn either inward or outward, or possibly both. The thing about drawing a distinction is that the value isn't on the inside, the value is on the inside, the outside or both... You end up with an infinite regress of distinction drawing. And in a sense, that is just what we as human beings do. I look at you and I look at you again, but you are slightly different but you are still Bill, enlarged, enhanced. But we go on. Well this a sort of way in which we do this. But there is another thing we can do. We can say, well, the

problem is, he is drawing a distinction and it is forming an inside and an outside. If you don't have an inside and an outside, you can't have a value for inside or outside. The only place you can have a value is on the mark. In that case it doesn't contain anything in space so it is a Moebius strip. And so I have been arguing we should think of distinctions in that way. What I now say is we use both of them because they serve different purposes… You can have a distinction that is its own value. Then it doesn't contain anything. It just distinguishes itself. If you imagine this thing, this circle or Moebius strip made out of a thin strip of cotton with a half twist all about it, and you imagine yourself looking down at it. You would never be able to tell. You can choose to have whichever way you want, depending on what it is you want from this. If I want the distinction to distinguish itself, that is, something which is not autopoietic, but it in a sense is slightly poetic, then it is a Moebius strip. If I want the psychology of regress and redefinition, Piaget's object constancy and so on, then I want it to be a circle. I have known about the two, I have known what I have said, what I hadn't understood is this complementarity, or choice. That is new.

Bill: That is also making a distinction between those two states.

Ranulph: That is what is in this column that I am doing at the moment.
Here is a new definition of cybernetics: Cybernetics is the art of balancing complements…

Bill: This distinction thing…to jump back to that. I love this approach to inside/outside… I love the idea of the continuum of space somehow.

Ranulph: Well if it is a Moebius strip there isn't any space at all. You don't even have to have a continuum. Even better. Nothing there [in a ghoulish voice].

Bill: But of course Moebius strips always have space. It is just very thin.

Ranulph: No, if you describe it mathematically it has none.

Bill: Mathematics…we start to have all of the different nodes that are playing in. And somehow mathematics is something that has meant quite a lot to you?

Ranulph: No,

Bill: Or just as another language to help clarify things?

Ranulph: When I was at school aged about 4—nursery school or kindergarten, I was very bad at math and I fell over one day, grazed my knee, and said to myself: "Tomorrow I will be good at math." And tomorrow I was good at math. I became very good at math. Extraordinarily adept. I could pass 15 or 16 year-olds exams at the age of 10. Without being taught. I would invent. I would invent proofs for things. My maths master would say "I've never seen a proof like that. I cannot find anything wrong with it... No one has ever done that before." Then I went to another school and I had a maths-master and I simply could not understand him. He spoke the wrong way and was quite rude to me as well and I just lost the whole thing. Heinz used to say that I had a mathematical brain, but if I don't have the technique, I can't do it. Lou Kauffman and me talking together is quite fun. Lou gets to sit down and be the mechanic for me, and we do this. Because there are insights. But to say that math was really important to me would be wrong. Math is something I am uncomfortable in operating in but I understand quite a bit about what it is about and what matters. Quite a lot of the beauty in it. I think the thing that is most important to me is beauty. I like beautiful ideas. I'm not interested in being useful.

Creativity

Bill: I was digging around and found a paper that you did about creativity, and beauty: "The Value of being Unmanageable: Variety and Creativity in CyberSpace."[18] The cyberneticists in general didn't talk about creativity or beauty so much.
Ranulph: I think creativity is terribly difficult to talk about, partly because whatever you say, someone will say you are wrong. I think creativity is best left as something that those who know, recognize.
Bill: That's a nice definition.
Ranulph: I think that this has something to do with variety. You know there is that old romantic notion of Beethoven drilling his fist into his forehead, you know...the tormented soul, and all of this sort of thing. And maybe some people are like that, but I think very few. I think people can gain creativity just by being open to more stuff. People are always giving you presents. All you have to do is be there, and land something that you have not heard of before. Misunderstand it in their own way, and show you a different way of doing your own thing. I'm off...that's it. I see that as being something you can do for creativity. You can increase the variety that is available. And it is not very difficult because the variety my brain can cope with and the variety of everyone else's brain in the universe, is just incomparable. All I have to do is shut up a bit. And let someone suggest something to me.
Bill: Did Ashby write about creativity at all?
Ranulph: Ross W. Ashby wrote about intelligence amplification.[19] And then there was someone in that cybernetic serendipity book [*Cybernetics,*

---

18. Glanville (1997).
19. Ashby (1956).

Art and Ideas] called Ali Irtem. And Ali Irtem wrote about happiness amplification.[20] It was basically the same thing, basically simplify—only concentrate on one thing at a time. Just do that and then you will have more variety to deal with it. Instead of trying to be happy on one hundred levels, be good at being happy on one. That was it.

Ashby—The Law of Requisite Variety and the Black Box

Bill: What would your definition be for the law of requisite variety?
Ranulph: Well I think—hum… The law of requisite variety states, that for one system to control another in a facilitative way, it must have at least as many states as the states taken by the system to be controlled. And then I would add to it, if it is a second-order cybernetic system, it must have exactly the same number of states. And that is it. And I think that is clear and simple and obvious and I do not understand why Ashby wrote it for me the way he did, opaque… It is the only thing he has written that is intentionally and willfully opaque. Why? But there we are.
Bill: Let's jump back to the black box also. I know you are very much enamored with this black box theory. Is that correct?
Ranulph: Well I don't know if it was a theory. What I liked—why, I think I theorized it probably. What I liked was Ashby saying maybe the black box was a model for everything. What I like in there is the notion that you are working from ignorance and not from knowledge. What I like is the idea that I can go up to a door, turn a handle, and it opens, and I have no idea what is going on in there and I don't need to know. And it's that, because there is an awful a lot of nonsense about having to open things up and see what is in them, and in my opinion, you never open them up. I open this up and I still have to open what I opened up again.
Bill: And it is an infinite regress again.
Ranulph: I think so.
Bill: You can never get at the thing in itself. You can only point.[21]
Ranulph: I am not even sure there is a thing in itself. I like the lack of certainty. I like that we can live in a world in which we are ultimately ignorant…which we cannot know. And we can work in it (positive tone). To be able to do that is a trick of absolutely supreme genius. And it is what humans do all of the time. I just think this says "we are amazing"!
Bill: [Laughter]
Ranulph: And I like it a lot.

---

20. Irtem (1971).
21. Cf., Casti (1994).

Interview Next to the Water (and Ranulph's Favorite Mobile Coffee Shop)

Bill: I love just sitting at the water, I don't know, there is something about the different states of the waves and the light…

Ranulph: Me too! And we can do it again. This is a waterside city. It is the only island city left in Britain.

Sitting Place by the Water. Courtesy of Aartje Hulstein and the Glanville Archive, University of Vienna.

Stone Sculpture with Vista. ©Bill Seaman

The Thesis Has Two Different Names—Ranulph's Writing and Curiosity

Bill: You did all of this reading, you did your thesis, and this was the Object theory, is that right?
Ranulph: The thesis has two different names: the official name for the University which I can't remember, and it has the name I gave it—which they rejected,
Bill: Which I love, yeah—it is a great name...
Ranulph: I did quite a lot of those—*The Nature of Fundamentals Applied to The Fundamentals of Nature.*
Bill: This language play seems like a way that often enables you to express ideas very clearly...through these titles.
Ranulph: Consciousness and so on. It is, it is just a machine. It just keeps adding 1. I should say *ein*, I am always thinking about saying it in German.
Bill: Where did this love of this language come from?
Ranulph: I have no idea. Lou Kauffman says that you can read my papers as literature. I was never considered particularly good at language in school. I think that writing tightly came from the academic stuff.
Bill: Reacting to Gordon?

Ranulph: A bit of poetry and tightness produce what I write. And I do spend really a lot of time on it. I sent a quite well organized transcript off, the lecture I gave in Tallinn. They had done it quite nicely they'd certainly reduced it…they put it into shorter sentences which had a nice completeness to themselves, also a bit of rhythm in them. I had done a good job. I had re-written it 6 times. And not because it is desperately wrong.
Bill: It this kind of re-writing a way of thinking through something?
Ranulph: Well, sometimes yes. I write much more slowly than I used to. I suppose I am more fastidious…I have always used the writing as the research. It is very rare that I am writing something that is already sorted out. What I am doing is writing to find out.
Bill: Inquiry is central to all of these different perspectives in a way…
Ranulph: Yes.
Bill: A lifelong, ongoing sense of inquiry.

"The Most Curious Person I Ever Met"

Ranulph: Peter Maxwell Davies who was until just recently "The Master of the Queen's Music," Britain's senior musician. He is an old friend of mine and I hadn't seen him for many years. I went up and saw him in Orkney which is where he moved to. And Max is very gay and extraordinarily musically adroit, and talented, what have you. He used to invite me down when I was younger to a place in Dorset, I'd take my work down, and he would take a look at it, and talk about it with me. It was the lessons that I got in music. And I asked him, why did you ask me? I'm not the pretty one, I'm not gay, and I am not the great talented musician. You had all those other guys who knew much more and were much more talented than I. Why on earth did you pick me to invite down. And he sat there for about 5 minutes and he turned and said, "You are the most curious person I ever met." I think I am cursed with curiosity and I think Max saw that and found it interesting, because of course someone who is curious is helping him too…and I went up to Orkney. There was a question I had been wondering about for 30 years and I could finally ask it. And he just sat there and shook his head and said—"You know that is the sort of question…" I asked what relationship there was between theme in different movements in a symphony or a sonata—was there one, how come these different tunes get put together? He said "I think of it as a very very deep relationship…it is very difficult to find. But what is interesting is that nobody else ever asked me that… Nobody has said what is the relationship between these things."

Bill: One of the things I have been thinking a lot about is this multi-perspective approach to knowledge production[22]—that we can draw from all of these different kinds of areas. If we look back at your curiosity there is the musical curiosity, there is architectural curiosity, there is philosophical curiosity, and there is maybe cybernetic curiosity if that is such a thing…
Ranulph: Well I just think I am curious about many things.
Bill: Or about everything?
Ranulph: Well, for instance I'm not really interested in biology. I'm not very interested in plants. All of my relatives are gardeners, and farmers and they do all of these sorts of things. And I did none of it! And I have a splendid aunt that remarks about how she always wanted to get me interested in the stuff and I just simply refused. I just wasn't interested.
Bill: But you are deeply interested in ideas, I think.
Ranulph: Yeah.
Bill: In general.
Ranulph: I think I like ideas…
Bill: And elegance somehow. Simplicity and elegance, but not reductionism
Ranulph: [driving up and shifting topic] There she is! Wearing orange [And there was Aartje, smiling and happy to see us].

Some Talk About Music and Sound Art

Bill: These are your archives over here?
Ranulph: Archives?
Bill: Downstairs. Isn't that whole front room filled?
Ranulph: The front room is just a store at the moment. This was evacuated quite a lot in the summer. Thanks to Mrs. Wife and Mr. Son. That's the carpet for the new room. We bought that last Christmas in Turkey, in Ephesos.
Bill: It's nice. It is very cozy.
Ranulph: It is a lovely room. We don't use it. I can put you on to something. This is so rare. It is rare in two ways. This is my valve amplifier and it has been played twice in 30 years.
Bill: Valve amplifiers are supposed to be the best sounding.
Ranulph: It is so good. I had forgotten. And this is a record which has some considerable warping on it. It cost me $150.
Bill: [whistle of dismay] A hundred and fifty! I still have my Technics turntable.
Ranulph: This is a Pioneer.

---

22. Seaman (2014)

Bill: I have a new album that I gave for the two of you [Aartje also present]. A year ago we released a double vinyl, double CD and DVD. With 7 hours of music.
Ranulph: This is it—La Monte Young—46 minutes on one side of this. [record starts being played in the background]
It is the ultrastable sound wave generator. And I heard this in 1975 and I have been unable to find it. And suddenly we got lump sums for pensions and I found it. It cost $150 and it is warped. I could have paid $500 which wasn't the case. I thought I could probably miss the first few minutes.
Bill: Doesn't he have a permanent installation in NY somewhere?
Ranulph: Yeah, Tribeca. 75 Church St. 3rd Floor.
Bill: Have you gone there?
Ranulph: Yeah, many times. Ted [Krueger] has been there. I took Ted there and Ted just went WOW. You have to go there for 3 hours minimum.
Bill: Do you know Phil Niblock, 224 Center St.—Experimental Intermedia Foundation?
Ranulph: Just up the railway line, near Poughkeepsie there is a big gallery. It has a whole set of Richard Serra's pieces, very big… It has a very particular sound installation that goes off on the hour. In that corridor with MASS MoCA and things like that.
Bill: Near Hudson NY. My friend has a gallery there. Phil Niblock does these pieces, either influenced or they were colleagues… He'll do 24 hours of these drones and then he shows movies that he makes, that are very beautiful that just go on and on and on…
Ranulph: There is a French woman called Éliane Radigue who was doing it in the 1950s who I just came across. There is a lot of this sort of stuff but in the end you don't need too much of it. This I first heard when I left my first wife. And now I've got my last wife.
Bill: How many wives did you have…?
Ranulph: Just the two. She is my last wife. I had a last girlfriend but that is not her.
Bill: OK. [addressing Aartje] And were you married before also?
Aartje: Yeah. That's the best way to do it.
Ranulph: You learn that excitement is more fun than politics.
Aartje: Most definitely.
Bill: Are these your Finnish models?
Ranulph: They are children's toys. There is a set of them. I've always wanted to put them on a set of shelves. I don't think they will go downstairs—it is a pity…

You know about standing waves. Just move your head and listen to them. Slowly!

Bill: Yeah, there is a beautiful piece that Alvin Lucier did with dancers where he would just slightly shift the standing wave and the dancers would find it again, then go still. Very simple but very elegant.

Ranulph: I think those guys do very interesting experiments. I am always uncertain where I should draw the line between music and sound art.

Bill: Yeah—it is hard to say.

Ranulph: But I think there is a difference. A lot of what I am interested in is sound art. One of the nicest pieces I ever heard was two vacuum cleaners in Amiens Cathedral. Boy were they good.

Bill: This guy, Jim Pomeroy, who died quite young, was very influential to me and he made a piece called "On the Ladder—The Beat Goes On" [southwestern accent]. He had two industrial vacuum cleaners, and he had two pipes, and they were set in the *blow* mode, so they were blowing over the top of the pipe making it a wind instrument. Then he put liquid in the top, with a tube that flowed down to the bottom (and was attached to the bottom pipe). What would happen is the pitch—one would be going down while the other one would be going up in pitch, creating this incredible play with beat frequencies. It was so palpable and beautiful and more sound art than music I would say.

Ranulph: Sound is something which we have played with in a very contrived way. And it is nice to have people playing with it in a less contrived way. You know—not clarinets and violins. Not these strange sound making devices.

Aartje: Bill—for you, when you speak about music it seems to be very much connected with this movement, these people doing this, or something being done at the same time. And I think for Ranulph it is much more the sounds itself. Is that true?

Ranulph: Yes.

Bill: Yeah, it could be. I've always had this image, sound, text interest.

Aartje: It just comes out and it is there.

Bill: It was only later that I allowed myself to just make music. Always before—everything had all of these elements. Too much maybe… But this new album, each starts with the line of a poem, and then you have these musical pieces (s_traits[23] with John Supko). The idea is, when it is finished there will be 300 of these poems. And you can play them on random basically.

---

23. Seaman & Supko (2014).

Ranulph: I have a copy. I'll see If I can find it—"As If."[24] I wrote a paper. This guy took it and broke it down into bits and set it as a set of hip hop songs.
Bill: [laughter] Who was this?
Ranulph: The name is Richard Hule
Bill: Do you play music, no? [to Aartje]
Ranulph: You can sing.
Bill: And you are working on a new sound piece now, is that right? [La Monte Young standing wave still playing in the background]
Ranulph: Yeah, well the idea is to… Max and I will talk about it and then rehearse it as part of the RCA film. I don't know when I am going to have time to write it… Poor old Max is not well. His cancer has come back and I think he is having an absolutely miserable time. There is my funeral piece. I'm writing a piece to really upset my relatives at my funeral.
Bill: [laughter]
Ranulph: They've never bothered to listen to a thing of mine. While this time they are just—going—to have—to. [Ranulph delivers this line slowly with a Cheshire grin.]
Aartje: They won't come if you tell them!
Ranulph: I don't propose making it terribly pleasant. [laughter]
Bill: You are dying to have it performed…
Aartje: [chuckles]
Ranulph: Yep.
Aartje: He likes this sound [speaking of the cat].
Bill: It is almost like purring.
Aartje: I think that is what involves him.
Bill: Do you know Terry Fox? He is a west coast performance and sound artist and he scored a piece for the purring of 12 cats. [actually 11—entitled "The Labyrinth for the Purring of 11 Cats"[25]]
Ranulph and Bill: [make purring sounds]
Aartje: [speaking of the cat climbing on Bill's lap] It is very unusual for him to come when somebody has just arrived. To sit in the room like this, and he came when the music started.

---

24. Scientific Fly http://www.scientificfly.de/sf_asif.html (retrieved June 8, 2016). Listen also to Scientific Fly http://fm4.orf.at/soundpark/s/scientificfly/main (retrieved June 8, 2016)
25. Fox (1977).

Cat. Courtesy of Aartje Hulstein
and the Glanville Archive, University of Vienna.

Bill: I grew up with Maine Coon cats and a dog, and they were all very friendly, and would sleep with each other, and so on.
Ranulph: You are a very gentle guy, I am sure that he senses it.
Bill: Yeah (quietly). I think so.
Ranulph: Whereas I'm not!
Bill and Aartje: [laughter]
Ranulph: You sense that don't you [Ranulph addresses the cat]. You are waiting until my eyes go white and then you will run away. So, if I get really angry I lose all of the color in my eyes!
Bill: Seriously!
Aartje: Yeah—and you don't want to see it.
Ranulph: I tell you, leave the country
Aartje: It doesn't happen often. I think it has happened twice in 20 years.

Bill: Its funny, the sound [Le Monte Young in the background] you sort of get used to it after a while. Then you don't hear it. Then it comes back again.
Bill: There is another sound artist that does architectonic sound pieces. His name is Max Neuhaus.
Ranulph: Yes, I know Max Neuhaus. It was Neuhaus I was thinking of at that gallery. Devo?
Bill: [Laughing] I don't think it is Devo. You know Devo is that band.
Ranulph: Devo was a band amongst other things.
Bill: He may hear pitches we don't even hear… [concerning the cat]
Aartje: That is also possible, yeah.
Ranulph: We have an invitation to Vienna; we have an invitation to Tallinn; we have an invitation to India and Stockholm; so, 500 for insurance. We don't really need it in Europe. We need it for the ferry because they insisted.
Bill: Are you taking a ferry up the coast? That's nice.
Ranulph: To Bergen.
Bill: With your car?
Ranulph: No. Up to Kirkenes, and I would love to go see the memorial to witchcraft [Steilneset: Witch Trial Memorial[26]], which is right in the north of Norway. Peter Zumthor is the architect, and so is Louise Bourgeois.

Steilneset Witch Trial Memorial 2011. Copyleft.

---

26. Zumthor, & Bourgeois (2012, January3).

Steilneset Witch Trial Memorial 2015. Copyleft.

Bill: I don't know this piece.
Ranulph: It is extraordinary.
Bill: I like their work very much.
Ranulph: If we had gone one day later…we could have scheduled it…but we can't. It is a five hour drive over the mountains with snow, each way… I'm not driving 5 hours each way under those circumstances to go and see this. So then it is a plane, but there aren't very many seats. On Sunday there is only one flight and it is late in the evening. Then it doesn't work for getting back. The next day the timings of the flights are all wrong. If we arrived one day later we would have been able to get the morning flight, and back in the evening…
Bill: If I couldn't go I would buy the book. My studio—you've seen it before—it is wall to wall books, now 3 layers deep.
Ranulph: Do you know the Sir John Soane Museum in London?[27] [See a detail from Sir John Soane Museum under Creativity in "Composing Composing," p. 270.]
Bill: I have never been there.
Ranulph: You should go—it has fold away pictures…layer after layer of pictures. It's fabulous. And in the end, it just becomes a home, and you look down. It is the most extraordinary room. We've got too many pictures and don't know what to do with them.
Bill: Do you have an attic?
Ranulph: We've got a roof but it is full too.
Ranulph: It has my architecture portfolio!

---

27. Sir John Soane Museum, http://www.soane.org (retrieved June 8, 2016)

Bill: All of the models.
Ranulph: No no no. A few drawings, many of them falling to bits because they were glued together.
Ranulph: The Architectural Association wants to shoot them. I am going to lend them the portfolio then send it off to Vienna.
Bill: And Albert will keep it.
Ranulph: They've got my archive.
Bill: They do, oh well that's good. He is a very wonderful person to take care of it [Albert Müller] … to keep archives. I've met Albert only briefly. I spent a week there and he gave me everything and I went away with about 10,000 pages [from the Pask and Von Foerster archives]. Which I am very happy about, and also the images from Von Foerster. All of those beautiful funny ones he used in his books.
Ranulph: Here is something that my boss quotes me on. We have meetings and so on and he keeps writing things down. I always wonder what it is about. And in introducing the program at the RCA to the students the last thing he said was "There is no book of spells but there is magic." That was me!
Bill: That's nice.
Ranulph: Yeah it is, I thought: "That's really good, I'll try to remember that one."
[La Monte Young still playing]
Bill: [talking to Aartje]: As a boy, when I would ride across the lake, we had a 2½ horsepower motor, which basically made almost this exact sound. And I would sing with it.
Ranulph: [Ranulph singing with La Monte in the background] Yeah.
Bill: Just like what you are doing…
Ranulph: I had a friend and he just died in Australia, and had an espresso machine. And it just does the beginning of Holst's "The Planets." Bumm BAAAAA then I would always sing the tritone. Dummmm BAAA DAAAA. Richard would look at me and say "What are you doing?" and I would say "Planets." Oh yes, how do you do that? Tritones are very difficult to sing.
Aartje: But you are right, this has some kind of quality of a lake and an outside motor.
Bill: I can almost see the wake.
Ranulph: Where was this?
Bill: In Maine. Every summer we went to this house in Oakland, Maine. My grandmother was the Dean of Women at Colby College and she was a concert pianist—she was very influential to me. She was a very interesting

woman. We always went to this summer place. There were little boats and sail boats so I had a nice childhood of always being at the lake. [East Pond]
Ranulph: Wonderful. Swimming.
Bill: Swimming and sailing. I had a sunfish and I would go out for hours and hours. And the windier the better. If it were so windy the whitecaps were there I would just fly across the lake. Very fun.
Ranulph: I did that for some years in Finland. A couple of summers ago I was asked to look at some project work that was done at one of the Estonian islands. And I ended up having a summer again. If I talk at this pitch you can hear… [low voice with the drone]
Bill: I get the resonance… I took the ferry from Stockholm to Helsinki, I don't know if you have done that…
Ranulph: YEEES.
Bill: And I sat right at the front part of the boat, on the bow, and it is like you are flying because you are very high up and there is an archipelago. You are so high up and there are all of these little islands and this giant ship passing through.
Aartje: There was a sauna at the front of the bow last time. So we had the sauna sailing in and it was fantastic.
Ranulph: I think the last time I did it was… I had damaged my back. And I had been all of the way around Europe on Interrail and I had damaged my back. I discovered they had a physiotherapist on board, so up on the 12th deck I had my back dealt with and he was playing music and it turned out to be—there was a very cultish influential DJ named John Peel.
Bill: I know his work.
Ranulph: And John Peel had a once a week a Finnish radio show. And the last thing he did was he played Cpt. Beefheart doing "Mirror Man" from the 8th of June, 1974 show, in London. I was there. It was 2 days after my son was born. Right up there, looking out over the Baltic, the sun coming across the boat. A very pretty "Mirror Man," a very short one.

Circling Back to Objects

Ranulph: OK, You want me to try to do Objects?
Bill: If you want to. Or we can do it tomorrow…
Ranulph: The thing about second-order cybernetics, one of the things is that if we say that the observer is different, the observer is always present, so the observation that each observer makes is different, you are left with a problem: How do we know we are talking about the same thing? It is really a serious problem. And people say we'll we talk about it. But I don't know

that you and I are hearing the same sound. So we have no way of actually knowing that we are referring to the same thing. Without this, second-order cybernetics is itinerant. What I try to do with the theory of Objects was to invent a structure, which if we assumed it, and it is only an assumption, it is not a truth, it would allow us to think that we are talking about the same thing. Which is what we want, and it is the best we can do.
Bill: So in a sense recouping objectivity?
Ranulph: No. Absolutely not. But recouping the belief that we are talking about the same thing. Now, I am convinced that none of the other guys [in second-order cybernetics] bothered with this at all. Heinz came up with his Eigenfunctions[28] and what-have-you, and those don't solve it. All those do, is that they are just a machine that produces a self-producing value, but what does the value mean to me? And anyhow, they are sort of peculiar because there is no observer in them. Why did he bother?
Bill: Why do you think he was—?
Ranulph: I think he had emotional difficulty leaving objectivity. Gordon deals with conversation. In a conversation you build things in parallel but they never really communicate. Maturana doesn't even try. Glasersfeld isn't interested in this question anyway.
Bill: And why do you think these guys weren't interested in the question?
Ranulph: I don't think it occurred to them as being significant. I think they were confused in their thinking. I don't think they pushed it far enough. But I never really got to talk to them about it. OK. Now my job is to design the structure which allows me to think that there is some thing, whether there is or not, this is not about truth this is about construction or as I now call it *composition*. I don't observe anymore, I *compose*. I am interested in a universe made up of things that are observable. That is what second-order cybernetics is about. Now what I want to know is what is the minimum condition under which something can be in a universe of being observed…observable—what is it? There is one minimum condition. It requires one agent, if it required two agents it wouldn't be minimum.
Bill: You tell me…
Ranulph: No, no, you tell me.
Bill: Self-reflection, self-awareness?
Ranulph: Go back in terms of the word *observe*. The least condition for being observed is to observe yourself. It only implies you and it brings you into being. You observe yourself. It requires nothing other than yourself. It doesn't rely on anything else being there beforehand. And it brings you into being in a universe of observation.

---

28. Von Foerster (2003b).

Bill: Hard to say that that would be very scientific though.
Ranulph: I'm not trying to say that it is scientific.
Bill: OK.
Ranulph: I wouldn't waste my time trying to say it is scientific. Science doesn't deal with problems like this. Science is absolutely incapable of dealing with problems like this. One of the great weaknesses of scientists is they don't realize this.
Bill: This is probably why cybernetics had a hard time getting off of the ground…
Ranulph: Scientists are very lazy about the conditions in which they work…the assumptions they make, and they are also very dogmatic about it. Anyway—this is perfectly scientific in an older understanding of science pre-Descartes and Newton. If this thing might be me in both observed and observing itself. Actually, there is not one of them but two—both subject and object. How do you act so that you can be subject and object, but not two—just one.
Bill: I call it a subject ↔ object unity.
Ranulph: No, that is not how you can do it, that is how you name it.
Bill: I think you can't avoid it to be honest. You are it.
Ranulph: No. That is a bit of an easy way out.
Bill: [laughter]
Ranulph: That's not very scientific.
Bill: No, OK. But I am an artist… [laughter]
Ranulph: It doesn't allow you to be sloppy. Well, how about if it switches roles: subject object…subject subject? What is that?
Bill: An oscillation.
Ranulph: Another word for it.
Bill: A dialogical situation,
Ranulph: [makes the motion and sound of a clock pendulum]
Bill: A clock.
Ranulph: What does a clock do?
Bill: Clock of the self?
Ranulph: No, what does a clock do? We say—
Bill: It makes the illusion of time.
Ranulph: It tells time. What has happened now? We have this one thing, it changes between subject and object; in doing so, it creates time. Creating time means there are two worlds and it is in only one of those roles at any one time. That means that the other role is empty. If I am busy observing, I'm not being observed. Something else could come and observe me. Because there is a slot so I can be observed.

Bill: Just playing devil's advocate though, can't you, for example if I am doing this, [staring directly at my arm movements] I am being observer and observed simultaneously. I am not oscillating.

Ranulph: I don't think that is true, not in the sense we are talking about here. Really, we are talking about the very simplest type of mechanical things. Mechanisms. Because cybernetics is interested in mechanisms. Cybernetics does machines. There may be another way you can describe it. I don't mind… I describe it in the way I like to describe it because I can… I have this thing, I have this moving, I have empty slots, I have other things that can have a look in the empty slots. Now, suddenly I have a way of relating things. I can observe other things and it's obviously different to observing the same thing. When you observe it, it is you in that slot that I can observe it in that slot, I look through that window. We can both do it at the same time but we will both see differently. The relation will be different because we are different. Yet, we have this structure which allows us to believe that we are observing the same thing. And we have that because we have given it the autonomy of generating itself. It comes from itself, it doesn't come from anything else. I didn't make it. Now, say I am here and there are some of these things and they are all viewable at slightly different times, I can view this one, then I can view both, then I can only view this one. What would that be? Imagine it as a second. $A…B$:, I'm observing A, I am observing both, I am observing B. What's that.

Bill: The intersection of the set? [Ranulph showing Bill a drawing]

Ranulph: No the intersection is there. This is the union of the set, so this is an inclusive all. Now I've got logic. I can say this stands for this and all of these differing sorts of relationships. Now I can do almost anything. What I have produced is a system that accommodates absolutely everything, is as simple as can be, and allows us to believe we are talking about the same things, whether we are or not. Since we can never know, who cares. All we really want to do is to feel that we can say we are. That's our actual task. Really very very simple… Because…well, it has taken me forty years to distill it since I wrote my PhD.

Bill: What would you say about this… If I make a pun that has multiple meanings.

Ranulph: No—things don't have meanings.

Bill: What do they have? [laughter] Identities? Self-identities?

Ranulph: You give it meaning. Meaning is from you. Of course there are puns. Things where I can construct two different meanings, sometimes going uncomfortably together and sometimes not. This is all very sophisticated stuff... I'm doing the least you can possibly do. I'm really lazy. Yeah?

Bill: [laughter] You can attest to that. [said jokingly to Aartje]
I'm just wondering if we put a pun into this system how do I know that you and I are sharing this concept of the pun.

Ranulph: We don't know that, but we can believe. It is a construct to allow us to believe. Not a construct for knowledge in the sense that "I am certain." But, you know, everything is an object so we can both talk about a pun. Here is an Object called pun. Here is an Object called blue, here is an Object called Danube. Together I have Blue Danube. [Ranulph characteristically starts humming Blue Danube.] S. has arrived at the space station.

Bill: And I got that reference too. [this was a sound pun related to both the music in itself and the music used as soundtrack for *2001 A Space Odyssey* by Stanley Kubrick].

Ranulph: 2001.

Bill: That's exactly what I knew.

Ranulph: The question is, don't ask me to do the building, ask me to show you that there is such a thing...as you can make a kit like this, which is made of less than nothing, which will satisfy the requirement. That's all. That's all I have to do. Now we can have second-order cybernetics. Now we can have multi-authors, we can have plural mathematics, and all of the rest of it. Before that, you couldn't.

Bill: People were just assuming.

Ranulph: People were lazy. And people were very lazy about second-order cybernetics. They didn't sort it out. Heinz changed Margaret Mead's request to consistency into a thing about observing...and observed. They were lazy. [takes mic and holds it close to his mouth and whispers]—They were lazy.

Analogue vs Digital in Recording and in Bodies

Bill: It's flipping along there. [the recorder] Isn't that a beautiful...this has twice the fidelity of a CD if you want to set it to that level.

Ranulph: Uh huh. But a CD is actually very clean. And listening to that vinyl reminds you. [listening to Messiaen earlier in the day—*Quartet for the End of Time*]. You know what that is.

Bill: Analogue vs. digital?

Ranulph: But you know what the difference is.

Bill: Yeah.

Ranulph: What?

Bill: Infinite vs finite. Or discrete vs—

Ranulph: At every point in an analogue system there is an infinite amount of analogue information.
Bill: Yeah—I am very fond of analogue systems. I think of us, as human beings, as mixed analogue and digital systems.
Ranulph: No, we are not digital; we are discrete. There are on and off but they are not digital which implies a uniformity of unit.
Bill: Well the synapse firing is a bit of a uniformity.
Ranulph: I am not sure that synapses really fire. I mean, this is an enormous construction. And it is things built on things, built on things, built on things…and I am not sure how much evidence there is, really for any of this. But of course, if I say that to a scientist they'll just say I'm silly.

The First PhD Thesis

Bill: In this beginning time, you did this thesis, which Gordon asked you to do. You got this fellowship through him. So what was the next step after that was turned in? Where did you go or who were you working with?
Ranulph: When it was first turned in it was turned down by the University Examination Board, who said [Ranulph in a slightly cockney accent] "We don't have a category for 'things' in the library." That is why it has its pretentious title.
Bill: After your thesis was denied, you renamed it.
Ranulph: I renamed it.
Bill: Re-submitted it and they accepted it.
Ranulph: Yep.
Bill: And after that what was next?
Ranulph: I was examined.
Bill: OK. And who was on your committee?
Ranulph: Gordon Pask and Heinz Von Foerster.
Bill: That is a kind of interesting thesis committee. [Laughter]
Ranulph: Gordon and Heinz couldn't be there at the same time.
Bill: You had, sort of, two different days or something?
Ranulph: No. Gordon wrote some message to Heinz and Heinz flew in and I collected him from Heathrow. Bantered about a bit and then he sat at my dining table. Then he went through the thing and corrected the English. Especially the spelling. He didn't ask me any decent questions. Then he said, "What are you going to do with it?" And I thought, "What on earth do you mean?" Are you going to publish it like this, are you going to make a book, are you going to write lots of papers? It never occurred to me. And Gordon was really a very naive academic. He was the last person you

needed as a professor to teach you how to survive in the academic world… Heinz said, "Publish lots of papers," so I did. And he took us out to dinner. And the next day I was summoned—he had to write the joint report. And he decided that I could have the same joint report as his uncle, Ludwig Wittgenstein. With the word Glanville substituted for Wittgenstein. I was sent to the British Library to get the quote right, this comes from G. E. Moore? And it goes "Mr. Glanville's, PhD may or may not be a work of genius, however that may be, it is clearly up to the standard demanded by this University, for the award of the degree of Doctor of Philosophy."[29]
Bill: Nice.
Ranulph: I had this signed by Heinz and Gordon. I don't think I have a copy of it.
Bill: Can we get the Wittgenstein letter again?
Ranulph: No, I know it by heart.
Bill: But there is something nice about objects too. But now we just made an Object of it.
Ranulph: The funny things about *object*, the word, is that it means subject too. It is a word that is completely ambiguous. A pun if you like. My objective in using the word *objective* was to confuse people. The object of Objects is not to be objective, but to look carefully as if through an objective, with the intention (or objective) of not objectifying…and so it goes on.
Bill: And it reminds me a little bit, and I think you did have an interest in Zen—these Zen Koans.
Ranulph: Yeah, yeah, yeah.
Bill: Because I think at times you use that strategy, a kind of confusing strategy that forces somebody to make a jump—
Aartje: To take a leap.
Ranulph: And sometimes I write a really clumsy sentence for the same reason. Before studying Zen, men are men and mountains are mountains. While studying Zen all is confused. After studying Zen, men are men and mountains are mountains. The difference is, that in the first case, the feet were a little bit off the ground. That's dear old John Cage.
Bill: I love that Indeterminacy album.
Ranulph: Have you got the CDs of him performing it?
Bill: I don't.

---

29. G. E. Moore's comment on Wittgenstein's *Tractatus* reads slightly different: "It is my personal opinion, that Mr Wittgenstein's thesis is a work of a genius; but, be that as it may, it is certainly well up to the standard required for the Cambridge degree of Doctor of Philosophy." See Monk (1991, p. 272).

Ranulph: Well they are Folkways. It was a double LP, with Cage and Tudor playing the piano and orchestra parts. [Ranulph mimicking the sweet voice of Cage] "One day when I was still living at Grant Street and Monroe, Isamu Noguchi came to visit me." He spoke rather like that…
Bill: It's a charming voice…
Ranulph: Heinz said I should publish lots of papers…and I did.
Bill: You passed. You got the stamp of approval.
Ranulph: I wrote some papers.
Bill: And what were your first papers about?
Ranulph: The first paper is "What Is Memory That It Can Remember What It Is."
Bill: Again—these playful linguistic titles.
Ranulph: Well it was also that Heinz had a paper, called "What Is Memory That It Can Have Hindsight and Foresight as Well."[30] So it was a little play on that. And I said that Objects were a form for memory, which is just to say they keep on going. Heinz said of them that what I wrote was the first calculus for Piaget's "Object Constancy,"[31] which makes it this sort of—his Eigenforms are also an attempt to make Piaget sorts of things and it makes it a bit disingenuous for him to claim that his had nothing to do with mine because he stated mine was the first example with anything to do with Piaget, and then chose to ignore it and I always felt very bitter about that.
Bill: Well one wonders also about his language plays, because these came after your thesis.
Ranulph: Yeah, well… People used to write to me and say can you invent a title for me.
Bill: I like titles, I think they are very important because they lay out a kind of field—a way to approach something.
Ranulph: I haven't had a fun title in a year.
Aartje: "You've Become Too Serious"—that is the title.
Bill: …Then you were in the thick of things. And you visited the states and visited…did you go to the Biological Computer Laboratory?
Ranulph: I visited the Biological Computer Lab as a student because it was closed by the time I finished.
Bill: I don't know my dates.
Ranulph: So I went to the BCL.
Bill: When you were there did you meet Maturana and Varela?
Ranulph: No, no, they weren't there; they were in Chile.

---

30. von Foerster (2003c).
31. *Piaget's Model of Cognitive Development*. Retrieved September 27, 2019 from https://www.massey.ac.nz/~wwpapajl/evolution/assign2/LO/piaget.html

Bill: Who were some of the other people you visited while in the States?
Ranulph: Well, I didn't visit people. I went to conferences. I went to George Klir's NATO Conference in 1977, which had the big public thing on autopoiesis and I did something in that[32] and then I went back for a couple of ASC conferences.
Bill: And the Macy conferences were done by then?
Ranulph: They were done in 1952. I went to one of the Gordon conferences (1984) but I was banned from them by Ernst von Glasersfeld for being rude to my professor.
Bill: Seriously?
Ranulph: Whereas my professor and I had actually agreed on this. Which was that I was going to savagely attack him…
Bill: Intentionally?
Ranulph: And I stormed out, serious, attacking Gordon and I was banned. Ernst really never had a sense of humor, at least not of the absurd. Anyway—
Bill: And you did have some conversations with Ernst before?
Ranulph: Yeah, I had several conversations with Ernst towards the end of his lifetime.
Bill: And that is a very nice title: "The Importance of being Ernst."
Ranulph: A bit obvious isn't it. And then (in the late 1970s) I was just a little old architecture teacher. And I used to ride in the train down from London every morning.

The Second PhD

Aartje: But when did you start doing your second PhD?
Ranulph: Almost immediately after I finished the first. Because I realized I had been doing all of this stuff. I thought I would get it properly tested. It was supposed to be a masters. Because my first degree was a doctorate, I was going to do it backwards. The only doctorates I am working on now are all of my students.
Bill: It was a masters but it got turned into a PhD?
Aartje: That was about space…
Ranulph: *Architecture and Space for Thought* was what it was called.
Bill: Did you have a playful punning subtitle or something like that as well?
Ranulph: Yeah… It is actually a very nice piece of work, it goes against everything that is happening in environmental psychology at the time.

---

32. Glanville (1978).

Bill: You were in the paper-writing mode, you went back to school to do this second degree. What is the upshot of your thesis? The title kind of says it all but what was your approach?

Ranulph: When I went to architecture school, I never really studied architecture at all. But it was very typical of the AA that it would invite the best, but also the "worst" students back. Because the worst students... I am not talking about people who were without talent, I am talking about the people who were difficult, who refused to do things. And the reason you invite them back is because they are the people with passion. These are the people you want teaching because they are going to bring new passion, new ideas.

Bill: And new approaches, and you were there.

Ranulph: I was invited in as one of the difficult people.

Bill: A dis-architect.

Ranulph: Not in those days. Nowadays it would be architect!

Bill: First it starts as an architect...then...

Ranulph: And I ended up with these very very talented architecture people and I really didn't fit in very well.

Bill: But they liked you...or they didn't, or they tolerated you?

Ranulph: More than that. Some became very good friends and some were very different. I discovered that architects talk about space and I had no idea what they meant by this. I started by doing some experiments with first year students about developing a vocabulary that might be shared in order to describe spatial experience.

Bill: Nice.

Ranulph: We used slides and we used Kelly grid techniques[33] and what I discovered at the end of this was that actually, no one used the same words, so you would have a limited vocabulary, you would show slides, you would say what words were appropriate, and none of them used the same words.

Bill: And were you reading some books to inform things.

Ranulph: Well yes, I was reading the environmental psychology stuff and some stuff on space but there wasn't very much of this stuff around then.

Bill: Was Christopher Alexander[34] somebody you looked at, or you didn't like it so much...

Ranulph: Well, Christopher Alexander wasn't doing this sort of thing. Christopher Alexander had really very little interest in space.

Bill: This beautiful Bachelard book[35] is a nice one I think.

---

33. Informational grids are used for the comparative analysis of multiple topics.
34. See Alexander (1977, 1979).
35. Bachelard (1964).

Ranulph: I'm not very fond of it but that is because I am not a very symbolic person. For me saying—oh the head is the attic. Sorry, the head is not erratic. I thought, maybe the problem is words... I did what were called blindfold surveys and I would take people into very unusual spaces, blindfolded, then they would go out and draw. And go back in and try to understand it.
Bill: To try to understand it physically, phenomenologically.
Ranulph: And what was interesting, when you finally took them into the space—of course! But doing all of this work and drawing meant nothing to them.
Bill: Merleau-Ponty come in there anywhere?
Ranulph: No. But that is partly because he is such a pain to read. I didn't need any of that stuff. I was actually doing experiments. Lots of people were pontificating but I was out there with the people. So we moved to drawing. My idea was that—the problem was about serialism. I don't understand the space by adding this bit and this bit and this bit. That you understand it as a whole, you see the whole thing instantly, and you feed detail into it. So I took people into other unusual spaces, blindfolded and then I gave them 2 seconds, 5 seconds, 10 seconds...to look at it, then they went out and drew it. And then we had an exhibition. And the question was, as someone who has been through this experience, as an example, how long a viewing do you think this represents. And it made no difference. There was virtually no correlation at all between someone that had seen something and how long people judged the drawing they had seen. Now of course, some people drew well and some people didn't.
Bill: Some people have that knack for the fast drawing that is very beautiful.
Ranulph: They weren't constrained to fast drawing, they could draw for as long as they wanted.
Bill: I see.
Ranulph: Then I started using this as a design device. I would show them something very quickly, a space from 4 corners, and get them to draw it and get them to *force* detail into it. When they weren't sure that had to think about it, to pick and choose and put detail into it. And then go back and compare their drawing with what they saw, and say that the stuff in the drawing is part of my mental model—drawing what is not there. They did this comparison and they took this stuff out and they said—these are things that I think are in the spaces, not that are in spaces, but these are things that I would put in there.
Bill: Project into the space...that's nice. A kind constructivist approach almost.

Ranulph: And I had them design with these—
Bill: Design with the things that arose out of the difference.
Ranulph: With the *wrong*—
Bill: Yeah—I like that.
Ranulph: And what they designed was an Object. I was at Saint Paul's and they were asked to design something to go into the church, that somehow reflected the totality of the church. This went back to an old project of mine called Wittgenstein's monument...
[Break in conversation]
Bill: My mentor and teacher just died who was at MIT—Otto Piene. I had seen him four weeks before, had lunch with him, and talked about all of these future projects.
Ranulph: What did he die of?
Bill: He had a heart attack. He was 86, I think.
Ranulph: There are some extraordinary people around, and many who have been overlooked in really bad ways. It is as if one name can do for an entire movement...an entire generation. And that happens too much in the world of art. It happens mostly in the world of "agent." God preserve us from them, et cetera. But this manufacturing of the new sensation to be forgotten in three years. The selling of the completely vacuous. And the attempt to be spuriously original. Then not. They are just spuriously empty.
Bill: He [Otto] had some big shows recently but they were things he had more or less waited for, for a lifetime. In the US he made these inflatable works—as well as fire paintings and ceramics that weren't really shown in Europe. In Europe, he was known for paintings and light sculptures. I think it meant a lot to him. He went to the opening and died just after.
Ranulph: It is wonderful to have something like that. Strangely it is what I am getting, I sort of think to myself most people are wearing blinders to this extraordinary body of work that I have done. Now there is sort of a queue of people saying can I come and talk to you about this. And I am getting endless ones on the Internet—you can't do this.
Bill: Thanks for taking the time to talk to me.
Ranulph: No, no, no that's not it. And I am not asking you to say anything like that. It is a pleasure! Absolute pleasure. You are such a nice man and it is so nice to do this with you. It is also very flattering because you are not exactly someone short on achievement.
Bill: Did you publish a bibliography of all of your papers?
Ranulph: Yeah, it is on my CV.[36] I have just updated it.

---

36. See Appendix 1 of this book for Ranulph's full CV.

Bill: If this becomes a book, and I am hoping that it will, is that something we could put in the appendix?
Ranulph: Sure. I'll just send you the latest version.
Bill: I was thinking that it would be interesting and that this glossary of terms from your thesis would be another interesting appendix entry, possibly. I was reading back through our initial list of things to discuss and physics popped out. So you had studied physics when you were younger?
Ranulph: At school. My school leading subject included physics. And you know we specialize in this country.
Bill: I do this kind of multi-perspective approach to knowledge production. Whereas, you have many different perspectives where you focus on the specificity of a particular one or find these analogies in pairs…as opposed to doing this kind of overarching kind of list.
Ranulph: A lot of what I have done has been about making room for multiple views. I think Candy quoted this at the ASC [American Society for Cybernetics] conference. What I am more interested in is "creating the area in which people can design pitches on which games can be played, according to rules other people have designed, under the control of other people, with people interpreting rules on the pitch and then being watched by other people." So I am interested in the minimum you can provide that allows people to direct themselves.
Bill: Is this a dialogical approach?
Ranulph: No, no… I'd would say it is minimalist, I used to write a lot of music like this… It's kit music for people to make their own music, not to make mine. I do two different sorts of things: I set up situations for anyone, and sometimes I use those situations or I don't bother with them. But if I do, I am the anyone. Sometimes I am setting things up to make it possible for anyone to do some creative act and sometimes I am the person who is doing them.
Bill: And that was really your strategy in how you ran the conferences.
Ranulph: Yeah. I have always been interested in making possibilities for others to express themselves, and therefore, from frameworks that allow many different points of view. Actually, I am working all of the time with multiple points of view. I think I am always interested in multiple points of view, and bringing them together, but before I bring them together, I am interested in the framework that supports them, in other words, the framework that allows multiple points of view to exist. Very few people do that. I'm not interested in sort of bulldozing them into some overall uniform view. I am perfectly happy to have two completely conflicting views, present in the same place at the same time.

Bill: Does this in some way relate to Stafford Beers's *team syntegrity*?[37] Or is it a different kind of thing?
Ranulph: No, I think it doesn't relate to Stafford in two ways: One is that here is the imposition of some completely random and arbitrary structure and you know, I am sorry—he may have seen the dodecahedron as a wonderful device, but in the end it becomes a sort of mythical insistence. I really detest all of that magic number stuff. It is also one of the reasons why I don't like Bucky. I admire him, he did fantastic stuff, he is not really a systems guy; he is a systems engineer. All his views of engineers' delight doesn't come in to fill a world. But, he is dominated by these things that are sort of special, and things that have magical properties for him, or come to have that. One reason is that structure which I find just ridiculous really—in that it is not even very convenient. And then there is the second bit which is—I always feel with team syntegrity that they are trying to come, not to an understanding of each other and then build something that would contain all of this, but to a conclusion. To me it always feels as though there is a sort of logical consistency in all of it. And I may be wrong about this, because I have to admit… Stafford's work doesn't most interest me.
Bill: I was just reading it and it didn't draw me into it as something—
Ranulph: There is something almost hectoring about this stuff. You know after the business in Chile, he retired to this little hut in Wales. He used to appear once a week on the late night religious show on television. And he would be there with his Hindu chants and he was sort of selling himself as an all-knowing mystic. It was an incredibly difficult time for him, I don't mean to be unkind in any sense, but I wonder if everyone else would have reacted in that way—and needed to. One of the other things, that Stafford and I had an absolutely wonderful meeting. We didn't really know each other was the truth of the matter and I asked him to do a couple of things and he couldn't. And then in 1978 at the Symposium For General Systems Research Silver Jubilee conference, which was in London, he and Gordon were doing an anti-conference, a conference against conferences. They asked everyone to put something on a card, there was going to be lots of computers number crunching as you would sort of expect. Brian R. Gaines asked me if I was going and I said, "No." They said "Why not?" "It is ridiculous, it won't work, there is no interest." They said, "What about Stafford?" I said, "Stafford is a wanker." They wrote this down and put this in as a card for me!
Bill: Oh jeez, that's not so—

---

37. See Beer (1994).

Ranulph: I was then taken by Elizabeth Pask to meet the great Stafford who was holding court—you know he was sitting on a throne in a throne room, in the Russell Hotel, and he got up and he said "Ranulph we meet and he hugged me and he said now what is this thing, you didn't write it." I said, "What thing?" "A card that says Stafford is a wanker." I said, "Yeah, you are right, I didn't write it." "Then why did someone put it in there for me." "Because I said it!" [big laughs from both Glanville and Seaman]. "How dare you! How dare you talk about me like that?" And he never met me and I just looked at him and said, "You know Stafford, you greeted me with a big hug and embrace and that is very presumptive. And we have never met. You don't know me either." So, Stafford and I met—two large—and I think we both drank a good bit that day, two large drinking alpha males and it was terribly funny. And of course we made up.
Bill: The first meeting with him and the first meeting with von Glasersfeld also was a scandal.
Ranulph: Yes, yes I suppose so, you know he was a very proper man and Ernst and I became very fond of each other, eventually. And I did eventually manage to explain to him that the whole thing was set up. That Gordon and I—and in fact Paul Pangaro—had gone through this and set it up and the reason for it was this and this… So I admire certain things Stafford did, and I admire a certain sort of courage—of realism. I feel for him in the loss of the Chilean project. I have been to Chile. I was there the day they turned off Stafford's life support.[38] I went to the Allende mausoleum while they turned it off. In the end, there are fantastic ideas that never really quite worked—I'm not sure that the followers can make them work in any Staffordian sense. We have team syntegrity, which was bought up by Malik who has done good things; he extended it but he doesn't have the grand vision that Stafford was going for. But I may just be wrong.
Bill: Well, it doesn't pull me in the way Pask or Von Foerster's or others did.
Ranulph: Well the thing about Stafford is that it is all sort of loud and political and real world. My thinking is not of that sort. It might be interesting if I talked about some of my student projects because some of them were truly extraordinary, many requiring enormous nerve and a vast streak of Dadaism, which remains hidden in me.
Bill: I think it pops up all the time.
Ranulph: Yes, small naughtinesses. Anyway, I could start with the most outrageous thing I ever did as a student—apart from never doing anything on time or properly—the most outrageous thing I ever did as an architecture student was in my second year. In my first year the AA (Architecture

---

38. Anthony Stafford Beer, September 25, 1926 – August 23, 2002

Association) brought in building technology and science. And we spent our first year learning about bad science by technologists. We were examined; I cheated and got distinction. At the end of the first year the man who had brought these guys in—the principle of the AA—got sacked. Not because of this but because he assumed he had the authority to sack one of the fourth year masters. He was the chairman of the AA for 21 years. In the second year we had a sort of stand in head. A man called Dr. Otto H. Königsberger. And he ran the tropical architecture unit at the AA. I knew Dr. K. because my aunt had been at the AA, so there were all these very senior people who knew me at the AA and kept an eye on me. I was very lucky. At the end of that year the AA had to find a new head of school and also had to join Imperial College. The government had decided that the RCA could survive as an independent unit but the AA couldn't. The best tech university in Britain was the Imperial College and it was a terrible place for the AA. We were having a building built for us—by Dow Chemicals—in the middle of the Vietnam War. You can imagine the scandal, the student riots to come. The AA, which was unruly and difficult—it wasn't pretty. There was a big hunt for a new head chairman. I, as a second year student, applied. I was put on the shortlist by everyone at Imperial College. They didn't realize that I was only a second year. I had this really futurist vision about stuff which would be on film and you would go and watch in a little carrel, recording stuff and indexing it, there was an enormous amount of available tech and also just people like Gordon Pask. It was really a very future and tech-y looking thing. But anyhow, that was probably the most outrageous thing I ever did as an architect.

Bill: This is already after the Internet shopping?

Ranulph: No that was all the same sort of thing. We were all asked in our first year kinds of things, I did them all really badly, because I didn't understand what was being asked and why. People said, "You've got to learn to think with your pencil" and I'd say, "Why? I've got a perfectly good brain." But you know as an artist there's something about touching the medium that changes what you do as much as anything else. I had a first year where I did really crap stuff. The third year we had to develop a nursery school or something and I discovered Piaget, and I said this is outrageous, we're providing for 12 kids, for half a day, for a borough in London that has 180,000 people and has an extreme shortage of nurseries—this doesn't help. Furthermore, it pays no attention to how young children learn. My tutor was a wise man named David Bernstein, who was a really significant architect in Britain and he knew a lot of really important things. One was to keep the student waiting. Every week he'd say, "Well go away and draw the thing." I said, "I'm going to provide a sort of system build factory type thing. I'm going to put a hundred and fifty people on this site." Dave said, "Go away and draw it." I'd come every week and I'd have nothing ready to show and he'd keep saying, "Well, go away and draw it." I'd have nightmares about this thing. It had to be hexagonal. After about 6 weeks of this he said, "Come in Monday night. Around 7 we came in and he started drawing my hexagons as boxes and I said, "You can't do that"—and he said, "We'll make them round later, let's just get something down." He did the whole thing and it was very complex. It would never have worked but it was an amazing response. He showed me by simplifying the thing, by concentrating on what was truly important rather than getting caught up in some detail and trapped by it, I could begin to design this thing. He did it by driving me to the point where I was so desperate that I would accept anything. I will always remember that lesson. About ten years ago, I got to take him out to lunch and thank him. Then we went on and we were set to do a supermarket. In Britain at that time you could make free phone calls by going to a public call box and instead of putting money in and dialing, by tapping the bar at the same speed as the dial rotated—so you had phone hackers just sitting there and hitting the bar. You could make long distance calls and everything. I knew that you could send these pulses down the line, so I knew it was possible to send data signals by dialing, so I decided I would design a remote shopping system so you didn't have to go to the supermarket.[39] The idea was everyone would be given a catalog and you could look up what you wanted and you'd dial its code, then you would pay with a brand new thing—a credit card. The first credit card appeared in Britain in 1966. We had ATM's on the wall, you could go and get money from the bank. I got into Zambia by being able to demonstrate this. This was magic. Anyhow, I designed this system with a warehouse, and trucks and tippers who would drop things off. Then it would compile everything and charge you and send it to you. My idea was to have a little motorized hot air balloon fly it over. This was the drone delivery stuff we are beginning to see just now. I designed it fully and brought it down to Gordon and he talked about it talked about the organizational diagrams and then I was told that it didn't matter that I had designed this thing because I was supposed to design a building—so I designed the most unpleasant building I could. It was made of unpleasant plastic clip on panels, cast off of all the local houses. It had some very cunning detailing, which wouldn't have worked, and it was drawn in 5B pencil, on tracing paper and it was then inked over with the first

---

39. Ranulph described it as "Internet shopping, in 1970!" This project got Glanville speaking to Pask, which changed his life.

felt tipped pens in orange red green and blue, so the drawings were as unattractive as the design, and I had to plan it. The way I planned it, was I threw darts across the room at it. They said you have to have two really different plans so I threw really different plans. I always did the minimum that they wanted, and it was always awful. I produced the most awful and unattractive projects. But conceptually they were radical.

Bill: I was thinking that you were always interrogating space and interrogating ideas and context—maybe that's not the right word.

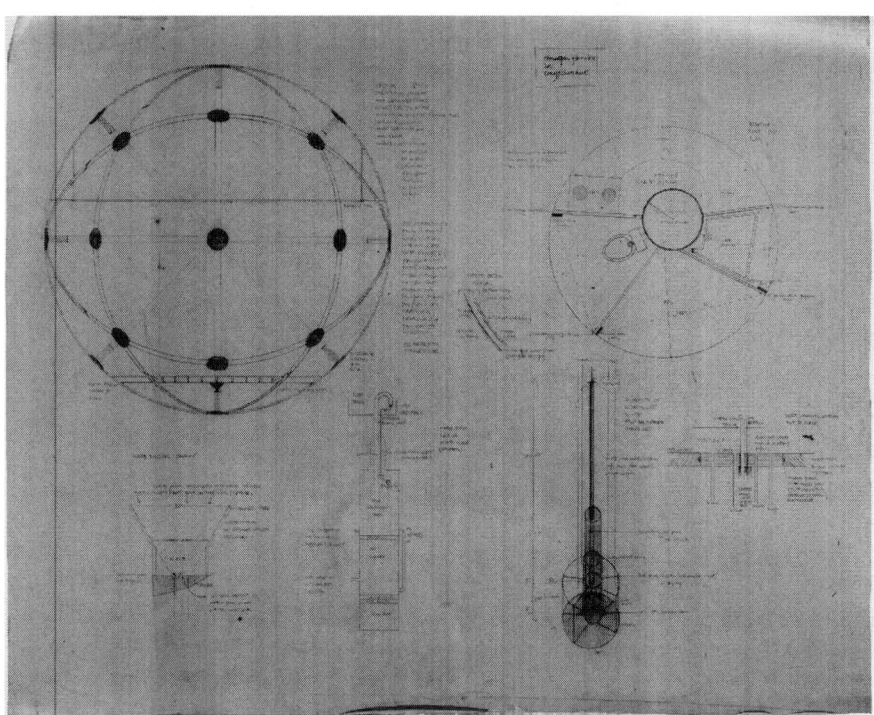

Year 2, Cray Fishery_1. Courtesy of the Glanville Archive, University of Vienna.

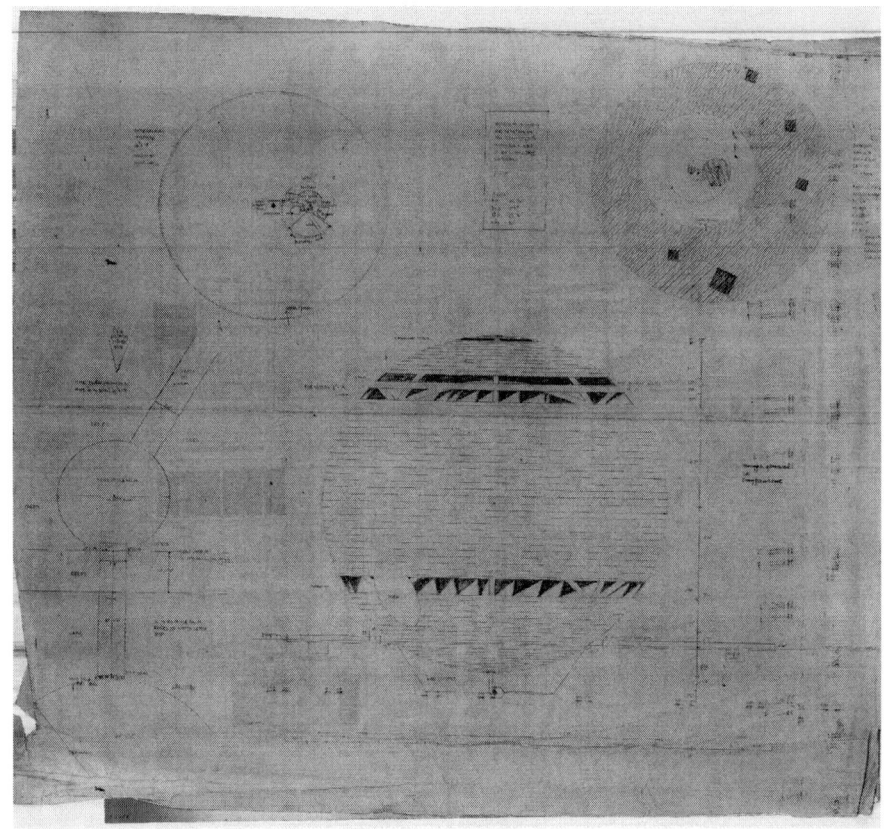

Year 2, Cray Fishery_2. Courtesy of the Glanville Archive, University of Vienna.

Ranulph: Well, no. I just questioned. And I did it in the first. We were asked to design a hut for a cray-fishery in Sweden, and I'm saying there's no sight, nothing. I designed a floating house, because they had to go round the pond. But my floating house consisted of a vast sphere with a floatation collar on it. With flexible joints so it could roll. With waves or wind, it had everything in there concentrically organized. The floors would rotate as well. You had to go and empty the cray pots every four hours so I needed something that would wake people up. You slept on hammocks, up on a rail. The thing tilted, they would swing around and wake you up. It was described by the year master, he said, "This is actually the only serious answer—that he has made a boat and it is waking you up." I was very well known. I did incredible presentations as well. I did one where I pretended to be completely dumb. And the jurors did nothing. I could hardly speak and I read some fake Koans. At the beginning of the next semester, I asked if I could be on the jury because I always complained about how bad they were—we got the same jurors and then I was back there twice as bright. I completely messed up all of the roles.

Bill: You talked about how you always wanted your students to work to the highest level so this was the beginning for you. You did it on your own.

Ranulph: The art of being good is to expect a lot of yourself. Not to be lazy—to be rigorous. There's a wonderful quote—this is not quite accurate, but I got it from an obituary for an architect named Denys Lasdun. He said, "Our task is not to give our customer what he wants, but to give him what he didn't know he wanted until we gave it to him."[40] I think this captures the arrogance of the architect that is actually a deep humility, which is deeply concerned with a visionary aim to make people's lives better than they could have imagined. I would always try to do that—I would look at a brief and rewrite it.

Bill: I was reading your paper last night because you were questioning this whole idea of *form follows function*.

Ranulph: Yeah but that's easy to do now. When I was a student that was different and it was the basis of the scientization of design and design methods. The people who were involved with it almost all gave up. People like Chris Jones, just became mystics. People like Horst Rittel, invented wicked problems.

Bill: Do you have a short phrase that you say to people like "form follows function?"

Ranulph: *Function follows form*. What I have is just a slide where I usually just move the words across each other very slowly.

Bill: What is a wicked problem?

Ranulph: That comes from Horst Rittel. He ended up at Berkeley but he was the mathematician at the Ulm School of Design in Germany, which is where design methods got developed. This sort of attempt to make design scientific. He did set theory…

Bill: What are design methods?

Ranulph: After the Second World War, there was a feeling that science and its active agent, technology, would solve everything. I grew up in that, and I remember the world's first nuclear power station going online. A lot of stuff to do with flight and space flight. We all of the sudden developed all sorts of wonderful things. The man who was the chief engineer was in fact my cousin-in-law. Well my cousin was married to his daughter. We got the tilting trains, the levitating trains on mag rails from Shanghai. Well, we were

---

40. See *The Guardian* 28/11/2003, "Not what the client thought he wanted but what he never could imagined existed."

doing that in the sixties but we let all these things go. So there was this view that we needed to science-ize everything. That science would do everything. I remember seeing the news reels at the time that said, "unlimited cheap electricity," and of course this wasn't true. There was a feeling that design and architecture were inferior because they didn't have proper scientific theory. And there was an attempt to scientize the things without realizing that design wasn't—science is about recording what exits, supposedly—and design is about making new. It's about changing the world, always. The smallest thing, even a new matchstick, it changes the world, whereas science attempts not to change it.

Bill: Although cybernetics is a science—

Ranulph: No it's not. It's not a science and Wiener didn't call it a science. It has been misnamed. Wiener's book does not mention science in there. But because of that era people wanted to make it a science. Gordon described it as the art and science of the defensible metaphor. The point is that it was presented as a science and I think Wiener made a massive mistake in publishing *Cybernetics*[41] before *Human Use of Human Beings*.[42] Because *Human Use of Human Beings* puts it there as a way of acting, a sort of art, a sort of humanity, a sort of philosophy. It's full of concern for social consequences, improvements in society. In contrast, the cybernetics book turned it into technology—control engineering. I think he made a terrible mistake. There was an attempt to scientize design, to make it systematic, methodical, rational, blah blah blah. And it produced supposedly all that stuff that Braun, the German company, produced, so that is all seen as the product of stuff. I, as a student was taught that this was the way to design. I never understood anything about design until I left the AA and started teaching. Others were more clever and learned to design when they were students. I used this scientific approach, I designed buildings from set theory, I didn't understand the element of delight: "What's delight got to do with it?" So I was trapped in this and it was only much later that being trapped by it I realized that I had completely missed the point. The peculiar thing was that I was writing music which was about delight—what it sounded like. I had an artistic practice that was doing this.

Bill: You had this German approach.

Ranulph: It was what was being taught at the AA.

Bill: Was Luhmann?

Ranulph: No, Luhmann has nothing to do with design.

---

41. Wiener (1961).
42. Wiener (1950).

Bill: I know but it's interesting that there's this German sensibility about philosophy and publishing that did work.

Ranulph: Well yes, German and Austrian friends were astonished although I come from this English, pragmatic, experimental, realist tradition, and although I'll do things like experiment clearly philosophically, I still have parts of their idealist tradition. They always found my work very confusing. Horst Rittel left Ulm closed and it was quite clear his program wasn't going to work. Very important people in the world of design had been there—Thomas Maldonado—but this stuff wasn't going to work. Okay, people left Ulm, it closed down, Rittel ended up at Berkeley, and he started talking about things called *wicked problems*. Which were problems that you can't solve. He had ten reasons which have now been worked down to six, but essentially they're full of self-contradiction. They're incompletely described. He introduced this to West Churchman[43] and gave it the name "wicked problems."[44] The opposite would be a tame problem. And with this, there was some realization that the tidiness of science was not likely to succeed. We had a moment where we invested vast amounts into the technology to make it work without checking all the funny things happening around the edge—forcing the atomic bomb, radar, et cetera. Then we began to discover that it wasn't really quite like we had thought.

Bill: What was the new approach that was born?

Ranulph: Well it was the realization that systemization only helps to a certain degree. Of course there are things that can be systematized and of course functional means functional. Let's get that right; it's not against the things that can be done mathematically in terms of proximity and in sets. It is that which has been left out, it's the thing that really matters and it's the thing that is really hard. It doesn't come out of a description of things operating together—equations. What it comes out of is a process of ignorance, a process of not knowing that what we are doing is we are playing around and eventually we come up with something that is wonderful—and this thing, this solution, tells us what the problem was. Instead of the problem defining the solution like in science and technology, designers and artists have solutions that define what the problem was. We don't know what the problem was until after we've found the solution. This process is not articulate, and it is not step by step, or predictable—it's about making a mark in a bit of paper and being surprised. It's not quite what I'd thought I'd drawn, something else pops up. It affords me something extra. That's what Rittel did, he introduced the idea of the wicked problem, and

---

43. Churchman (1967).
44. Rittel & Webber (1973/1984).

there are lots of people who want everything to be more scientific, they try to define problems out of the equation and what they end up doing is defining them out. Defining them out is not the same as facing them. There's a lot of low-grade cheating that goes on in design research. Most people are quite thuggish and impose themselves on others.

A Conversation with Ourselves Through Pencil and Paper

Bill: Do you have a title for this new kind of architecture that arose?
Ranulph: It's not a new type of architecture. It's how architects always worked, it's how artists work. You make a mark with a pencil on a piece of paper, or a bit of code on a screen. It is this conversational business that goes on.
Bill: *Conversational* would be the term here.
Ranulph: Yes, if you ask architects they'll say yes of course, that's what we do or if you ask artists. There's no question about this. You say to people, here is what is at the center of design. What we do is we have a conversation with ourselves through paper and pencil. They say: "Yes, I make a mark, and it surprises and inspires me. Then, I make another mark. All the time it's giving me ideas I didn't originally have."
Bill: But yours would be a little different from Pask who has the conversation with the client.
Ranulph: That's a completely different thing. Pask was talking about a way to better communicate between architects and clients. Most architectural briefs are very boring—most clients have no practice in writing briefs and they're using, they say, "I want a bathroom." You say, "What's a bathroom for?" "Oh, I want a toilet." "What's a toilet for?" "Oh, I go there to shit and piss." "Okay, what's the room you go to when you're really upset and want to be alone and lock the door?" "The toilet." "So the toilet isn't where you go to piss and shit, it's a place of refuge?" "I'd never thought of that!" "So you want it to be a place where you can lock yourself in and cry—lock the world out." Everyone accepts this as a proper behavior. Well what about reading… We describe it in terms of some basic biological functions but we don't actually describe it in terms of the wishes that we invent for them, the wonderful things we can do with them. It's fantastic. Gordon was giving a chance for people to explore. I tell you something, you give it back and I tell you where I think you've got it wrong. But he did also refer to the drawings on a piece of paper as sort of a conversational thing. I don't think he got much beyond that. I would describe conversations with myself—on paper—as the key, central outing of designing or of all artists. It's very rare

that you get someone who just thinks of something and does it, and it's right. What they do is they think of something, they go to do it, and it becomes different.

Bill: Did you do a project—a kind of Dada approach to this design?

Ranulph: No, I think it would be incredibly difficult to do because you would need to be present everywhere for such a long time. For me, the proof of what I say is the response the design has given it. That for me is enough.

Student Projects in Music

Bill: Oh yes, and you were telling me about your student work.

Ranulph: Yes as a student I did all these concerts, and they basically followed these two quite different paths. A time in English and even American avant-garde music there were two very different schools. One which is probably better known, so indeterminacy and sort of randomness and the idea of the liberation of sound—the idea that humans impose their will on sound which meant that we only listen to certain types of sound and certain sequences and we shouldn't do this. Whether Cage's stuff has any musical value or not, that thinking clearly has an enormous value. And the notion of sound material that is available for musical use, ways of using it has changed massively. Although we can put it down to technology starting with the wire recorder I think in the end, it's not so much the tech, it's the conceiving of music which suffers from the imposition of human will on it. In England, we had a school that was thought of as being sort of *Cagian* but as I looked back on it much later I came to the conclusion that it wasn't really Cagian at all. It was the English tradition of folk music, of music which people took part in, where everyone could try to play, stuff which you'd play in the pubs. Cardew was one example. He was the one who looked most Cagian, and was a graphic designer so that Cardew's scores had remarkable visual qualities. But there were a lot of other people like John White and Howard Skempton, and those people have ended up in a rather small little backwater. In fact a lot of that stuff happened here in Portsmouth at the Art College. So, there was something called the Portsmouth Symphonia which would play classics.

Bill: And Scratch Orchestra?

Ranulph: The Scratch Orchestra was different,

Bill: And the Penguin Café Orchestra?

Ranulph: No, the Penguin Café Orchestra was a different sort of thing. It was nice and entertaining. The Scratch Orchestra were people who just showed up and played so it was like the Portsmouth Symphonia. And there was another one, I forget what Cardew and John Tilbury called themselves, not when they played as a group but they played together. I did things with Cardew, premiers of pieces. At the same time, I did stuff with the Serialists and the Total Controllers and I did electronic music, which is pretty determined. I did stuff with Peter Maxwell Davies and Harrison Birtwistle in the "Fires of London."

Bill: Do you have those on tape, or albums?

Ranulph: No, I'm sure I have bits somewhere, there is a version of one of Maxwell's pieces where I did an electronic part which he later replaced with a spoken part. But this stuff went out on the BBC. I was also writing my own stuff, I had an improvisation group and I think we were the first group in the world to improvise with live electronics. We did things like play at the electric garden, which became the perfume garden, which was sort of hippie place around 1967. Cornelius Cardew and I did La Monte Young's "Studies in the Bowed Gong." But we would record and mix and process sound during the course of the evening to make a piece out of it. And Stewart Beetee and his dancers would improvise to this.

Bill: It seems analogous to the architectural technique of finding the thing through the process.

Ranulph: Well in a sense it was essential and self-referential, but I wasn't doing cybernetics in those days. But it shows I was using that kind of thinking—and that Gordon Pask was right—I knew a lot about it I just didn't know that I knew.

Bill: Do you know the "Punkt" Festival? They do this live remix stuff.

Ranulph: Well yeah, this was a long time ago that I did this, I think I can claim precedence. [laughing] It's really because the AA was such a wonderful place. Everyone was busy transcending architectural limits—space stations and walking cities what—so…

Bill: It'd be cool to incorporate one of these pieces.

Ranulph: One of them is on the Internet. I should find them and process them, somewhere on those tapes. And there's a lot of stuff that exists as manuscript—and never was played.[45] So there was that sort of stuff and that meant that I was very interested in sound and sound in building. For instance I was asked to provide some music to go with a futurist exhibition and I did. One of the tricks was, I couldn't afford mixers and all this clever stuff, but by using really cheap and bad speakers, they all had very different frequency responses. So I could produce a 12 track tape by having really poor loud speakers that were all different. Sound, I think we were using sound in the mid 1960's, the first sound artists, as a virtual reality. We were

---

45. Ertl, Korn, & Müller (2016).

composing new sound spaces using this stuff. Long before the stuff got visual—virtual reality done in sound. Just stereo was the beginning of VR. When it came time for me to do my final two years at the AA I went back and did the sort of compulsory building stuff, and produced quite an interesting little shop. It was a small corner block in Oxford street and it was supposed to be a sort of shop and I turned it into a small version of the Finnish franchise shop, and we had to have a foreign restaurant there so I put in something called a "Moo Cow." There were two or three of those because the most foreign food you could get in England was English food. Downstairs there was a disco which I think my tutors were right—there wasn't close to enough headroom but… It was a strange building it had lifts on the outside and things which hadn't been seen but in Finland. I did this history thesis, which one was required to do, in which I…what I had discovered about Finnish architecture and language was that if you look at the way Finnish language is structured—the grammatical structure and the mindset—they're mostly to do with being at places, not to do with movement. They're about location, so they're locative rather than transitive. The way that Finnish works as a language, unlike European languages, I found that these were paralleled in the foci. For instance, Finnish words are designed so that you can integrally add on parts and they have ways in which they change their shape in order to bind this extra bit in. It's not like German where you just bang another word on there, there's some actual integration that goes on and it reshapes the words. There are rules as well, only certain types of vowels can go together. I found a lot of stuff that was about being Finnish words and concepts being available for expansion in a particularly integrated way. And I showed how the farm houses were built like this and if you understand how, you could see a very common thread where the houses in the west seem to have extremely large exterior courtyards, but very different from the ones in the east which are great monolithic blocks. But there was a way of interpreting them that showed that the way which these were organized and put together and the way the language is organized and put together showed similarities. I wasn't saying that the language made the architecture or the architecture made the language, but if you put the two together you can find something that they had in common, and that they could both be seen as reflecting this deeper commonality.

Bill: It seems like this analogy ties back to the Messiaen-Klee[46] relationality you articulated.

---

46. Glanville (1966).

Ranulph: Yeah, it does, and a lot of what I've done is finding things that can be held between two so that they become one.
Bill: It's making a whole but its making a new whole because it's finding some kind of relationship.

Finnish Architecture Photos, Ranulph Glanville. Courtesy of the Glanville Archive, University of Vienna. (Collage by Seaman)

Ranulph: It's looking at something deeper that's behind it. And a lot of people disagree with me but I would have to say that almost all of those who disagree never bothered to learn Finnish.

Bill: Was this a thesis?

Ranulph: It was my AA history thesis. Vastly too long, with something like 500 colour slides. So I toured around Finland and found farmhouses and buildings, back when Aartje and I were there this year we went to one of my favorites. It was wonderful to see it again, I hadn't found it for forty-five years or more. Then I went back to the AA and I did this building, and this history essay, and we got to the final year where you do what you want. I was interested in ways in which electronics could affect the environment so I came up with a way—with electronic music and mixers. So, there were several projects. One was called "The Suitcase Secretary" which is the portable computer. It was a suitcase which had a cassette recorder in it, you could make notes on, a diary, an address book, a notebook, so you took this around with you and you could dictate into it. It's just a portable office with recording capabilities.

Bill: Were you friends with anyone from the English pop-art scene, Allen Jones? Or—

Ranulph: No no, I didn't know any of them. I did know some of the kinetic people like Takis. But no. I didn't know artists or rock stars or Eno or Ferry, I only knew serious people doing serious composure. It was a different world. But I did know a few people going into the avant-garde jazz scene. Jazz has never really been my sort of music but I knew Roland Kirk and Albert Alier, and I knew Ornette Coleman, and John Coltrane.

Bill: Did you know Michael Snow at all?

Ranulph: No no. He did that wonderful slow zoom in *Wavelength*. In a sense it was preceded by Antonioni's *Blow Up*. It was full of AA students, all the wackily dressed characters were students from the year after mine. The years before and after mine were both interesting, mine was not. The year after was people who were sort of media people. The year before were people who were wild architects like John Frazer—really wonderful stuff. Many of them have become absolutely first rate academics. Almost all of them became teachers because that was the way to keep the vision alive because it was clear we wouldn't be building the things we were dreaming of.

Bill: Can you talk about John Frazer a little more, I think he is very interesting.

Ranulph: Well John always said he was interested in the computer because he wanted to be lazy. He just wanted something that would do it for him. He was given a project by a school he'd been to, which was to cover an area to provide a sort of covered play area, and he came up with this idea of a space frame. It was the idea that you could do this if you planted a seed and let it grow, so he planted a seed and let it grow on a computer in planned section and elevation and he produced this roof—I don't think it was ever built but of course later on things that he produced were built. John has always seen computer aided design, not as the engineers see it, which is an automated drafting, but as something that would save him the bother of having to design anything. He did some very remarkable things with other people of course. Cedric Price's generator, which came within a gnat's breath of being built, was a holiday resort for a paper company. What was interesting about it was that it was a lot of boxes that could be moved, and they moved to break habits. John provided the computing for that, and the habits, and then there was a wonderful thing he did with Walter Segal who was a Swiss émigré architect who lived in England a long time, and he was interested in participatory architecture—architecture that was designed by individual ordinary people. John provided this really extraordinary thing which was a plug board, and he had bits of furniture which included walls and all sorts of things. These were all modular so you could pick them up and plug them into this plug board and build a 3d model of a house. There was a WC, so you could plug it in and put a room around it, and by plugging it in it was automatically drawn, and all of the services were automatically calculated and provided. Not by drawing but by making little models.

Bill: Who was this?

Ranulph: The architect was Walter Segal. He is the only architect who's got a street named after him in London. John did this in Northern Ireland with his wife Julia Conner. And he's done stuff like that since. He did a futurism thing for this might be Groningen in Holland which was a whole set of very arbitrary ways of imaging your understanding of the city, which people could play with in order to develop future plans for the city of Groningen, and they could do it on pub tables in bars. What I'm never quite sure of is what happened to all of this stuff when he left for China and became head of school in Hong Kong.

Bill: Yeah, I met Frazer in Hong Kong and got his book there. Did you collaborate with him?

Ranulph: No he was always a year ahead of me. He and I knew each other but I think we both had a sort of charismatic effect on each other so I was always very impressed by John's charisma and his character and his intellect. I think he was probably much the same with me so. It's just, we behaved in different ways.

Bill: Yeah, technology didn't take the spotlight in your work. You were much more interested in ideas.

Ranulph: Also, I became less and less interested in being the head of school. There was a time when I would've loved to be the head of school—I applied, second year student—but I stopped being interested when I realized it wasn't an educational position but administrative. That's a little bit about John. But I don't know what he's done in the last, say, fifteen years. Whenever I've seen him lecture he's talked about old stuff. I think he got sidetracked by the administration and going to other places, QUT [Queensland University of Technology] and Brisbane.

Ranulph: Let's get back to this project: There was the suitcase secretary—and then as well, there was a listening wall. What I proposed was that in a terrace, there should be microphones placed in the walls, all the way through all of the houses—and these would randomly select and record what was going on and mix it separately on a set of tape recorders that would gradually build a history so things would fade away. The purpose of it was to stop husbands beating up their wives—I didn't announce that but that was the secret purpose of it—it was not to produce this rather tiresome thing, an endless drone of noise but rather it was that you would hear, and then begin to fear that you might be overheard, and others would hear you, screaming.

The Environment Modified by Sound

Bill: It's kind of a second-order cybernetic—the observer observing themselves in a way they wouldn't have—

Ranulph: I just see it as a bit of therapy. And this was not made, it was un-makeable. The thing about using analog, it's essentially un-makeable, because you just don't have the control for it. Mark Burry from RMIT [Royal Melbourne Institute of Technology] did eventually make some sort of version of this in Birmingham which is an enormous wall driven by pistons, that adjusted its shape It was done by him and Mark Goulthorpe of dECOi [Mark Goulthorpe's architectural design Atlier]—there was a Bartlett student who, forty years later, produced this masterwork and I went up to him and said yeah I did that in 1971.

Bill: You could probably do that with a Mac now.

Ranulph: Yeah these things are not difficult now. There were a couple of other projects as well.

Bill: These are more in the line of architectonic installation in a way.

Ranulph: Yes, they were uses of electronics to modify the environment. There was another interesting project from when I was in my third year. We were given a monastery to design which seemed completely stupid but later

I realized it was quite a good project. As I decided I'd do this instead of the project I was supposed to do so I did a plug in monastery—the plug in because that was the thing, and the plug in monastery was sort of Zen, and when you got to the top of the thing—having been put through all of these obedience classes—your food would appear and disappear randomly, the WC and the rest of it—you would be ejected out into the landscape on a little helium balloon with a little engine on it, carried by the wind and eventually you would crash, potentially being killed but surely being reborn.

Bill: Was Lebbeus Woods someone you were interested in?

Ranulph: No, well Lebbeus was my age or younger, but no he was just at University of Illinois, he did those rather dark drawings and he's continued doing them, that's what he does. We were doing great—not visually great, bits of hardware but completely—bits of life. We wanted to change everything. There were two other projects that I had…oh yes, I did operas. One with a guy dressed in white, in a white box, with every light in the theater on him, sort of croaking this nonsense text that was repeated and repeated. But of course he got more and more confused so it wasn't repeated, it was stuck through ring lines and modulators, belted out at the audience. Something totally incomprehensible and of course that was the purpose. A very Beckett image.

Bill: Do you like Robert Wilson's work at all?

Ranulph: I don't mind Robert Wilson, I'm very fond of Robert Ashley's *Wolfman* [musical work], I remember a project with Cage and Ashley in London, sort of standing at the urinals next to Paul McCartney and having a piss before he went in to get some culture. *Wolfman* was truly shocking. But yes, *Private Parts*, there were several of those operas—one of them was commissioned for the opening of channel 4 on TV.

Bill: He did a very nice series of conversations with other composers.

Ranulph: I think it got overlooked in a way. I think of the cult minimalists who were just dire. Steve Reich is much better. I like Reich but I have little time for Phillip Glass. To take advantage of a quite clever rock star who's going through a really bad time, to make a hit record—it's unforgivable.

Bill: Was this the Eno stuff?

Ranulph: No no, it was Bowie—*Low*.

Bill: *Low* was the album that made me want to make music. It was so different than anything else, and I was very moved by that. Then that period with Eno, *Before and After Science* and *Another Green World*, I liked all of that stuff.

[pause]

Bill: After this whole series of projects, what was the next stage? You started to work at an architecture firm?

Ranulph: No no. Then I went off and did a PhD in cybernetics.

Bill: This was the pre-Paskian moment?

Ranulph: Yes, I met Gordon in 1967 when I was doing the supermarket project. I met him again in the four years that followed. Two of those years I worked, one of them in a school as an art master, one of them in a couple of architect's offices that we spoke about yesterday. After I had gone through my final two years at the AA, then Gordon said, I've got you a scholarship to do a PhD, and I said oh, okay.

Bill: Maybe we should talk a bit about this wonderful book project that you did?

Ranulph: Yes, one moment. Let's order some food.

Bill: I have a paper, it's called "Interflow Architectures" and it's a list—you don't like lists—but I'll send it to you.

Ranulph: What I don't like is lists that are constructed as arguments. But there, one of the great damages done by PowerPoint, the belief that if you make a list of things sequentially people will understand it but that's not the case.

Trojan Horses. Courtesy of the Glanville Archive, University of Vienna.

Ranulph: So *Trojan Horses*[47]
Bill: Yes, where'd the title come from?
Ranulph: Where is the conference…
Bill: Oh. Troy. [laughing]
Ranulph: And it's a nice offering that's full of subversion. What I think is a great sadness about conferences and particularly, I've tried to put on these conversational conferences so that people are actually learning things not just being told things. The big problem with conferences is that people go away and they don't do the bit where they do the learning. What I'd hoped for all of the conferences was to have a meeting at the end in which we would schedule some sort of outcome that would capitalize on what had been learned. What happened after Troy was that people stayed and people were really quite serious, someone made a movie in two days, and we planned this book. There were these three or four groups that people fell into and I refused to do things. People were endlessly asking me to decide, and I would just go and get the next round of coffee. It was a book which was supposed to develop what we found in the conference and take it further. I suppose that it began to have the qualities of cybernetics of cybernetics in the sense of Heinz Von Foerster, that it had drawings, it was witty, it had comments on things, and it was very multifaceted. And there were four editors I suppose. First the picture section had interviews of people, done by Claudia Westermann and Aartje. Then there was a section—the German group—which was Tom Fischer and Candy Herr, and Tim Jachna who's a third generation Polish American who actually speaks very good German and spoke German with those two. Then there was the Methods Group. That was also edited by Michael Hohl and a Danish guy. But Michael is 110% reliable, if he says he'll do something, he does it. He put together the method bit. Then I did the bit on the cybernetics of cybernetics competitions—which we had already found a way to publish but I thought it'd be nice to have it in that book and publish a paper—and I wrote an introduction for it. And that's how it came about. I think that one or two people had too much power in it, and whose ignorance of cybernetics comes through by their choice of only one person to have as a hero. That's very typical of the ASC. There are people who you know: "I've got my hero—Herbert Brün or whatever—"I don't need anyone else," and they don't have the courtesy or the intelligence to look wider. Of course, they also make sure that everything is very backward looking and Herbert Brün has been dead for a long time, and if your hero is going to be Herbert, then you're not looking forward. What you're doing is this endless supporting of

---

47. Glanville (2013).

the dead old white men. It's an enormous weakness of the ASC, and of the whole field.

Aartje: It was very difficult to get a balance and not to be overpowered. To say, "No, I also want other people in there." It was extremely difficult to say no to anything because I was married to Ranulph, it always seemed like I was defending that part of the cake and that's not it because for me that was a different way of thinking and—to me—pretty important.

Bill: At the conference in Washington, there was this big timeline, and of course all of these heroes and when they died—and I was saying, "Why don't you add when these new members come in, instead of going out—

Aartje: But they are not really interested in that.

Bill: But their books are still incredibly rich, just to play devil's advocate.

Ranulph: They are, but those guys played up to cult status. And they made no room and no connections and they handed over to no one else. They didn't care about the subject, they cared about themselves. You look at it and it's really quite disgraceful. The only one of those guys who really pushed students was Gordon, but he wasn't very good because Gordon didn't know how to promote his students. I mean, I got Gordon all of his jobs at the end of his life. Gordon didn't know how to do it.

edition echoraum

Bill: This is an interesting place to talk about the other book projects you published through echoraum? You did how many books for them? Three?

Ranulph: I did eight. I can't remember all of them, there are three that I did which were my own work but before that there were books of collections of things like, *Trojan Horses*, *The Importance of Being Ernst*, *Gordon Pask Philosopher Mechanic*, and I can't remember but I think there were a couple of others. I contributed to Karl and Albert's one—*An Unfinished Revolution*. I think I'm the only person who contributed two pieces, two chapters to it.

Bill: Now you have this wonderful *Black B∞x* series.

Ranulph: Well that's the three, yes. The third volume was the one published first because it was easiest to do. That was everything I had written about in *Cybernetics & Human Knowing* coming up to the end of 2009. Everything stops at 2009. There's actually about 5 years of production since. Then the next one was the abstract theory stuff, five sections, and the problem with it was that they were in such a rush for production reasons and finance reasons to get the thing out that they didn't execute the corrections—there are lots of typos. And they sent it to me while I was in Toronto along with *Trojan*

Horses with two days to proof read. It got churned out and in some respects it is actually more confusing. Then the new one is coming out now, and that's being much more carefully done. Each of them, there was an introduction to the whole thing, and an acknowledgement, there was an intro in each volume, there was an introduction to each section, there was an abstract provided for every paper, even those that didn't have them, and of course, every paper was re-edited not in the sense that I rewrote stuff—unless it was really confusing, nor that I corrected silly errors that I'd made—but I read it through two or three times and did correct it and, God it was an awful job. It was horrible.
Bill: When does that come out?
Ranulph: It will be published when we go to Vienna, because I'm going to sign all *The Black B∞x* in black, invisible ink, on the inside, so that no one can see it, signed and numbered so that they're all unique as limited edition art work.
Bill: How much is it going to cost?
Ranulph: I don't know, what I would do is write to Karl and Albert and say I'd like one. But please buy it. There's a guy in Belgium who I've worked with for a long time—a mathematician—who is astonished to see this stuff. He and I only meet through design research. He wants to put together a book to go with the three books, essays by other people on what it is I've been doing. Which is nice. Of course I won't see any of these.
Bill: I think the idea is to have a positive mindset: You're going to see all of the books and the other book and we're going to continue to meet for the next five years. I told you the story about my grandmother, right? She was fifty something and she went in and they told her "You have terminal cancer" and she saw seven different specialists, they all said this. She had a radical mastectomy, and a bit of her lip was removed, and she went on to live to 99 years of age. I try to keep the other perspective.
Ranulph: Cheers.
Bill: And I hope what we're doing now becomes a book someday.
Ranulph: What I'd like to do is go out and write some new music. I have two pieces that need to be written.
Bill: I think the music has played a much bigger part than I ever realized.
Ranulph: I think I have, it's like I have a mathematicians sensibility, but not the skills, and I think I have a musicians sensibility and not the skills. Writing music is for me an extraordinary process and an extraordinarily slow process. I reckon it takes a vast amount of time to get a minute's worth of music.

Aartje: But with all of the books coming out, you haven't had time for the music.

Ranulph: Right. I still have Joe's piece to do. This is a piece for my original best American friend who died of cancer a number of years ago. He was ten years older than me and was a small Sicilian who joined the army. He did lots of military music and was in Alaska and I told him my first Rastus joke, all sorts of things like that. He was always in search of an ethic for himself and became a Jew. This Italian Sicilian becomes a Jew, and he was at the University of Illinois with Lejaren Hiller and Herbert Brün. Then he went to Iowa and found himself a wonderful woman to marry. They were part of the Iowa contemporary music band that came over here and Joe was just wonderful. He was a very deep thinker, very intelligent. He and I got on extraordinarily well and I wanted to write a piece on piano but I never did it before he gave up playing piano and then died.

Bill: What's his last name?

Ranulph: Dechario. He had a buddy from the town he came from, Girard, Kansas, whose father taught music. Tom was an outstanding trumpeter and we'd been staying with Tom in New York, a house out in the countryside in Pennsylvania. Girard, Kansas, was the home of the little red books—which were these slim primers, short books on Aristotle, everything to do with knowledge—and there was this great social publishing venture there. So, you know, large parts of the state were very socialist like me—all of those town community things. Barbara asked me what I'd like as a memento of Joe and I got the little books. Carefully wrapped in some newspaper, very special things. I'd wanted to go to Girard but I never made it.

Portsmouth

Bill: For many years you've been hoping to be doing all of this. Do you want to talk a little bit about your PhD supervision?

Ranulph: Well, just up that road there, opposite of here in Portsmouth there's a man called Geoffrey Broadbent who was a very important figure in promoting things like design methods in the 1970's when they were big things. He was head of the architecture school here and when the AA didn't really have room for me anymore, Geoffrey employed me. He says that he employed me as an irritant, (and I've become very fond of Geoffrey in a funny way—we go out and have lunch now) and so at the end of the 1970's I came here, Portsmouth, UK, the last thing I'd ever have expected. I used to commute from London. First of all I needed a job, secondly this was regarded in the public sector—in public education—as being a school most like the AA, and thirdly it wasn't so far from London that I would completely lose touch with my son. I was here for 19 years and as times changed I discovered all sorts of things here. I discovered that academics could get paid to go to conferences. No one ever told me those things. I had no idea. I did things here: I ran first year, I did the dissertations. Mostly people here were jealous and small minded. What they used to say was well, he's always off on some free trip—and that was because they had never been to a conference and they didn't know what you did and what hard work it was. They didn't understand how I would compose trips out of this and tiny bits of money. They always got angry that I was still working at the AA. Obviously, I wasn't working properly there, but it was a condition of my working here—they made it a condition that I went on working at other places, and the associate head of school, hearing all these complaints about me, carried out an informal survey which showed that I was in more days a week—earlier and left later—than any other member of staff. All of these people with all of their complaints—but it just meant that I didn't have friends on the staff. I was very much alone. I worked with people, some of them were very nice, but mostly they had no interest in me. They didn't like what I did, what they wanted was ordinary projects—design a small house—whereas I'd ask people to design a corridor, or design a fairground. As time went on and conditions got worse and worse in public education, I was treated worse and worse and I was eventually kicked out of architecture and put as an associate in a new media center where they had absolutely no interest in me at all and I basically supposed that I could've stayed doing that so I got a salary—I got some money to go to conferences—but at that time I bumped into *this* woman and she wanted to move over here, and when she moved over I was offered early retirement and she said "Take it, I can earn enough to look after us, you're absolutely miserable doing this job," so I did take it.

Bill: And you've been working just as hard ever since.

Professor of Odd Jobs

Ranulph: I had a good friend at the time and he had started working with the University of Hull, and he went over to help people develop their research. I thought, "I could do that!" I went off one summer to see my brother who was quadriplegic and brain damaged and my mother and I would go take him on holiday, but my mother was beginning to suffer very severe memory lapses and dementia so Aartje came with me to help. We took my brother on this trip, I sent Aartje back and said this time I can go on. This time I don't have to go back. I paid for me to go and see my aunt in New Zealand. Then I managed to get myself a visiting fellowship in the Chinese Hong Kong University, and I managed to stop off in Melbourne where Leon van Schaik and I had started teaching, so we'd been working together for a long time. I supervised his PhD, the first I ever did, and I had the strange experience of meeting people and I wasn't allowed to ask them anything. The condition for being a doctoral and masters examiner was that you hadn't met the person, and it was deemed that if I hadn't asked a question, but had only answered questions, I hadn't met them. I had this really weird weekend. Then sometime a bit later Leon said, "Come here for three weeks, come and do some examining, and what would you like to do for three weeks? Would you edit our journal?" and I said, "Absolutely not." I said, "I'll come and sit in a room and talk to anyone who wants to talk," and I did. At the end of it, Leon and I had breakfast and he said "This has been sensational, this has been the most extraordinary success. People needed someone to talk to, we never realized this. They like coming to talk to you about their work, or things at home, whatever it is, they just liked it. We'll do it again." I said, "No, we won't." But we agreed on a minimum twice a year. I started going to do things in Melbourne and working as a casual visitor, who was sort of an ombudsman, a supervisor, and I'd be given project—asked to sort things out—so I really was a professor of odd jobs.

Bill: That's another nice title!

Ranulph: I would meet people, after a bit I'd say, "Can I come to your school?" They'd say yes. Having seen me do what I do with reviews, which is generally quite good, I tend to sit there silently with my eyes closed, then as Mark Briar says, then he would ask exactly the right question—what everyone was trying to say. I did that, and gradually more people trusted it. At the moment if I were able to go back, I would have posts at Monash at Melbourne University, and at Newcastle who've recently been my strongest supporters. Then the Bartlett asked me if I would go and do a pre-doctoral program and would I look after these real problem students because I had experience. Gradually little things built up, these people in Belgium started talking, I'd go over to the head of school about research and gradually as things developed I'd get pulled in to do teaching, little projects, workshops, supervising. It was strange.

Bill: That's probably about five?

Ranulph: Probably, there's the Royal College of Art. I'm actually a professor at the University of Buenos Aires in Argentina.

Bill: Seven.

Ranulph: I'm a professor at the Catholic University of Leuven.
Bill: Eight.
Ranulph: Imperial College.
Bill: Nine.
Ranulph: Yeah well, people say you know, how do you get this? How do you do all this? And I say well, once you've got one, it's just easy. It's the first one that's hard—the one that you believe is an achievement, that is really difficult to get and it is hard to become a professor. But once you become one—once they offered me a job at the RCA, started talking about it "Would you like to be this?" I said, "Yes, I don't do any jobs less than professor." I said, "Yes, I think I've done this enough, I know how to do this." It's been an extraordinary life because it's been full of new challenges and interest and fantastic people and wonderful places to go to, and always a more wonderful place to come home to.
Bill: And you've had some amazing students like Ted Krueger, Ben Sweeting, Thomas Fischer.
Ranulph: Tom was not officially mine, but he was actually mine. None of his advisors put up with him so they dumped him on me. And he is to go back to Hong Kong. A lot of people in Australia though.
Aartje: Do you know Craig?
Ranulph: Craig Bremner.
Aartje: But it's very funny because for both men, in a sense, you have very much formed what they think and what they do.
Ranulph: I do ask them all at the beginning, "Are you prepared to completely change your mind about everything you believe?" If they say yes, then we can go. I'm not saying they will, but they might.
Bill: Ted had the seeds of that way of thinking, but he's completely clear about how he goes about it.
Aartje: It took a long time. Lots of visits.
Ranulph: Ten years, we would meet once a year in Melbourne, once at RPI [Rensselaer Polytechnic Institute], and once here.
Bill: My CAIIA [Center for Advanced Inquiry in Interactive Art] stint was three and a half years. You know, it forever changed me. I'd began reading in a way that I had never done before. I had one college class—I didn't read in high school, I was a TV kid—but CAIIA got me critically reading, mining ideas. Mike [Punt] really forced me to ask, "What is your question?" "What are you trying to say?" "How do you frame that?" So he actually helped me in how I approach the PhD students I now work with.
Aartje: But it is about changing your way of thinking, and being able to make a logical argument, being able to go back and knit together and see if

there are connections. That makes an enormous difference in the way you think. One of the things you often say, you don't use an idea unless you've introduced it. So you have to do that.

Bill: We've talked about this multi-perspective approach to knowledge production—what are the methodologies you use to draw on? How do you make that work? I know that we talked about analogy and metaphor.

Ranulph: I don't use the word *methodology*. Methodology is an inflationary term used to describe something quite ordinary, which is a method—make it sound grand. Methodology should be about the generalized study of methods. But, I don't think there is a method. The more I do it, the more I realize that I'm completely intuitive and I say what comes to my mind at the time and I have learned that saying that is alright. If I propose a project to a student that's just come to me with something they should do, it's the right project. I just say, "For whatever reason, whatever way, however it is. I happen to be someone who is able to listen to you and to respond in a way that can produce something." Craig Bremner, said that I was the best listener he'd ever come across. Which is not natural at all. He said, "What you do is you present something, and then you find everything that I say back in response is somehow appropriate and sensitive and helps clarify and take you along." I suppose that's good. It is a gift. I had a letter from one of his doctoral students who I met twice and the first time he bussed down from Sydney to Canberra and Craig said, "You can meet Ranulph," and Ian said "What does he do?" and Craig said "You know what goes on. You speak, he listens, he tells you what you were going to say." This guy sat there and started off. I'd looked at his PhD draft and it was crap, as they are. It's okay. And after forty minutes and not being through the introduction I said look, I've already read this, I don't need you plodding through this. It's not very good. Let's find out what's really going on, and I then complained about certain things that he'd done. And he left absolutely furious, and sat on a bus steaming, and he got home and sat down a little bit quietly and he said, "That guy just said everything you need to hear. He wasn't wrong. Shut up. Act." Very often, I think that I don't understand or hear people at all so I listen to this stuff and say what am I supposed to be making up, I can't hear any of these words, what's going on. And then somehow, I always think when I've heard someone speak that I can't remember what they said, but if you get me to begin I'll give an extraordinarily accurate summary of what they said. There's a little ha-ha there. I see the ha-ha and think I'll fall into it, I just have to step over it.

Bill: One person who comes to mind is Bernard Scott. This idea of teach back. He seems like a very important thinker to you. There are the big heroes of course that you don't like to think about as heroes, but Bernard seems like a true hero in an inverted way.

Ranulph: Bernard's history is very interesting. He studied psychology, and was on a sandwich course and was placed with Gordon Pask. And he thought Gordon was fantastic, and indeed he was. He went back and worked in Gordon's lab and became director of the lab, he was the person who designed the experiments, and he provided a lot of the thinking that goes into conversation theory. If Pask had been just a little bit more honest, it would've had a few other names in there, including me.

Bill: I think the paper really shows it.

Ranulph: But an enormous amount of the shaping of conversation theory came from his students and his employees, and Bernard was that. And Bernard went on working with Pask. Eventually they ran out of steam with each other and Bernard went back to education rather than education and technology, became an education officer improving things, and then became a school teacher for a bit. He sort of disappeared from all of this stuff. He got his PhD and vanished. Then Bernard went through all sorts of things.

Bill: [to Aartje] The funny thing about Ranulph is that when we first met, I thought he hated me. I don't know why, your style sometimes is very terse, right to the point and finally at one conference I said "Ranulph, do you hate me?" He said, "On the contrary."

Ranulph: People often think that. I really don't believe that you should hide these sorts of criticisms behind pleasantries. I think a lot of that happens and I think it's very damaging. It leads to people believing something is okay, when it's not. I'd prefer to just go straight in, boots kicking, and do it. And just be very, very direct. Occasionally I won't, but occasionally when I'm not, I will be savagely indirect. I have something I call walking on water—when you walk along and keep asking questions, they give answers that are more and more ridiculous. Until you get them somewhere where they're contradicting themselves—you let go of their hand and you walk back on the water and they sink—and they look at you with despair and horror. I'm terrible at therapy. I'm just not good at it.

Bill: As you gave the definition of *cybernetics*; it's the balance of complements [in a circularity].

Ranulph: You've agreed to take my list of publications and break it down into elements and put it into a spreadsheet, which is fantastic. It's such a saving not to have to do that. It's up to 2014.

Bill: Alright, I think we've covered a lot of the things we wanted to cover. Can you think of anything we missed?

Ranulph: There was something you wanted to talk about.

Bill: Ah, your project, but we can do that later.
Ranulph: We've talked about supervisors who've had a big influence on people. The point of supervising is to get people to a place where they can stand on their own and not sink in the muck. Anyhow, one of the things that was interesting about Gordon was that Gordon was very inspiring as a person and a thinker, but he wasn't terribly inspiring as a supervisor. He did give me books to read, he read things and talked about it with me, but that wasn't the way he opened up worlds. He opened up worlds because he was working on this stuff, came in and talked about it—
Bill: Well this is the idea of learning through practice.
Ranulph: And Bernard used to come sometimes and I was absolutely intimidated by him. We both thought the other was awe inspiring and we had no time for it. It was strange. I think Gordon's heart was always in the right place although sometimes it got very dislodged. His own problems were very considerable. But as a supervisor, he wasn't particularly good. If I wrote something, he'd read it, take me to tea and talk about it. And that's good enough, but the sort of person who can get in there, tweak your thinking—he didn't do that. I did that. A large part of it was coming across rigor as something—there's a sort of obsessiveness in rigor. Rather than seeing rigor as being something which kills you. For many people, you say *rigor* and it's like saying *test* to an artist. They just don't want to know about it.
Bill: *Rigor-mortis*
Ranulph: Yes, exactly. And that's not so, we're just asking for higher standards. All artists, when they do work, do it to extraordinarily rigorous standards. Even Damian Hearst.

The Glanville Name

Bill: Your namesake…
Ranulph: You understand the people who came here in 1066 weren't French. They were Vikings. It was the Norsemen—hence Normandy. My family lived there at that time. And they lived in a little village, which was called Glanville. Which was named after them, or they were named after…and Ranulph de Glanville, one of us, came across in 1066. We had come across before as well. Aartje and I were invited over to the little town where the Vikings first landed 1100 years ago to take part in the 1100th celebration of the arrival of the Vikings. We strolled around this town and went to a special service in French and English, at the Cathedral. We actually stayed in the bed and breakfast in Glanville, which of course

excited the owner of the bed and breakfast. We were given a reception by the mayor of Glanville, and all sorts of things—it was really fun. Totally silly! What a ridiculous thing to do. But you know, someone has got to do it. People don't get to do this so why shouldn't I?

Bill: Have you gone back there again?

Ranulph: Well you know I have been to Glanville several times, because there was a very nice couple. They sort of chased me. She moved to Glanville and she was dealing with La Patrimoine, which means culture. Tairan… It is the whole thing about these sorts of things, very French. She was doing a history, restoring the church, and all of those sorts of good things. She looked up Ranulph de Glanville and discovered there was another one. Not the famous one she knew of, who wasn't the one who came over with William the Conqueror, the one who wrote the *Book of Common Law*. He set up the English legal system and the English Constitution. Ran the country when the king was abroad crusading. She was very excited to find another Ranulph Glanville. Asked if I was any relative, and clearly fixed us up. We had this interview comparing the old one and the new one and this sort of thing. We went over and visited them. It was all part of the same daft business. And I say you do it because it is totally silly! But, why not? But you probably don't do it again. I doubt if I will be around in another 1100 years.

Bill: Unless you come back…

Ranulph: Yep.

A Contrary Friendship

Ranulph: My old buddy in Sydney died just a couple of weeks ago. His name was Richard Wingate. He was a medical doctor. He and I used to drink prodigiously here in Portsmouth. He discovered how not to drink rather before I did. I always thought he was much older than me…but I was older than him so it was a big surprise. I used to have the run of his house in Cronulla, and the run of his house out in the countryside, and Mudgee. I had this home away from home. He was very generous, and he was very very private. To stay for months like I did was very unusual. We were very unalike each other. He was really a Tory. And converted to Catholicism—a believer, and all of this sort of stuff… And his walls were covered with guns and knives and pictures of ships. And he was very military…and he thought it was terribly sad that we lost the navy, because it was nice to have a navy. And I am sort of a floppy little liberal with slightly socialist leanings.

Bill: How did you meet him?

Ranulph: Drinking. A pub around here. Yeah yeah. It was just extraordinary. It was chalk and cheese but it was very good chalk and cheese. I did the eulogy for his funeral, which I showed him a year and a half ago. What's the point of writing a eulogy if the person never hears it?
Bill: What did he say?
Ranulph: Oh, he was pleased. I did revamp it a little bit before he died and I didn't show him that because...he lived a very difficult life with lots of very big problems and there was a sort of way, despite of being very privileged, he was also very disadvantaged. And he didn't like this being said and I don't think you can appreciate the extraordinariness he achieved unless you include this.

Architecture Student at the AA, Assorted Architecture Offices, Then Abroad to Finland, and the Introduction to Cybernetic Concepts, and on to the PHD—Object Theory...

Bill: So March 1968 you worked in an architecture office. Was this at Edgington?
Ranulph: Edgington, Spink, and Hyne.
Bill: And what kind of work were you doing there?
Ranulph: Well, an architecture education in the UK is not like it is in the States. It is a full time, five year program. It's not a masters you do with something else. In those days it was broken into two halves. Three years and then you would go out and work in an office for a year. And then two years and then you would go out and work in an office for another year and then take the final exams. You would then get a diploma or masters or whatever, at the end of the five years. One of which was in practice so it is more like six. So I spent my time at the AA, not doing architecture. I spent three years putting on concerts, and every holiday I would have to catch up in one way or another with the architecture. I have to say I left the AA knowing no more about architecture then when I went there. Which was nothing. [laughter]
Bill: Did you have a supervisor?
Ranulph: We had wonderful teachers—very famous architects. My tutors included Richard Rogers; I knew Norman Foster because he and Richard had an office together. I used to go there... I knew people like David Bernstein. A lot of very good architects. Tutors. And I was very lucky. The AA at that time was very open and liberal. They took the position and they said this at the end of my first term: "You'll never be an architect but we are good for you." And they were right. But you could do that in the 60s.

Bill: More like learning through osmosis when they showed their own work or just through talking?
Ranulph: It was the 60s. Things were really very different. We were all much too arrogant. Interested in our completely new way of doing things.
Bill: And were the concerts chronicled anywhere. Did you keep an archive?
Ranulph: This was before the time when people kept archives.
Bill: You just did it.
Ranulph: I might be able to find some old programs…one or two were recorded. Anyhow. It came to be the end of three years and I hadn't passed the intermediate exam, and quite rightly. I went out and spent the 1st year teaching Art, at a boys school. And then I went and worked in architect's offices, one of which was Edgington, Spink, and Hyne. They were reasonably competent. Very ordinary. They did little jobs for the Queen in Windsor Castle, because they were in Windsor. This sort of thing.
Bill: Were you building models?
Ranulph: No, I was asked to have a go at designing a village hall, which I did. But not the way they wanted. It was OK. What was interesting was to go there knowing nothing about how to make a building work, and ending up, because I'd been to the AA where they taught you how to learn, being really very good, very quickly. And then I went to Finland for six months. And worked in a chain of offices run by young people. They would each house me for a month and I would work on something.

Finding Analogies

Bill: Is this different from the time when you were going there to study the relationship between Finnish language and its architecture.
Ranulph: No—this is around the same time.
Bill: Because that sounds like a very unique kind of approach to architecture.
Ranulph: Well, yes and no.
Bill: But it didn't come from any of them did it?
Ranulph: No, it came out of me. It brought great astonishment to the Finns. I was actually the first architect who had shown a general interest in trying to make some sense of the whole of the Finnish farmhouse. People used to go around quoting me.
Bill: And is the material from that period available? Is there a finished written work?
Ranulph: Of course there was an AA history thesis. I suppose I have a copy of it somewhere. But there were about 500 illustrations and they were color

slides. I think those have fallen out of order. But I think the thing which was interesting was... I mean most of us have thought that language somehow limits us. We are trapped in language and this sort of thing. Benjamin Lee Whorf for instance did all of this stuff about Fire and the Hopi Indians. The notion that language limits and shapes was not new. It was not popular at the time, people being post-Chomsky structuralists. But it wasn't all together astonishing. For me, when I began at the AA we were asked to do an analysis of a painting, looking at its proportions, and I pooh-poohed this idea and the art master, Paul Oliver, who was a great authority on the blues, said to me, "Well OK, if you think it is rubbish, go away and prove it." So I did, I went away and I measured and I found out that it wasn't rubbish at all. What happened was related to the way that Klee had adjusted his grid. This was very much like the way Messiaen put together his rhythms. I said, "Look – a similarity." I was finding analogies like that all of the time. That was what I did.

Bill: And that is very much in the spirit of cybernetics.

Ranulph: Cybernetics.

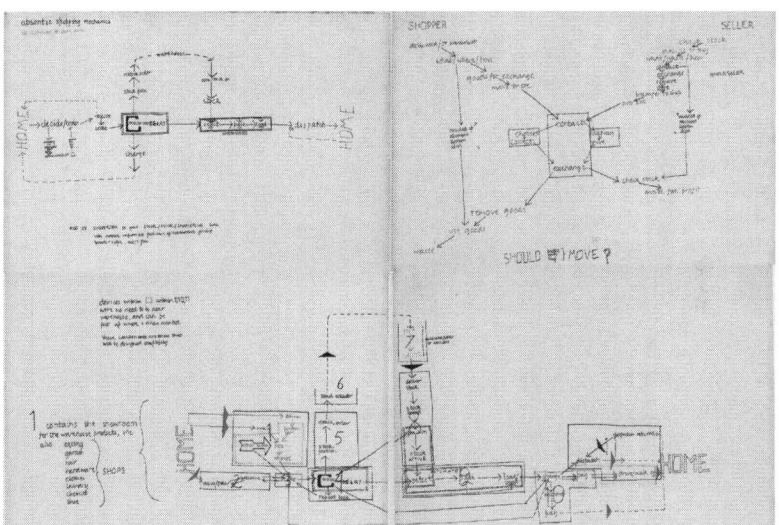

Internet Shopping, Ranulph Glanville. Courtesy of The Glanville Archive, University of Vienna.

Bill: And Gordon Pask...the use of analogy and metaphor, the making of models, exploring relationality...

Ranulph: I did this. Well it is a world of relations rather than things. Cybernetics. But I did this stuff, and then went back to the AA and did my final two years.

Bill: But now in this kind of expanded sense.

Ranulph: Yes. I was a bit more architectural, because the concepts had vanished while I was away, and I didn't have the energy to start it all up again. I had done that and it is hard work. I took the architecture rather more seriously having worked with architect's offices. And then at the end of it people asked me to go and teach. My final design project was something which was so far ahead of the times…it was actually Internet shopping, in 1970!

Bill: And you do have drawings of that.

Ranulph: Somewhere I have some diagrams.

Bill: Just to go back for a second. You already had this beautiful set of relations between image/painting—music, language—architecture. They were all beginning to talk to each other, maybe? Or did you keep the worlds separate a bit?

Ranulph: I tend to keep things separate until they insist they're together. I am very dubious about—synesthesia doesn't work for me, to start off with. I am not synesthetic. I think there is an awful lot of nonsense talked concerning the relationship between sound and vision. I think about those sorts of completely arbitrary and bizarre parallels and connections. It is fine that people want to do this for themselves but don't pretend that these are somehow or other essential and important things. What they are about is how we think. And that is what I am interested in. That is why I teach.

Bill: So when did the second-order cybernetics come in? The way I think of it is, that you were already doing it in the way that you thought about things. For me that is how it was. I was already making circular causal systems, not knowing the first thing about Pask.

Ranulph: Well Bernard Scott, who you probably know, always remarks that everyone else needs to study to do second-order cybernetics, whereas I just do it. I think there is an element of that and I think maybe it has to do with a more artistic and less scientific background.

Bill: With Otto Rössler, his endophysics[48] is basically a second-order cybernetic idea… where one makes a model system where the observer and the world are together, and then you become a super-observer of this. [And in my own virtual world generating system.]

Ranulph: Gordon was presenting that in 1970. I was sent by a fellow student to meet Gordon. He just said you need to meet this cybernetician, you are doing this project, and you really have to talk to this guy.

---

48. Rössler (1998).

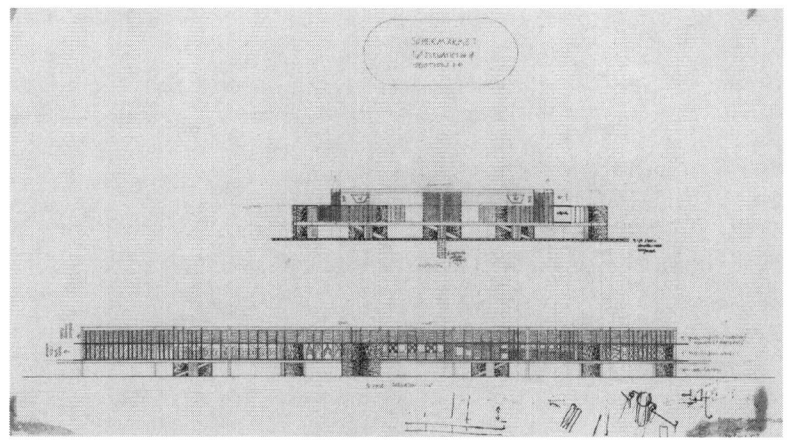

Brent Shopping Market. Courtesy of The Glanville Archive, University of Vienna.

Bill: This is the pre-Internet shopping project?
Ranulph: Yes—so the supermarket project. I went to see Gordon, and explain myself, my image of it is, that it was incredibly incompetent. I didn't know what I was saying. I was lost. And I took about three hours and Gordon summarized it with acute clarity in three minutes.
Bill: Nice… He got it. It clicked.
Ranulph: And I said to myself, it is not just that this guy is very clever, it is also that there is something in this approach and I want this. I would go back and see Gordon ever so often. And when I finished at the AA when I scared all of the jurors out of the room, so they never looked at my final projects, so they couldn't fail it… Gordon came up to me and said "Oh, I've got you this scholarship; you are doing a PhD with me." I didn't know anyone had a PhD. What in, I said. "Cybernetics." I said, "I don't know anything about it." Pask replied, "Yes you do, you don't know you do, but you do."
Bill: That's fantastic. The fact that he approached you is so exciting.
Ranulph: It was good for his career. He had a scholarship, and he didn't have anyone to fill the scholarship. And I could fill the scholarship. And that is why I got it. If it hadn't been for that, he might very well not have bothered. I have no idea.
Bill: That is a pessimistic reading I think [laughing].
Ranulph: Maybe…maybe. You know, one just doesn't know. Anyhow. This wasn't second-order cybernetics.
Bill: This was still first-order cybernetics.
Ranulph: But it was changing. And Gordon would do his seminars, and he would start talking about this new cybernetics. How it worked and why. I

just found the ideas fascinating. I was supposedly building circuits that would do things in the environment, but actually this was much more interesting. I became a bit of a philosopher.

Bill: It was the network of cybernetic concepts…and I read the review that you wrote of his book, the little book, *Cybernetics*[49]…and that it is one of his best books…was that your introduction to the set of concepts in that book…?

Ranulph: No, I didn't read his stuff, I read Ashby. I had that book. I read that book when I was asked to write something about him and I thought, I bet you, if you look you will see all of the themes from Pask already in Ashby. That is a really interesting thing to do. No one has done that.

Bill: All of the themes in that book…

Ranulph: Yes.

Bill: Yeah—I think they are. I think he touches many many of the same concepts. I like that book very much.

Ranulph: It was edited by Heinz. That is why it is so precise.

Bill: With an introduction by McCulloch, which is also nice. There is a beautiful quote … I can't paraphrase now but it is really lovely. [In the Preface to Pask's *An Approach to Cybernetics* Warren S. McCulloch stated:

> This book is not for the engineer content with hardware, nor for the biologist uneasy outside his specialty; for it depicts that miscegenation of Art and Science which begets inanimate objects that behave like living systems. They regulate themselves and survive: They adapt and they compute: They invent. They co-operate and they compete. Naturally they evolve rapidly.[50]]

Bill: You came in. Did you meet Ashby?

Ranulph: No, I just missed Ashby. Anyhow, this was the time that cybernetics was changing. The relationship between models and Objects, "Is the model of a model an Object?" "Is the Object the model of the model?" So, all of that was coming up—"What is the role of the observer?" And Margaret Mead in 1968 said we need to look at cybernetics consistently, all of this stuff. Around 1973, I was quite interested in this way of thinking. I was doing stuff, writing notes, et cetera.

Bill: The book you were reading was Ashby's *Introduction to Cybernetics*. It is a very nice book.

Ranulph: Well Ashby is far more advanced than people recognize. What I most liked in there was the stuff about the black box. That was fantastic.

---

49. Pask (1961).
50. Pask (1961, p. 9).

Bill: The notion of the black box was also very influential for you. Around that period when you were reading the Ashby, what else were you reading?
Ranulph: I spent a year reading 3 or 4 books. One was *Plans and the Structure of Behavior,* one was *Introduction to Cybernetics*.
Bill: Who wrote *Plans*?
Ranulph: George A. Miller, Eugene Galanter, and Karl H. Pribram—a bunch of psychologists. Slightly cognitivist psychologists. At the time in the States it was very rare.
Bill: Very new.
Ranulph: I can't remember the others, but I read these things in a very determined way. I would do every exercise at the end of the chapter. If I made one mistake I would do all of the exercises again.
Bill: This was a self-imposed—
Ranulph: I got the basics pretty right.
Bill: It took me a few years. In the beginning I had a rough idea. It wasn't until reading across all of these different people's ideas that I got it.
Ranulph: See the great thing about being involved in cybernetics—oh, around 1970—was that you were inventing the thing. In a sense you didn't have to read anything. It was all going on. You lived it. You discussed it with him. Heinz came by, Maturana came by...
Bill: It was a very heady time!
Ranulph: I had a really easy way in to all of this. And others didn't have an easy way in because they couldn't grab it. But they potentially had the easy way in because they were there. It was all being taught.
Bill: Now your Object theory, that came much later?
Ranulph: No. That was my PhD! 1975 was when it was examined. I wrote it in the summer of 1974. 15 hours. What I am doing at the moment is—I write this column for *Cybernetics & Human Knowing*. And I am currently writing what will probably be the last column I write. And it is about wholes and parts, and it is about society and the individual, and these sorts of things. And of course it refers to Objects. Then of course, I shall re-write the Objects stuff...without reference to the old stuff...just as a new, you know 40 years later, probably much more clearly, much more tersely, with context, why it matters to cybernetics, what the situation was around the time it was written. I'll do that.

The Architecture of Delight

Bill: Somehow the wicked problems remind me a little bit of the Zen Koans.

Ranulph: Yeah, they have that same sort of structure of being without resolution. Or—if you're going to resolve them you have to find a completely new standpoint. A new way of conceiving it I suppose.

Bill: A couple of times we've come to this term *dis-architecture and anti-architecture.* Are those terms that you use or was that me projecting onto things?

Ranulph: Well, I regarded myself as against… The more I've had to do with architecture, the more I've come to admire it and like it.

Bill: It's almost like this circle. What was dis-architecture becomes architecture. If you use it in this manner

Ranulph: I grew up without any understanding of the quality of architecture and what I've learned through my involvement in education and design is about Vitruvius's third quality that design must have, which is delight! That delight, is the difficult one and also the important one. Any old engineer can give me the other stuff but delight is difficult. I am interested in that side of architecture. I had this wonderful time when I was going to RMIT, when I met a lot of good architects who were sort of obsessed with the artsy side of architecture and with the experience of architecture and getting their Ph.D.'s through practice. I began to see what they were getting at and began to feel I could actually do this, although I had done little bits of conversion and that sort of thing. They really hadn't had much quality—but the stuff we have going on in this house for instance at the moment, I think has plenty of quality. That comes from having been invited in to that program which helped to set it up and all the rest of it. And having practiced for a good time, and having learnt to appreciate things that I had missed before.

Bill: The architecture of delight!

Ranulph: Yeah, I mean, it's a good word. It's not beauty, it's not pretty, it's not nice, it's not lovely, it's not any of those words. It's a word that suggests some sort of elevation.

Bill: Are there any architects you know of who explore this concept?

Ranulph: I think all good architects do. It's pretty trivial. I would have to admit an enormous admiration for Alvar Aalto who was, I think, the best of all the modernist architects.

Toppila Oy, Sulphate Pulp Mill, Oulu Finland, 1930-1933.
Photo: Aino Aalto, Alvar Aalto Museum, 1931.

Alvar Aalto's church in Riola. Courtesy of the Alvar Aalto Museum.

Maison Louis Carré, Bazoches-sur-Guyonne, France, 1956-1959, 1961.
Photo: Martti Kapanen, Alvar Aalto Museum, 1980s.

Paimio Sanatorium, Paimio Finland, 1928-1932.
Photo: Maija Holma, Alvar Aalto Museum, 2014.

Continental Philosophy

Bill: Where were we? Ah yes, Continental philosophy. Do you want to say anything about that?

Ranulph: I mean I just, I don't know why I should be un-English in the way that I think. And if I am, well then that's the way I think about the world and that makes me friends with a whole lot of people who aren't English then that's alright by me. I think that English pragmatic realism is rather silly. I'm sorry but if you don't have room for the observer then you don't have room for anything else. The failure to recognize the contribution of at least human intelligence in building the world is daft. And the pretense that somehow or another we can know without knowing everything about a—just come on guys, stop it. It's really good for the children to pantomime Christmas but it just isn't really good for thinking. Most of these guys are too clever to have been so stupid, but of course it's because they are so clever that they are so stupid.

Bill: Are there certain philosophers that you prefer, or reference?

Ranulph: No, I'm very poor at reading philosophy. I am told that I should go back to people like, oh all those German-mystic-romanticists of the late 1800s. I do occasionally read articles by…

Bill: Like Goethe?

Ranulph: I think yeah that's nice stuff but no I don't think there is very much in common between Goethe and me—what I'd sort of like is if that…

[Both Bill and Ranulph get up and move inside.]

[Poem recitation of "Jabberwocky" by heart by Ranulph]

> 'Twas brillig, and the slithy toves
> Did gyre and gimble in the wabe:
> All mimsy were the borogoves,
> And the mome raths outgrabe…[51]

Ranulph: What I think is wonderful about Jabberwocky is that he does manage to make an entire verse almost without a single word you know, yet it's completely convincing.

Bill: Yeah, you feel it. This notion of a felt meaning. Which might be what you're getting at with the architecture of delight.

Ranulph: Yeah, I mean the rational deals with some things but it doesn't deal with everything. The great mistake has been for us to insist on the rational and only the rational and it's one of the great mistakes of

---

51. Carroll (1871).

science—to say, "I wear really limiting spectacles, and what I see is everything. And If you don't see in my terms, then you're not seeing this. Not proper." Which is what people like Richard Dawkins have said.

Bill: What term would you use as the alternative to the rational?

Ranulph: Well you don't need an alternative to the rational, there's nothing wrong with the rational. Where the word *rational* is appropriate. Where it's sensitive and helpful, there's nothing wrong with the irrational, or tacit knowledge for instance.

Bill: I tried to write a curriculum with a tacit knowledge idea, which was fun.

Ranulph: The unspeakable curriculum.

Bill: Or poetic maybe is the other word.

Ranulph: I just think, I told you that new definition for cybernetics yesterday by balancing complements [think complementarity], and rational and tacit is one of those complements. They complement each other. You have something, rationale is great. Rational with the irrational is greater. If you want to talk about wholes, then you should talk about the two together, not as parts. The whole is the combined two. For instance, when we were in Bolton, we had acting and understanding. The point is not understanding or acting—which goes first? It is that you need both of them for each other and for a richer view. So I've come to think that because I happen to be writing about wholes at the moment. First of all I think there are only wholes. There are no parts. What we do is we compose wholes into the roles of being parts of other wholes. This is one of the reasons that *compose* is a very good word. Better than construct. We make these things, and then we make a new whole, which is these things together and you have to work out how we do that. One way we maintain those wholes is through conversation between people. But the way we make it ourselves, if my theory of Objects has some value, one of the things it does is it explains how we make new wholes and how they then continue to be whole for us in our universe.

Bill: I like *holistic* thinking, I think a number of cyberneticists have thought this way.

Ranulph: Systems theory is generally, or very often described with the whole being greater than the sum of its parts. Kurt Lewin said it much more accurately: "A whole is *different* to the sum of its parts." He was one of the original social scientists involved in general systems theory. He was also one of the founders of the SGSR, The Society for General Systems, which is now the International System Science Society.

Bill: Do you go to those conferences?

Ranulph: I have been going most years. I am sort of treated as a member. I've been building this loose coalition between the ISSS and the ASC, against quite a lot of opposition because there's a terrible history of the way the ISSS behaved toward the ASC so we have reason to be very suspicious of them. On the other hand the people who are running it now are a long way away from that generation and most of them are friends of mine. What I have discovered in the world of cybernetics is that I can do things that other people haven't been able to do because I know people and I can get things published. It comes down to treating other people as worthwhile and being polite to them, and remembering them. We get published in journals, we get conference proceedings. They were never published, they hadn't been published for years and years and they bitch about things not being fair, but there are two points of publication: One is that we all know we can read them (which we all know is rubbish), and the second is that it is what gives credibility to academics. They have to be published in recognized journals. And we've done well at that. We've done better than any other society in the world.

Bill: It [*Kybernetes*] is a very strict journal. I've never rewritten a paper so many times.

Ranulph: Well I want to say something about the way you write papers—

Bill: Yeah, badly.

Ranulph: Well, there's too much of the heavy scientist in third person and some of the reviewers got really irate about that last time. If instead of saying "Seaman this" and "Seaman that" you should play that down. One of my papers was written with a couple of linguists from Hong Kong Polytechnic is about the language in which people write up science. You'd be interested to know that those who use the first person most (of classic scientists) are physicists. It looks less like you are trying to impose this great Seaman way of doing things. Which is not what you're trying to do. I would suggest, if you are trying to write something up for cybernetics, you write in a more literal way and refer to yourself less in third person: "In my work I've done this, this and this…"

Bill: Yeah, I've done a huge amount of research for this one so I'm excited to write it up.

Ranulph: Well what I have to find out is whether someone, anyone, will take this on in the future because I'm not going to. This is last time I'm going to have something to do with it. And to be honest I'm going to be lazy about it so other people are going to do the work and I'm going to direct it.

Bill: I wonder if Ted would take it on? He's such a good thinker and writer and well read.

Ranulph: I don't know what Ted's been doing and I haven't had a good chat with Ted in ages.

Getting Tired

Ranulph: I was really enjoying not having to say a whole lot.
Bill: We did talk quite a bit today.
Aartje: I think that Ranulph becomes interested in the thing he wants to write about in his papers—we start including the area. You'll [speaking to Ranulph] do that when you're working, you'll do it at home. A lot of the time you're exploring the things that are being thought about so that it's always that pull back into the practice…

Listening and Impossible Problems

Bill: Some of the stuff that came to me had to do with the different methods that you use in mentoring, in terms of listening and getting people to articulate/express their ideas, and they're very different methods. So far I wrote down Zen Koan approach, logic (which is like Object theory), ideas, emptiness (so you put everything back on the person and they have to figure it out for themselves), humor as well as play. Are there any others?
Ranulph: I think we find all of those as intuitive. Yeah, I once wrote a paper on impossible problems. What I like to do is put people in impossible situations. Before I knew about wicked problems I was putting people in situations, and giving them strategies to complete this, continue this, contrast this. I think I also use the furious argument as well. Furious in the sense of fire, cold-eyed and merciless.
Bill: I think that that's the logic one, where you get directly to the point… This is how it is et cetera.
Ranulph: The logical argument about one thing, you know, developing something is different than the scalpel. I do that a bit. I do this especially with people who I think are bogus. I can be so unkind. Someone said that I'm one of the kindest people in the world except when I'm not. I don't suffer people who are manipulative and dishonest, and in the end I just won't speak to them. I will ignore them by just staring at them. It's when I've had enough of their attempts to pretend. People who say, "Well you have to respect Gordon." "Well, why?" "Because his work is great." "Well, so is mine, do you know any of mine?" "No, but Gordon's is great." "Well that's because you're too lazy to find out about anyone." "Well you owe it all to Gordon." "No, Gordon owes as much to me as I do to him." It's stupid.
Bill: I think it's unfortunate that that happens.

Ranulph: It's happened to a number of people. Varela had this problem with Maturana. I have an insider's version of that because after I wrote a review of the book containing an interview done by Bernhard Poerksen. Maturana wrote to me completely out of the blue. He hadn't answered when I asked him to contribute to a Festschrift for Heinz, but completely out of the blue he wrote me this long, long email saying what a wonderful review it was and how fair it was, and how carefully thought out it was, and what great work I did, and how Francisco had been my friend and how he'd worked with Francisco and where these misunderstandings had come from, what they were. It was a very open and honest message and since then we've been good friends. But I think the biggest problem of all is that it is very difficult for the gifted student of the gifted professor to appear equal and gifted and it's one of the reasons I've withdrawn. When I'm finished with a student I go away. I very rarely talk to Ted, and it's not because I don't love the man dearly. It's because he knows how much of what he did was in part insighted [derived from insight provided by Ranulph], and he needs time and space to be Ted Krueger, not my student. After a while some of them come back and one or two of them never quite leave. I think embarrassing is a good strategy.

Bill: I was thinking also about the humor, and about how many things are about listening. Listening to patterns and mimicking in a playful way.

Ranulph: Well one of my great books, not that I would take it to a deserted island or anything, is called *The Lost Art of Listening*.[52] By someone whose name I think is Michael Nichols. There used to be a good summary someone wrote on the Internet, it's out of print but it may have gone back into print. It's a fantastic book. It seconds Joseph Beuys's comment that there is no conversation if there is no listener. It's the listener who turns it into a conversation, not the speaker. This is something that we persistently misunderstand.

Bill: I think Heinz also said that.

Ranulph: Well Heinz said, many people have said this, he called it the *hermeneutic principle*, which is that meaning lies with the listener. What Beuys said is that there is no conversation without a listener. It is the listener who turns an utterance into a conversation. That's quite different to saying that the meaning lies with the listener.

Bill: A conversation is just two sided listening.

Ranulph: Yes, at least. The least you can do is a conversation with yourself, as both speaker and listener. That's what artists do.

---

52. Nichols (2011).

Bill: I was thinking I should interview myself, but what should I ask myself? You ask yourself things all the time, like why am I doing this, or why should I do this. Also, to come back to the books, you named about six so you still have about four more to go.
Ranulph: We'll have to look at the list. I thought we got up to ten?
Bill: We had the three about Beckett,
Ranulph: Glasersfeld, Pask, *Trojan Horses*.
Bill: Which Pask one?
Ranulph: The one I edited with Karl. It's called *Philosopher Mechanic*.
Bill: You mentioned the cyberneticist you were moved by, who died, Richard Jung.
Ranulph: Well there have been a lot of them that have had wonderful minds. I think what I like is not so much the cybernetics but more the properties of the thinking.
Bill: I was thinking that really, your architecture is about the architecture of thought. That the idea is much more important. I was beginning to think that you inverted the "form follows function" to "function follows form" but then I was thinking is it instead "function follows form, form follows idea."
Ranulph: I think it's this thing of form and content. It's dangerous to separate the idea out. You get people doing illustrations instead of art. There's an enormous amount of illustration being done, and being called art, and it's not. And there is an enormous amount of research being done, called project-based research but it's actually illustrative theory. I'm very weary of illustration.
Bill: But I like the idea of an idea that is form and content. It's embodied by both form and content.
Ranulph: Well what I'm writing at the moment is about wholes, and you could say ideas are wholes. What I do is I say all ideas, are wholes. There cannot be a part of a whole, but every thought you have is a thought in itself. Then to make parts, you assemble wholes together, in relation to each other.
Bill: That's what leads me to say you're kind of an architect of ideas that are form and content. Then you add these wholes and continue to create a greater whole. It's much more driven by ideas than by an aesthetic or by design methodologies, it has to be this thing that you find in the process. Is that correct?
Ranulph: Well I think I'm very interested in ideas. I think that when I build architecture, I think that little extension in the back of our house is an entirely unpredictable object. It came about through years of gentle thinking, and it has been knocked around in the process of building. It was never completely drawn. It only exists as itself and it's somehow, the form is

the content. There is no story. There is no point. It just is. It completely accommodates the content. And the content completely shapes the form. But I don't think that's what you're talking about when you talk about the architecture of ideas. It's always difficult talking to people who've done architecture when you start using it as something other than to create buildings or spatial thinking. You know, computer architecture.
Bill: Well I always think of architecture as a pun, and I like all of the branches.
Ranulph: Well I think of architecture as not being a building, that it's quite separate. The reason it's interesting to work in architecture as a teacher is because you can examine almost anything.
Bill: *Unpredictable Objects* is also a nice title.
Ranulph: Well *Predicate Objects* is also good because it's grammatical. Well if you parse a sentence or a phrase in English it goes subject – predicate. Predicate consists as a minima of the verb and the object. In a sense, objects are predicate. So *Predicate Objects* are just statements of the obvious. It's a definition.
Bill: So *Unpredictable Predicate Objects*?
Ranulph: Predicate and predict are just the same.
Bill: The next question: In the same way that there are books and influential people, are there architectural moments that stand out in your life? For example, is this an architectural moment that you find interesting? [Talking about the old water pump while driving by.]
Ranulph: Well yes there are fantastic things.
Bill: Is it that the architecture embodies a concept for you?
Ranulph: No, I mean I have one architecture book I like, which is called *Gothic Architecture and Scholasticism* which is by Erwin Panofsky,[53] and people say it's totally wrong [scholasticism] and I'm in no position to judge that, but what I am in position to judge is the parallel with what he does which is just breathtaking. This way of looking at a gothic building and finding in it, all these scholastic things, and seeing this system of thinking literally embodied in the building and that was for educational reasons. I just think that the construction of that analogy is wonderful. For me, that's my favorite architectural theory book. It's because it brings the two things together and makes sense of them and of course it reflects the mind of the person who does it. It's not a truth, it is someone looking at two things and seeing something similar.
Bill: I think you love these deep analogies that are each reinforcing the other.

---

53. Panofsky (2005).

Ranulph: And it's that same with Finnish language and Finnish architecture.

Bill: That and the "Klee and Messiaen." Are there other ones that you think are interesting, or that you use sometimes when you're teaching?

Ranulph: I endlessly come up with analogies, they're intuitive things I invent. I very rarely remember exactly what I've said. I don't remember my lectures either.

[Brief discussion about the boats and water nearby.]

Bill: I think that intuitiveness and listening and building is what's necessary about ideas and steering, in the cybernetics steersman sense, steering the conversation in a way that's always intuitive but very pointed coming out of a history and a large knowledge of many domains.

Ranulph: Well I think there's something else too, I don't care what the outcome is. Most people who teach are putting forward their own view. I will put forward to the student, 20 or 30 methods. I'm not interested in which one they choose, I am not partisan. What I am concerned to show them is that it's possible to choose something, and that there are good choices to be made and that if I was there I could make something. And that I can produce enough of these things in one supervisory meeting for them not to believe that they cannot do it. The fact that I can produce thirty and they can only produce one, doesn't matter as long as the one is a good starting point. My interest, and you can always develop ways to do this, my interest is to know that they do not follow my suggestions but that they make their own. If they happen to follow my method then I am usually a bit shocked. Ben was talking with Ted at a conference, about his exam, Ted being one of Ben's examiners and offering suggestions on how he moved forward, Ben said it was really helpful, and Ted looked at him and said, "And who do you think I learned that from?"

Bill: I have a similar method of teaching.

Ranulph: It's the mark of a decent teacher, that they're not promoting themselves. My job as a teacher is to promote the student.

Bill: This *Book of Notice* assignment is one way where people come up with very different things that they're noticing and looking at. I try to get them to feed that into part of their ideas. It's basically feeding them back to themselves.

Ranulph: But it remains a problem.

Bill: The other thing that's very different about my teaching, and this may be true about yours as well, but because I'm listening and responding at that moment, and I let the conversation go wherever it needs to go, some people think I'm not prepared for class, which isn't the case at all. It has to do with this navigation of whatever arises.

Ranulph: If someone says that to me, I say: "Well, I've spent a lifetime preparing. What do you mean 'Why haven't I gone away and read something?' Why should I?"

"Well you're supposed to know about it."

"No, you are supposed to learn about it. It has nothing to do with me."

"Teaching is a complete fraud."

"Well I can't teach you anything because I can't learn for you. You have to do your own learning, you have to learn the subject matter. I don't have to learn it. Why should I waste my time knowing stuff you're going to?"

"Well your supposed to show me how."

"No I'm not, you're supposed to show me how. That's called an examination."

Bill: It's a bit like a reading list, when I'm working with a Ph.D. person. I don't want to tell them everything to read. I want them to tell me the reading list. I'll add to it, I'll make suggestions, but it's not me saying this is what you have to read and this is what you have to know. They have to figure it out.

Ranulph: "Where do I begin?"

"Well, the beginning."

"How do I know where the beginning is?"

"You have a look."

"Well how do I know if I'm right?"

"You don't, isn't that nice."

"But I might be wrong, I might waste my time."

"You'll only waste your time if you don't realize that you were wrong."

Bill: If you're serious about something, there is no wrong way. It's just part of the process of going.

Ranulph: But I think being a good teacher at least at a university, is a matter of charisma. Some people can be silent, some vocal, but what works is their ability to convince people that they expect something good, and that they expect it from you, not from me. It doesn't matter how bad I am, nobody is interested in me, you shouldn't be interested in me. Your concern is how good *you* can be.

Bill: Although I feel like sometimes if they haven't made that thing, I feel like it's my failure sometimes if I haven't elicited it. I don't think it is but—

Ranulph: Well we're taught this thing, the teacher's responsibility. The teacher has no responsibility. I spent a year teaching in which I refused to answer a question. I said nothing. I showed up for every meeting and they'd say things like, "What's our project?" and I'd say "Hmm." And they would say "Well we need a brief" and I would say "Hmm." And they would say, "Well, you haven't given us a brief" and I'd say "Hmm"…

Bill: Was this when you were teaching at the AA?

Ranulph: No it was here in Portsmouth. It was the last teaching I did in Portsmouth and I'm sacked for it. The results were far too good. I discovered that students perform at what I convinced them I expected. The problem is that when I go, they get someone who expects much less, so they drop out.

Bill: I had an independent study student recently, very brilliant just in their sophomore year, and she had this idea of making a touch feedback system so you could feel a textile but through a virtual interface. We just started brainstorming, how we could do this. She went out and researched all these different ways to do it and we had a very strong semester and it finished, we were very close and I continued to talk to her, and she tried to find the person to do the next independent study and nobody would do it. They said "Oh, you're doing Ph.D. level stuff… you can't do that your second year." And it's the absolute opposite of how I feel. She's brilliant. Often, I'll do that with students, I'll say "Think, if you could do anything in the world, what could it be?" and then I say alright now scale it back to something you can do in two years. You end up with this 5-year research agenda.

Architectural Moments

Bill: Okay, back to these architectural moments. Are there certain buildings that stand out, that you'd love to visit?

Ranulph: Informally, Zumthor's[54] Pavilion 2011 at the Serpentine Gallery, three years ago. Just a purely magical building. That's where I got the idea of the waterfall coming out of the gutter of the extension of our house. The roof was a forty-five degree pitch, all the way around the central garden which was surrounded by gravel, and the water just cascaded down the roof, forming this sheet. The water going into the gravel, beyond which you could just see the garden. It was fantastic. I think there are a lot of very fine cathedrals. Salisbury Cathedral[55] on a sunny day is magical. There are a lot of individual buildings. Mies van der Rohe's pavilion in Barcelona,[56] the reconstruction of that is just astonishing. Gaudi looks cheap in comparison.

---

54. Zumthor Pavilion for the Serpentine Gallery, https://www.serpentinegalleries.org/exhibitions-events/serpentine-gallery-pavilion-2011-peter-zumthor (accessed 28 September, 2019)
55. Salisbury Cathedral, http://www.salisburycathedral.org.uk (accessed 9 June, 2016)
56. Mies van der Rohe's pavilion in Barcelona, http://miesbcn.com/the-pavilion/ (accessed 9 June, 2016)

I say that I'm looking forward to going and seeing the cathedral now that it's complete. The last time I was there it was still being built. I work with the guy who invented the Gaudi-esque shapes for all the stuff Gaudi hadn't done. His name is Mark Burry. He's at RMIT and just now moving to the University of Melbourne and changing the sort of stuff that he does, early fifties. Very interesting man. Lots of building.

Mies van der Rohe's Pavilion in Barcelona, 2010. Copyleft.

Zumthor's Pavilion 2011 at the Serpentine Gallery. Copyleft.

Salisbury Cathedral. © Antony McCallum: WyrdLight.com (with permission).

The Nave, Salisbury Cathedral. © Antony McCallum: WyrdLight.com (with permission).

Blur Building at Expo.02 in Yverdon, Diller Scofidio + Renfro.
Photo by Norbert Aepli, Switzerland, Licensed under CC-BY-2.5.

Wright's home in Oak Park, Illinois (1889). Copyleft.

Image of a favorite Cathedral from Ranulph's Scrapbook.
Courtesy Aartje Hulstein and the Glanville Archive, University of Vienna.

Bill: Vernacular architecture?
Ranulph: No, no not at all. I don't think cathedrals are vernacular. I don't think Zumthor or Aalto are. But still I think there is some beautiful vernacular architecture, and there is some beautiful ordinary architecture, and there is some beautiful art architecture.
Bill: Do you like the Blur-building that Diller, Scofidio, and Renfro did?[57]
Ranulph: I think the trouble with most of those buildings is that they're completely uninterested in anything else. They're just there, they don't relate to other buildings, they don't make a space around them, so I think they're a very poor form of architecture. I'm also bored with buildings that are clever shapes because in the end, a building is not paid for to be a clever

---

57. The Blur Building, https://dsrny.com/project/blur-building (accessed 28 September, 2019).

shape, except for in the rare case like Mies' Barcelona Pavilion. A building is paid for because it's supposed to be helpful and useful. That's what I want. I'm not against buildings being useful, even though I send my students follies because I don't want them to be dominated by functionality, but I think that a lot of these buildings being designed today are basically just done work.

Bill: This building in London—the Gherkin—it really bothers me.

Ranulph: Does it? Why?

Bill: It just feels like one that doesn't fit with the landscape.

Ranulph: I think there's another one next to it which is far more fancy, that one that goes up with a curve and that curve goes up with the sun and melts cars. But it's just an ugly building, and why is it there? Someone's got something that looks different. Well I don't know if we need more of that, and I think sometimes the contrivances are just ridiculous.

Ranulph: I'm not someone who obsessively chases around after buildings, I go and look sometimes. For me, architecture gives me a very wide field in which to think about the world and I like that.

Bill: Do you like the Dom in Cologne? This cathedral.

Ranulph: Mhm, but I can't divorce that from the first time I went on a long cycle trip at about 14, when I cycled through Europe. One of the places I stopped was the Cologne Cathedral and I climbed to the top of the tower. I went back a couple of years ago with my mother-in-law who, being a good Dutch protestant, did find Catholicism a little bit daunting but she had never been to any of these things before. There's one country in the world that speaks Dutch besides Holland and that's Belgium, and she's never been to Belgium. Well that's not quite true she'd been on a trip to Bruges.

Bill: Do you like Corbusier's chapel?[58] Is that one of your favorites?

Ranulph: No. Again, my generation of architecture students learned about these buildings from black and white photographs. Cheap to reproduce, as gray and gray, with a lot of bleaching out from high contrast. Ronchamp to me, is full of mysterious windows that somehow bleach into the wall. When you go there it is full of little holes which are slightly colored. So this entire image I have built is destroyed. In the case of the Mies's Pavilion, you have the opposite happening. There you have a building which was torn down and then only a few photographs, and the photographs were all programmatic of the modern movement. So they were all there to propagandize for modernism, form follows function, all of this sort of stuff. When you look at the building by itself, you realize it is the most sensuous

---

58. Ronchamp, Corbusier's Chapel, https://www.dezeen.com/2016/07/24/le-corbusier-notre-dame-du-haut-ronchamp-chapel-france-unesco-world-heritage-list/ (accessed 27 September 2019).

and sexy building ever built. It is just astonishing. The things it does with reflection and glares and bringing the outside in, it's just breathtaking. Fortunately, by the time I saw that building, which was reconstructed just in time for the Barcelona Olympics, the first time I went to Barcelona. And when I saw that I had already learned that form did not follow function. So I was looking for buildings which showed that to be a lie. This was the architecture building that supposedly proclaimed it as a truth, which had been used by the propagandists and actually showed something completely different.

Bill: Did you like Frank Lloyd Wright? Does he represent one of these modernists?

Ranulph: Yeah, yeah. I particularly like the family house in Oak Park. I'd love to see Fallingwater, I've been past in a trailer.

Bill: I've actually lived in one for a year, one of those Usonian houses[59] in Oberlin, Ohio and it was just fantastic. A really great memory. All glass, one side, and the other side just kind of had long windows, heating in the floor, views everywhere.

Ranulph: Again his is an architecture of solo prima ballerinas. What I like about the family house in Oak Park is that there's this view that Frank Lloyd Wright brought forth these plans and sections perfectly conceived with no modifications just truth. And here's a house he worked on, moved things around on, and you can say well at least once he didn't do that, had that occurred to you? Oh but he's perfect—No.

Bill: We were talking about Messiaen and that beautiful piece on the end of time,[60] and I was wondering if there was an analogy in architecture of this kind of playing with time, or time ending.

Ranulph: Well Messiaen architecture is not generally thought of as a time-based art. And music is necessarily a time-based art. In fact, architecture has a major temporal element because the sun moves round architectural objects and changes the way the light creates forms.

Bill: I think of them as time-based spaces and also you do this in the morning, and this in the evening, and gather by the fireplace…

Ranulph: But for many architects the construction in architectural space is really an abstracted geometrical thing. It doesn't pay much attention to lighting and this sort of thing. I mean it does later. There is an aspect of architecture which is about putting those blocks together right. I think one

---

59. See https://amam.oberlin.edu/flw-house
60. *Quatuor pour la fin du temps*. Retrieved September 27, 2019 from https://www.britannica.com/topic/Quartet-for-the-End-of-Time

can push these things too far, and I think also these sort of analogies are analogies we make for ourselves.
Bill: I was very impressed by the scale of the Great Wall of China.
Ranulph: Something I've never seen.
Bill: There are some beautiful architectonic sculptures, I like a lot of those artists as well. Do you like any of Bruce Nauman's work?
Ranulph: Well I like Christo a lot, not the funny flowers. I could do without the pink poppies. But the wrappings and the walls across landscapes are fantastic. Were you in Sydney when he wrapped the cliffs? Did you see them? I'm sure you'd remember if you saw them.
Bill: I saw the Central Park work. I've seen the movie *Valley Curtain* many times. Many of the smaller objects which I like, including his really early work, where he just has a wall or a storefront or barrels piled up.
[There is a short pause in the conversation]
Bill: I found this "Interflow Architecture" [Seaman's poem] text which, when I get on the Internet, I'll send later today.

To Not Be In Control

Ranulph: There was something yesterday that we were talking about which I thought was interesting. One of the things that I have found very powerful is not to be in control, not to be in charge. An example; there is no single job I have ever applied for, for which I have even been interviewed. The jobs I get, come because people like the former boss of where I work, he meets me in a dark room and asks what I do. I tell him. He says "Oh, will you be my research professor" and I say, "Yeah, alright." It was just like that. I have learned over the years, not being in control is better than being in control and that I'm better with influence than I am with power.
Bill: Can you give that beautiful definition that you gave yesterday about these two different approaches to control?
Ranulph: As I said there are two types of control, one is facilitative and one is restrictive. And in order for one system to control another system it must be able to see, model, act on, every state that the control system might be in. And if the controlling system doesn't have enough states to do this, then it will always constrain/restrict the system being controlled, now this may be a good thing. In extreme cases it becomes fascism. One person imposes a very narrow mindset on entire populations, in contrast there's facilitative control. Maturana gives the example of skiing. When you are in control as a skier, it means that whatever is thrown at you by the environment, you will be able to respond and accommodate. You have as many states as the environment

is going to throw at you. There's a lot of facilitative control, and sometimes restrictive control can be facilitative because it reduces the vastness of something but then we're getting into really unnecessary complications because they're really only extensions of the initial state. If your controlling system has less variety than the system it is trying to control, it will restrict it. If it has more, or as much as, it will facilitate it. We, when we live in the world, are faced by eight billion people or whatever it is, each one of them with as many mental states as any one of them has, and the notion that we can control these eight, or go unaffected by them is rather bizarre. Because the amount of variety that eight billion separate things has, is not times eight billion but to the power of eight billion. However, many brain states we have (call it a billion), so it's a billion to the power of eight billion. You and I couldn't even write that down in our lifetimes. We live in a world where there is much more variety than we have ourselves and we control it in all sorts of ways, traffic lights et cetera. But we are essentially not in control of this world. For cybernetics, this is not a good thing. But, I say "Well, if everyone in the world is offering me things that I don't already have, I'm growing richer by the second, thus I am transcending all the things I knew and the person I am, all the time." This is a way of enhancing my creativity. I give myself more options than I had or than I could give myself, by myself.

Bill: So is this a case of the law of requisite variety, trumping the control?

Ranulph: No I think it's the other way around. Being out of control is not bad. Because this is saying "I'm out of control, the world has all this over me..." that in a sense is like saying, "I'm better at the jobs that are offered to me than the jobs that I go out to seek." Or, "I'm better at influence than I am a power."

Bill: Like when you gave that lovely list of seven or eight schools that kind of one at a time, found you, whereas if you tried to get jobs at those same schools, it never would have happened.

Ranulph: Well in one or two I pushed a little. But eventually it was because I knew people there. Usually those same places then give you the resources you need without making it difficult.

Bill: Yes it's wonderful.

Ranulph: I wished I had had something like tenure but now I'm glad I didn't because I am the person I am, *because* I am everywhere.

Bill: Nomadic.

Ranulph: The other thing is that as a child I went to a Froebel Kindergarten, a real one with teachers trained at the Froebel College and that sort of thing. I regard that to being enormously formative. One thing I've picked up, I call driving from behind. You sort of sit back and let the students drive you along. Occasionally when they're not looking you reach out and adjust some things but essentially they're doing all the driving. I talked about this public, but of course then I realized it's just Froebel.

Bill: Similar to Montessori?

Ranulph: Many people think they're the same but not at all, Montessori is the fascist version of Froebelism. Whereas Froebel would say "Look after my flowers but look after my weeds also, for I have learnt much from both"—his dying words—Montessori is trying to *form* a child. Froebel thought if you force the child you create social and psychological problems. So for Froebel it's all created for the child and led by the child's needs. Whereas for Montessori it's all packaged into things. We may have some Froebel practices and play but the thinking, the structure is absolutely non-Froebelian. It is one of the great distortions that she pretended that somehow she was his heir. And of course, the middle-class who are a bit frightened that if kids are not properly trained they'll be no good, and these are all the people who read Montessori, not Froebel. This was in Windsor, where we will go and drive past Windsor Castle, I'll show you Christopher Wren's Town Hall where Chuck married Camilla in a civil wedding. The Queen didn't come to the wedding, because she is the head of the Church of England, which doesn't really have divorces. But I think it's important to encourage your students to look not for power but for influence. And not to impose but to work from the bottom upwards. That's how we'll get change. We won't get change from people imposing. We've been playing with that. There are people like dear old Stuart Umpleby who's still hammering on about this and he won't change anything. "If we can just get the really influential people…"

Bill: When I went to RISD (Rhode Island School of Design–Digital+Media), one of the reasons I went was because they let me design the curriculum. I designed it in a completely novel way, where every different department and its relation to the computer became a different kind of class. Master students could come in and take the class and they had to take a certain amount of the node classes but the people from the outside could also take the class. You had this completely interdisciplinary mix in every class and you could take three very different ones that would help you learn general re-combinative education. There were about sixteen of these classes. But the reason that it worked was because I went and talked to each individual department chair and convinced them that it would be a really interesting class for them. And that it wouldn't disrupt their current course, functioning more like a supplement. When they voted it in it was totally

bottom up and unanimous and they made it work. But what RISD did, which wasn't fair, was they had every other one of the departments pay for that class. So everyone was happy when they had enough money, but then when RISD had money problems my classes were the first to go. But either way, a lot of people benefitted from the classes.
Ranulph: Will it come back?
Bill: Yeah I think so. For me though, that was my baby. It was my creative initiative. At Duke I've taken a very different role. I'm back to my research. I teach two classes each semester; Experimental Interface Design and Experimental Drawing (with both digital and analog) but each time it's a new challenge like, make a drawing with light, or use an unusual material… We do this generative arts class that goes into image, sound and text. I team-teach that with someone in the music department (John Supko). And they let me do my "Body as an Electrochemical Computer" class where we read about autopoiesis and all these different things. People did neosentience[61] in whatever area they were studying—like design, furniture building, glass, architecture, and many other areas. So that was very fun. And also they can support me coming to talk to you and for these sorts of things.
Ranulph: Well, that's one of the things that I miss—always being properly funded. That's one of the things that I chose, because I consider them an outrageous waste of time and also being giant packs of lies and I don't wish to participate simply through packs of lies.
Bill: I've got two grants so far. Michael Lissack told me he spent three million developing his program, I've spent 25,000 dollars, mostly for equipment, being able to show it, programmers, but it's exciting. It's working. It may make more sense with your thing to work with a person who is close by.
Ranulph: Yeah, well they're always busy.
Bill: Or do it in two instantiations and see which works. Survival of the fittest.

Some Key Thoughts about Cybernetic Concepts

Bill: I found a few more things that we started talking about but didn't wrap up.
Ranulph: Well I thought of something worth saying, sort of key things about cybernetics. Cybernetics is interested in mechanizing. Cybernetics only exists because of error. It only exists because there is error to be corrected. So cybernetics depends on error. Cybernetics requires feedback

---

61. Seaman (2013).

so you can tell it is error. But feedback makes it sound like there is some force going in one direction, then a tiny little thing coming back which we can ignore and I think that takes away from the heart of cybernetics which is the importance of circularity and circular causality. I tend to avoid talking about feedback. It's an old word which has led to a lot of misunderstanding. It has led to a lot of people trying to pretend that cybernetics is a subset of physics, and Ashby amongst others says quite clearly that it is not. It's materialist information, it's not energy, and it's not governed by the laws of energy. It can only be placed as being governed by laws of energy if you believe that the physical has priority. And information is very nonphysical. I don't like to use *feedback* because it quietly suggests that circularity is not serious.

Bill: Can you talk about the black box?

Ranulph: Yes, it gives us a mechanism by which we can build what we believe is knowledge, without actually knowing what's going on. Now if you take physics, it sort of consists of endless attempts to find fundamental particles and they're always made up of other fundamental particles, and it's because we don't quite know how to work out what's going on. In fact, fundamentals are things you can't split so you can't see inside… There's this problem with fundamentals, and there's a problem with description too, which is that when you describe something, you take the thing, and the word, and you find an area of intersection between them. You say, "This is a tape recorder," but the thing is none of us get inside of those, so we don't know what that actually means. We don't know what's in there and if we break it open we find endless amounts of things in there and it all falls to pieces. It also means when we think things are fundamental, the moment we describe them, we've broken them. Because although they have this area of sameness, they must also have this area of difference otherwise there's only one. Physics has this problem because it's a representational science. It keeps looking for fundamentals but it's the very act of looking which splits the current fundamental. It's the very act of describing that does the same. Now, we don't know what's at the base of anything. We can go on and on and on splitting—so we can't base our knowledge on actually knowing what's inside something. We have no idea, and if we find something that's inside we don't know what's inside *it*. We don't know what it's made up of.

Bill: It's an infinite regress.

Ranulph: It's a potentially infinite regress because we don't know what things are made up of—in the sense that this has a resistor, and some wires.

Bill: But even at the bottom of our own thought, it's still a mystery.

Ranulph: Absolutely. What I like about the black box is that it gives us a model for building knowledge from complete ignorance. That's a fantastic thing. We don't have to claim it is true. We don't have to claim we understand what's at the basis. We just look at something which structures behaviors. A black box is something that doesn't exist, it's a thought experiment, it's not really there. You insert it in a mess of things, signals what have you and it converts some into inputs, some into outputs. You put something in through the inputs and you look at the outputs. Then you put something else in, and if you're being economical you just use whatever you just had as the last output for your new input, then you go round and round until you're convinced there's a pattern there. Then you say, "The black box, I understand what's inside it now." Well, you don't. You've looked inside, you can't look inside, and it's not there. But you have made some reliable knowledge. However, you don't know how long that knowledge will continue to be reliable. It's always open to suspicion and question and at some time it will stop working. That's why I really like the black box. Ashby gave the example of a child going through a door, turning the handle and the door opens. When you go to open a door, do you think about all those mechanisms? Ratchets, tongues, keys and springs? No. You say if I turn the handle the door opens. That's the level of it. It's not absolute fact, it's something that you've found out through using this model which allows you to handle your ignorance.

Bill: It makes me think of affordance theory—you don't perfectly know why you reach out for the door knob. But it's something that you can notice that you do.

Ranulph: Yeah, I think you can talk about affordance in this way, and that you can talk about most things. I think if we have a universal method of acquiring knowledge, it's the black box. I think it's extraordinary how people have misrepresented it.

Bill: Who invented it?

Ranulph: James Clark Maxwell. He invented it along with Maxwell's demon. It's very difficult to trace but I did get it traced by someone who can trace anything: Dr. Albert Müller. I also like various things about it—you can have fun with it. I don't know if you've seen any of these slides that I have but basically I show: you put in one, it becomes three, three becomes five, and five becomes seven. Then I say to people when I put seven in, what's seven going to become?

Bill: Nine?

Ranulph: No eleven, it's a prime number generator. But of course, if you'd said eleven I would have said no it was nine. That just makes the point. I have another where it goes through the numbers—first I get nine and I say, I better test this again, try it and then I get eleven and so I say, this is very strange. Then I put it in again and I get a picture of Ross Ashby and I say, I never said it had to be a number.

Bill: You're doing function follows form aren't you? You gave the form, then you decide what the function is after that.

Ranulph: Well I think that's using the word function in its mathematical meaning, not in its architectural meaning.

Bill: See, I always like these multiple meanings.

Ranulph: Could be fun to develop the relationship. There are other things with black boxes. First of all, the black box doesn't actually consist of just the black box. It consists of the black box and its observer as Ashby points out. He called the observer the investigator. It consists of that. Then you say, and I look at this and since this thing is black, the whole thing is black. It is recursively black. If you can make it white it is still black inside.

Bill: I was thinking about the white cube of the gallery, and the white cube of architecture—of minimalist architecture as having a funny relation to the black box.

Ranulph: Supreme minimalist architecture. But then there are things like, I have ventured that human intelligence has shown the ability of one observer to transcend the levels. To observe from outside and now establish that that's a black box. The intelligence is in its ability to go across levels.

Bill: This is second-order cybernetics?

Ranulph: I don't know. It seems to me that the black box is inherently second-order cybernetics. It's all about observing, or as I now like to call it, composing.

Bill: I had a teacher who did "composition in D" so it's Decomposition.

Ranulph: Not to be confused with a "construction in D" or even worse, when it's the French doing it it's a "construction indeed." [laughter] Shall we hit the road?

Mercurial

Ranulph: I think cybernetics is a way of exploring a very abstract world. An old colleague of mine, Leon van Schaik went to a conference and I presented a paper called *Five Cybernetics Friends*. It was just five things in cybernetics that I really like and I tried to bring them together. I sat there in the pristine space in the chair and turned around, wrote a few things, and Leon, at the end of it turned to Aartje and said "I don't know how he does this. I don't know where he learned to do it, but I have never seen anyone do it so well." He can just sit there and think, without reference, without humanizing or talking about artistic content, and I think that's quite interesting.

Bill: You move very fluidly across fields. When you need the mechanism then you call on the mechanism, and sometimes you need the emotion and the humor and the playfulness.

Ranulph: Yes, I think I am properly called mercurial.

Bill: Have people described you as a polymath in the past?

Ranulph: I don't know what that means. Mostly, something like interdisciplinary. Albert introduces me that way.

Bill: How do you feel about Albert's role in cybernetics? He's one of those people that's very quiet but has been very active in the scene.

Ranulph: I think Albert is very active. I think he will always be someone who's interested in talking about others and never about himself. He's someone that finds others fascinating but never himself. He would rather look at what others have produced than produce something himself. He is deeply studious, academic, widely read, well informed, all the rest in ways that I most certainly am not. Although I did get *something* right and he corrected me, and he was wrong. I was right. And I felt *so* pleased. Especially because it's something as German as German can get. It was the title of a Bach-Choral. It was a Bach-cantata that was quoted by Alban Berg in his violin concerto. The title of the Choral is "Es ist genug"[62] and Albert was sure it was "Ich habe genug"[63] which is another Bach-Choral and I'm afraid I was right.

Bill: Have you made analogies between musical thinking and cybernetic thinking?

Ranulph: No, at the moment I've been making certain analogies between cybernetics and design, not because I've been terribly interested to but because it keeps the pot boiling.

Bill: Were you at the Center for Design Research?

Ranulph: I don't think so, I was at the Center for the Study of Human Learning.

Bill: Were you in a design research initiative somewhere?

Ranulph: I've introduced design research programs. I'm not sure what you're thinking of.

---

62. From the cantata O Ewigkeit, du Donnerwort, BWV 60, http://www.bach-cantatas.com/CM/Es-ist-genug.htm (retrieved September 27, 2019)
63. BWV 82. http://www.emmanuelmusic.org/notes_translations/translations_cantata/t_bwv082.htm (retrieved September 27, 2019)

Bill: You gave me the name of two psychology books that were quite interesting that were the opposite of behaviorism.

Ranulph: Ah yes, George Kelly.[64] I hate behaviorism, even though cybernetics is a deeply behaviorist science. I despair of behaviorist psychology. It seems to me what's wonderful about psychology is individual difference. Not this sort of attempt to show that it's all stimulus-response without mentation itself. If that's the case, just kill us off. I'm always pleased to find a way to think about how we think and feel and act and behave which is not behaviorist. Especially when it comes out of America, and at the time where America was even more dominated by behaviorism than it is now. In the 1950s there really was only one voice that was an alternative to Skinner and his followers in psychology in the US. That was George Kelly. Kelly was a psychotherapist—psychiatrist—in a hospital somewhere in the mid-west. For Kelly, what was important was that we made constructs of things and that these constructs were bipolar. A construct wasn't a name, it was a line between two names. Actually, three names but I'm not going to go into that. So the name might be *warm* as opposed to *marshmallow*. These are not logical categories, they're not semantic categories, they're just things that we put together. We build large numbers of constructs and we connect our construct so they intersect with each other and include each other. If I say *sweet* and *water* and I might have another one which was *orange* and *curly*. I might think of this construct *orange and curly* and find that it went something like "orange ↔ all colors ↔ transparent ↔ bendy ↔ curly" now that *transparent* intersects with the other construct. Now you have constructs building together and providing ways of going through. Of course, you can find contradictions in structures, things occurring in two places.

Bill: It seems almost like an entailment mesh[65] but not quite—

Ranulph: No, it is very precise. Kelly's construct grid technique is quite sophisticated, it was one of the earliest things that was done interestingly on a computer and was done by Laurie Thomas[66] with whom I studied for my second Ph.D. Kelly used the eliciting of constructs and repertory grids, to find particular qualities. If there were extreme differences between things, the grid would show them. He and Gordon worked together and argued about learning and they both heard about Kelly and they were really interested. He was a psychologist.

---

64. Cf., AllPsych (website). (2018). George Kelly and the fundamental postulate: The Beginning of Cognitivism. Chpt 11, section 2 in *Personality Synopsis.*, Retrieved September 27, 2019 from https://allpsych.com/personalitysynopsis/kelly/
65. http://www.c5corp.com/research/entailmentmesh.shtml (Retrieved June 9, 2016)
66. See Thomas &Harri-Augstein (1985).

Bill: I thought of one other teaching methods for us to talk about. Symbol. The use of the symbol:
Ranulph: I tend to avoid symbols. I have absolutely no interest in symbols.
Bill: I'll give you two instances: One in your infinity title of your book and one in the infinity in Spencer-Brown…
Bill: What are the symbols? The infinity sign.
Ranulph: Well that's no different than using the number one. Infinity is a number.
Bill: And the Ouroboros?
Ranulph: Well that's no different than drawing a Moebius strip. Okay, I was worried you were going to accuse me of being obsessed with phallic symbols.
Bill: Oh no no! [Laughter]
Ranulph: Also Freudians concede that sometimes a cigar is only a cigar.
Bill: Did you read Michel Foucault's "Ceci n'est pas une pipe"?
Ranulph: I'm not very keen on second-half-of-the-20th-century French philosophers. I think they have much to be accused of, little to be credited with.
Bill: Who do you like from the period?
Ranulph: Ludwig Wittgenstein. You know that *Ludo* is the Latin form of "I play" so Ludwig—who was referred to as Ludo—was the player.
Bill: I like his *Zettel*[67] and *Philosophical Investigations*.[68]
Ranulph: I like his remarks on the foundations of mathematics.[69]
Bill: We've had that conversation in the past, I was interested in this idea of open-order cybernetics[70] with language and new kinds of technology which may change the nature of the observer. For example, if we have a neosentient observer we suddenly have another order to observation but you said—no it all collapses back down. Second-order covers the whole thing. I think Ludwig Wittgenstein said the same thing, that there aren't these meta-levels.
Ranulph: Well you can have as many meta-levels as you like but they collapse into two. Why would you want to go into having many levels it just gets you into a regress.
Bill: I think of it as a growth over time, of progress, as opposed to making a muddle so to speak.

---

67. Wittgenstein (1967).
68. Wittgenstein (1953).
69. Wittgenstein (1956).
70. Gaugusch & Seaman (2004).

Ranulph: That's sort of like saying "Here's something which we can draw a line—it's a whole." Next week I draw a line that's a little bit bigger. But in the end it's just the line that's around it.

Spirituality

Bill: Last night we were talking about baby Jesus and how he was being represented but I realized, we really didn't talk about religion at all…or spirituality. Is that something you'd like to talk about or just leave out of the picture?

Ranulph: I don't think it has much to do with my cybernetics. I'm not religious but I was brought up as a good Church of England schoolboy. I went to school every day, I was in the school choir. I am deeply imbued with Christian myth and Christian stories. I know the Bible more or less by heart, I know the common payer, the hymns—except they've all been changed now—and my ethics are essentially Christian ethics. Whenever I write about the ethics of cybernetics, you can look at them and say "Well, that's what Jesus said." I know I'm very touched by that region, as to the questions of the existence of a God or not—I find it a matter of profound lack of interest. Whereas most people would say I'm an atheist or I'm an agnostic, I'm an I-couldn't-care-less-don't-waste-my-time-ist. Which is of course the worst of the lot. I'm quite glad I have a bunch of religious people voting for me, waiting for me at the pearly gates, to see if I'll get in. But spirituality I see as something quite different. I certainly am interested with the mysterious and incomprehensible. Spirituality is something that we have lost. We have lost what we have meant by the word so we must start off with what spiritualty means. For me, spirituality is not watered-down religion, it's not the enactment of religion, and it's nothing to do with religion. That's the first thing. *Spirituality* is a word we can use to describe those things which are missing when we've used all the other words we use. It's not about the physical, not about the mental, not about the psychological, or the emotional, but it's about all those things you need, to be human, that aren't in those nice readymade packages. Now that version of it comes from Alcoholics Anonymous…

AA not AA

Ranulph: AA [Alcoholics Anonymous] says it is a spiritual movement and that it is not a religious movement. It then makes the terrible mistake like using the word *God* or some meetings using prayer and outrageous things that really piss me off. AA would say that alcoholics are people who have missed out on their spiritual development, which stopped the first day they took an alcoholic drink. And as a solid attending member of alcoholics anonymous I go along with that. And I go along with the fact that there are powers greater than myself and that turns me back to inference. When I found AA, I found a philosophy that did almost everything I wanted to do—and sometimes thought I wasn't doing—but it wasn't. Like my Sicilian-Jewish friend who I was telling you about yesterday, who found Judaism and saved his spiritual side, I found AA and it saved mine. AA is, I think, the most astonishing social achievement of mankind. It is absolutely extraordinary. My last two columns in *Cybernetics & Human Knowing* were about this. What AA has done (apart from saving alcoholics and providing a model for people who want to save others)—they have created twelve steps. Admit you are powerless over the substance…et cetera. The steps are not done one at a time, they're cubic. You do the first—and only two of them mention alcohol—the first one where you admit you're an alcoholic and sign on and the last one where you say I'll go out and help other alcoholics. One of the key things is that you give away what you have in order to keep it. AA has twelve steps, it also has twelve traditions. The traditions are not rules but they are ways of behaving which will enable AA to remain as it is. Remain as an anarchic organization.

Bill: When did it start?

Ranulph: 1935-ish, During the time it's been with us it has managed to remain true to itself virtually without distortion. Without government, without anything.

Bill: Was there a person behind it?

Ranulph: There were two people who had met, both had given up as absolutely unsalvageable alcoholics. They found that by sharing with each other they recovered easier. It was a way to maintain not drinking. After they had about 20 members who also hadn't been drinking for years—it was unheard of. No one had done this.

Bill: You've talked about it often.

Ranulph: It's the model on which the ASC [American Society for Cybernetics] should base itself. But they won't, partly because they're too lazy. AA expects everyone to play their part. Their part isn't particularly big but they do things like help put chairs away. So I think of AA as the only society that has managed to be and remains anarchic. It is without government, anyone can join it, there are no rules, you can't be thrown out, it funds itself completely, it has no stars, despite people who try to make themselves stars. A person on their first day is as valuable as the person who has been there for thirty years. I've been in AA now for 24 years and I won't

tell people how long it's been. I simply say, "I am Ranulph and I am sober today, and I have been sober for several days, one at a time." It's an AA thing, that you are sober today. It's very Buddhist in a sense. Many of the guys try to be Christians because the founders were. But they were very influenced by the Oxford Movement which became moral rearmament in the fifties. I think that almost every one of their insights is an insight that we find in Buddhism not in Christianity. For my friend who just died in Sydney, it was always a question whether or not he converted to Catholicism. He joined AA about four years before me and for him it was very important to keep his Catholicism separate from his AA. He and I would talk about AA in terms of Buddhism, but never in terms of Catholicism.
Bill: Did you study Buddhism?
Ranulph: Not really, not like Francisco Varela did who was a deeply-convinced, practicing Buddhist. In that film [by Franz Reichle] you will see him dying while meditating. He was the Dalai Lama's science advisor.
Bill: I wanted to ask you about paradox. It ties into Bateson's notion of the double bind, which is another area that somehow also became part of cybernetics.
Ranulph: Yes, paradox is only paradox if you insist on not having time. Classical logic pretends there is no such thing as time. But if you say "If A then B" and "If B then A," it's fine as long as they both happen one after another, but when they both happen at once you end up with trouble.
Bill: There's this book by George Brecht, it just lays it out for you. The way I see it, reality is more complex than language's ability to reflect it.
Ranulph: Well reality is not its description. I say the same for physics. When you describe a fundamental particle you've missed it.

The Longer Route

Bill: I never mind the longer route if it means seeing something beautiful. My wife disagrees.
Ranulph: Must not be an artist. To be an artist you must be interested in the journey, must like doing the stuff that's necessary to get there, not necessarily straight there. Of course, ever so often you can go straight there as part of the gesture…Moholy-Nagy's *Factory Paintings*.[71]
Bill: I do this thing I call "I Ching[72] driving" where I just go and go until I find something in the moment and then I'll come back and do the shoot. I did a piece when I was just out of college, maybe still in San Francisco, called *Architectural Hearing Aids*.[73] There were speakers all over the car and there was a person driving and when we came to architecture, I would

make a mix and play something. It was an hour and forty-minute tour. A very nice piece but it came out of driving just like we are doing now.

Ranulph: You see, where did that idea come from?... You were out doing things, not controlling everything but following your nose and it was spurred from curiosity. I think this is a basic source of creativity. Not the only one but—

Bill: I find I love music more than anything, but I didn't have a way to make a living doing it. Academia carries me. I could spend hours and hours at Ableton Live, composing, which I do more and more.

Ranulph: Yeah, at the moment I am still very caught up writing some papers. I have one for a Festschrift, I have to complete my last column for *Cybernetics & Human Knowing*... I have my ASC conference paper. I just felt well, I am saying goodbye as the ASC president so I should have a paper in there. I also have the reworked version of Objects. These are big undertakings especially because I don't have much energy. I find it difficult to start things and get the pressure on. It's very weak. I'm writing this paper about wholes and parts, and I would just love to write sections from scratch, and I've got bits and sections to finish, and I'm not getting anywhere. I just read it, make some minor corrections... With any luck I'll finish it up. We've also been having lots of visitors. I need to conserve my time and use it in a different way.

Ranulph: This church was built in 1052 and there is a very fine painting reproduced inside. The story is that St. Hubert came across the building after walking through the forest and so he got down and prayed and founded the church. Now these churches are all over and associated with spring. This one is unusual, there was a village here, as you look around this was mostly forest. It was never built to be seen and yet here it is and it's very subtly positioned. Normally, people would sight on top of the hill—this is sighted just below the top of the hill. It gives it a subtle distinction, which is very

---

71. Better known as *Telephone Pictures*. Shortly after joining the faculty of the Bauhaus art school, in Weimar, Germany, in April 1923, Moholy had *Construction in Enamel 2 and 3* made at a local enamel factory. He would later claim to have ordered them by describing them over the telephone, exaggerating both his distance from the manufacturing process that produced them and the degree of technological mediation involved. In doing so Moholy presented the artist in the modern age as producer of ideas rather than things. While sharing the same abstract geometric composition, the works use a mathematical progression to change its scale, highlighting the conception of the image as transferable data. Gallery label from *Inventing Abstraction*, 1910–1925, December 23, 2012–April 15, 2013.
    Informatic explaining telephone pictures on the Moma website. https://www.moma.org/collection/works/78747 (Retrieved September 30, 2019).
72. Wilhelm & Baynes (Translators, 1968).
73. Seaman & Hernandez (1980). *Architectural Hearing Aids*. Performance by Bill Seaman and Carlos Hernandez.

special. You should've been here when I took Louis Kauffman through the place—drawing distinctions within this building. Lou had heard me talk about this but he'd never seen me do it. Suddenly, "Oh yes, that makes sense."

Saint Hubert's Chapel from the Road. © Bill Seaman.

Saint Hubert's Chapel – Exterior Wall. © Bill Seaman.

Saint Hubert's Chapel – Door. © Bill Seaman.

Saint Hubert's Chapel – Gate. © Bill Seaman.

Saint Hubert's Chapel – Windows. © Bill Seaman.

Saint Hubert's Chapel – Ranulph (Tired). © Bill Seaman.

Saint Hubert's Chapel – Hymns. © Bill Seaman.

Saint Hubert's Chapel. © Bill Seaman.

[run into some people]
Stranger: Hello, are you from the States?
Bill: Yes.
Stranger: Where are you from?
Bill: Duke University, Durham, North Carolina. Lovely place you've got here. Do you live here?
Stranger: I live just down the road. It is beautiful. The manor house used to be between here and the old garden house you see down there. The old Idsworth house used to be between there. The house over there is now split up and lived in by those who support the church. As for the trains, this is the main route to London. The church is still full time.
Bill: Well thanks it was very nice to meet you.

Fresco in Saint Hubert's Chapel. © Bill Seaman.

Ranulph: You see the painting over there, and the version that has been put over there in the window, there is one significant difference, one very small difference that you have to look very hard to see. If you look up and left from the halo, you see an extra little building in there. There is a little chapel that has been added. It's a selfie. It was an attempt to capture aspects of life as they are in the twenty first century.

Bill: I was like you, grew up in the church. I was in a choir and a bell choir, and I would sneak in and play the organ. I have a little pump organ which I'm getting repaired... I got some wonderful photos this morning, the fog was just over the town [Portsmouth]. Sometimes just the silhouettes of people are beautiful.

Morning Fog with Monument. © Bill Seaman.

Ranulph: Oh boy, you're an impressionist at heart.
Bill: I am (among other things). I have a Matisse book that I keep close by. I return to it over and over again, there are a dozen of these paintings that I return to over and over, and Cy Twombly I like very much. I keep looking at him. [And abstract expressionism.]
Ranulph: Aartje's birthday present this year is the Blue Matisse cutout nudes in a big book, from the cutouts exhibition they had. It didn't have the presence of the first time I saw it in that gallery at the top of the National Art Gallery there in DC. It was just so astonishing.
Bill: That was one of the very nice things about the conference. I got to go see some of those galleries that I had never seen. This Rothko room—
Ranulph: Have you seen the one at the Tate?
Bill: No.
Ranulph: It used to be in the old Tate in a very post-renaissance room. It was fabulous.
Bill: I have a really beautiful book that Morton Feldman did, and it's about his relationship to the arts, the people he knew. It's filled with these different artworks including Phillip Guston. It's very great book.[74]
Ranulph: Did he know Beckett?

Bill: I bet he did. Did you meet him?
Ranulph: Yes. It was extraordinary.
Bill: You read his work when you were young?

Samuel Becket. Portrait Circa 1980. Copyleft.

Ranulph: Yeah—I went and knocked on his door—he was very friendly. He had a soft Dublin lisp and very long arms and he didn't know where to put them. I have some letters somewhere from him.

There's a famous story about me and Frances Yates, do you know The Art of Memory?[75]
Bill: Yeah,
Ranulph: She was apparently a very impressive woman, and I thought I'll get her to lecture at the AA, everyone's interested in her work. I found her and asked, and she said "I'd love to, why haven't you asked before?" So she came and we had a little walk through the place and I told her I was going to put her in the worst room in the place so I looked and said yup, these students are going to be absolutely intimidated and terrified so I want to put them somewhere where you cannot be seemed as anything other than disarmed. So I put her in this absolutely terrible room which was packed. The AA's who were serious intellectuals couldn't get in, they were furious with me. They came up and said "how dare you ask Frances Yates. We never had the courage, how did you do it?" I said—the telephone. And they thought I was disrespecting them by putting her in a small room but I

---

74. O'Doherty & Pellizi (2011).
75. Yates (1966).

wanted her to speak to the students, not stuck there posing for people like them. Some of them never spoke to me again. "He got Frances Yates, it's not fair." They could've done the same thing.

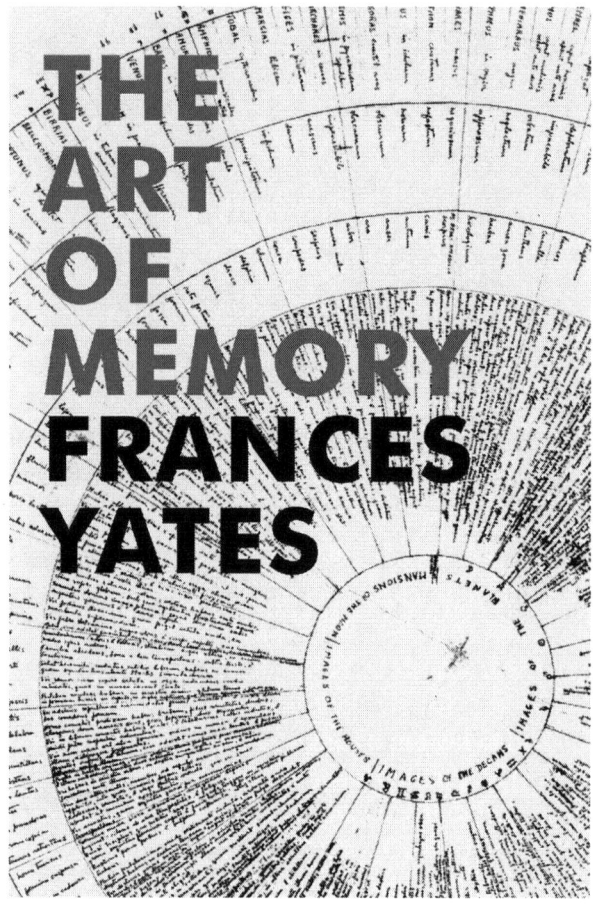

Cover of *The Art of Memory* by Francis Yates. Copyleft.

Bill: I've tried to do the same thing.
Ranulph: I was intimidated by some of these people but basically I was too ignorant to realize how famous they were.
Bill: Who were they?
Ranulph: Ernst Gombrich, Frances Yates, Isaiah Berlin, John Chris Jones, Jonathan Miller… I can't remember who else, a couple of logicians, some musicians—Xenakis.
Bill: You didn't record?

Ranulph: No it wasn't something we did at the time. I don't want to dissect it later. I was always doing it, I said you just pick up the phone and ask. They would look at the IAB [Interactive Advertising Bureau] for our conferences, and wonder how. All they have to do is call.

Bill: Did you ever like Pere Ubu?

Ranulph: No, particularly because I only liked the play [by Alfred Jarry[76]] and the band had none of that.

Bill: Duchamp was influenced by Roussel, did you like *Impressions of Africa*[77] at all? The author—Raymond Roussel?

Ranulph: Ah, no I don't know him. There was a composer with the same name?

Bill: Well there's this network of very interesting writers, like Huysman's. Did you ever read *Against Nature*[78]

Ranulph: I've never heard of those.

Bill: Shall we start heading back?

Ranulph: Sure.

Bill: I've always liked it when a building gets torn down and you see all of the history of the building next to it. Do you like Gordon Matta Clark's work?

Ranulph: Yeah, there's also the Rachel Whiteread stuff.

Bill: Yeah, I like the bookshelves, and the mattress. She did something recently that was just a colored soap bars, it was very beautiful.

Ranulph: I heard she had plotted out, day by day, her career for ten years, to make sure she was going to be a star.

Bill: Well she achieved it somehow.

Ranulph: I always think it's a pity, how artists become businessmen. I don't think the two things generally fit together very well. Lots of artists do very well and are very comfortable—I don't mean that. I just find that when you become so self-seemingly ambitious, the art may be left out.

Bill: Sometimes when I'm feeling down, thinking about if I'd stayed with my career and not done all of this other—PhD work, I would've had an entirely different kind of notoriety. Then I realize that's the stupidest thing I could think of—I am who I am, because of what I do. I think often you are secure but all the while part of you is quietly distrusting, at the level of your knowledge, why people have hired you.

Ranulph: Well the thing is, people have never really said these things to me so I've sustained it as a one-man act. It's very peculiar to find interest such

---

76. Jarry (1965).
77. Roussel (1988).
78. Huysmans (1988).

as there is now, and it's somewhat bewildering to think that you have to start dying for people to express this sort of stuff. They could have said something earlier. There are films the RCA is making now, someone could've done that earlier, I would've benefited from it instead of what it's actually doing. Taking a lot of time and effort at a time—

Bill: Well I think people who know you just took for granted that you were going to live forever and this wasn't the moment to do it—as you are still going.

Ranulph: I've had Steven Jones write back saying "I can't believe this, you are indestructible."

Bill: He makes me very sad but I heard he's stable now.

Ranulph: He's had some work done.

Bill: Was he in Troy for the mathematics conference?

Ranulph: Yes, he was very aggressively Steven-ish.

Bill: He's one of those people where I just pick up where I left off... I'm coming back to Rotterdam in about two months but I don't know if I'll have any money to come back again.

Ranulph: Well I'm looking forward to the end of next week when I just have a lecture to prepare, I always tell myself it will be easy.

Bill: You could just do it off the cuff?

Ranulph: Well people expect these slide shows—but I don't make slideshows like most other people do. I don't stick up pictures of stuff. I make slides of animations. It's a conference on systems and design with a lot of big name people there.

Bill: Did Bertalanffy[79] show up?

Ranulph: I didn't meet him, but Richard Jung worked in the office next to him—and to Bateson. No, I never met Bertalanffy I'm afraid. I think he died before I got interested. Where was he? Edmonton or Calgary.

Bill: Any other systems people you were friends with?

Ranulph: I knew a lot of them. It's a small world.

Bill: Michael [Lissack] is a very sharp guy.

Ranulph: I'm not sure about the future of the ASC.[80] I'm not sure it should have one.

Bill: I mean it's a bifurcation—either it's going to make it or not.

Ranulph: What I would not like to see is the things we've worked very hard to establish, just left to rot. If you decide not to do them, fine. If it's a wise

---

79. Ludwig von Bertalanffy (1901-1972), creator of theoretical biology and general systems theory (GST).
80. American Society for Cybernetics, http://asc-cybernetics.org (retrieved September 30, 2019).

sensible decision, okay. But to just not bother with things that were hard won—and let them disappear—would be a shame.
Bill: Ted talked to me about putting my name in there, I was very tempted but at the same time that's a big-time commitment.
Ranulph: Yes it was a big-time commitment for me as well. Plus, everything was falling to pieces and I needed support for writing my paper. I think it's better to have people who have an ASC membership to establish a more expert cybernetics connection. It could be a very nice thing.
Bill: Let's give it five years and see.
Ranulph: I was president for six.
Bill: That's the question I've asked myself: "Is this the conference that I want to have?"
Ranulph: I keep reminding people that the ASC is a democracy. It is actually very much a society.
Bill: …There is this younger group that I like very much.

Family

Bill: Did you have very supportive parents growing up?
Ranulph: My dad died when I was eight. He was a disturbed Irishman, and he came to the conclusion that he was not my father—even though I looked just like him. He was determined of this, I didn't realize it, so he was a paranoid schizophrenic, not believing my mother and he wouldn't believe any tests, so it was just… My grandmother died when I was sixteen of advanced dementia, and she was the only grandparent alive in the whole of my life. My father was fifty-six when I was born and sixty-four when he died.
Bill: What about your mother?
Ranulph: Oh yes, she was supportive but what we discovered a little bit later was that I have a brother who she judged as fragile—so I was okay and didn't need any…
Bill: I see, he got all the attention—
Ranulph: Yes, and then he had this terrible accident in Seattle and received even more of it. But the more I think my mother was just astounded—thinking where could you have come from, how I could possibly have…—and I think that was a compliment. I think she was truly astonished. Like when I travelled at 17 years of age to Paris by myself to meet Samuel Beckett.
Bill: My parents never quite understood what I was doing. Even now they'll say, "Now you're a videographer?" I think my dad wanted me to go into

advertising or something… He was very altruistic and spent his life fundraising as the director of development for several institutions. Money was always essential to him and didn't mean anything to me. I was never driven to make lots of money. I was driven to do art and explore.
Ranulph: I was never driven to make money either and I am always astonished by how much I did end up making in the end.
Bill: Yeah, I've always just followed my nose. Duke has always been very good to me as well. Sometimes I get a sense when you talk about things—and I get this from myself as well—that you're very secure in the things that you do, and then there's somewhere in the back of your mind, this doubt.
Ranulph: Well the doubt's the thing that keeps it moving.
Bill: Something Otto Rössler and I were talking about—this "tragic laugh" where things are so intense and you don't quite have a way of dealing with it. I think that's sometimes what happens with you.
Ranulph: Yes and when I'm nervous I get very funny. When I went in for my first day of Chemo, I had the entire ward laughing. Now I just go in there quietly. "Never, never, never…" [Ranulph sings lyrics from Ian Dury's *Clever Trevor*].

The Most Cheerful Author Ever

Ranulph: This morning we talked about the importance of being out of control.
Bill: We also talked about black boxes, beautiful architectural moments, humor and tragic humor, Samuel Beckett. We went to the church…
Ranulph: I think Beckett is the most cheerful, positive author ever.
Bill: I was going to say what is this sort of cheerful streak in you?
Ranulph: The Irish in me. I do my best not to use my favorite Irish accent when in Ireland.
Aartje: Did you talk about meeting Messiaen?
Ranulph: Ahh, he never asked me.
Bill: Who are all of the important people you knew that you would like to talk about?
Ranulph: I did meet Messiaen and he didn't speak English. Frances Yates was sweet.
Bill: I tried to talk to her but she had already passed.
Ranulph: Well this is what the RCA is doing—they're saying "We've missed videoing so many people we should have videoed, we're not going to miss out any more."

Bill: Did you meet Boulez?
Ranulph: Yes, he was very interesting. But he's very intense, and he's off doing the next thing.

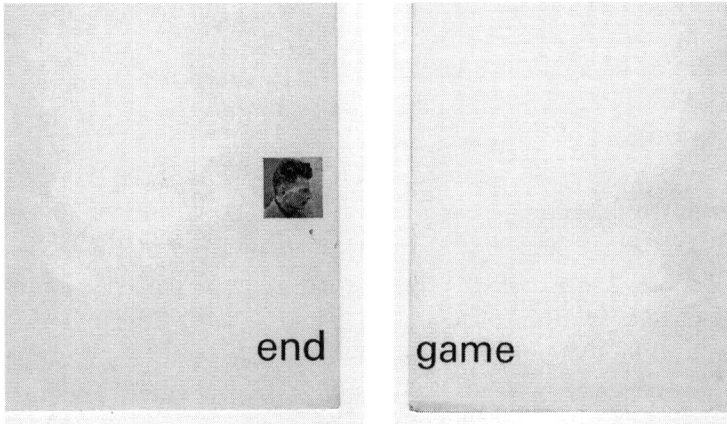

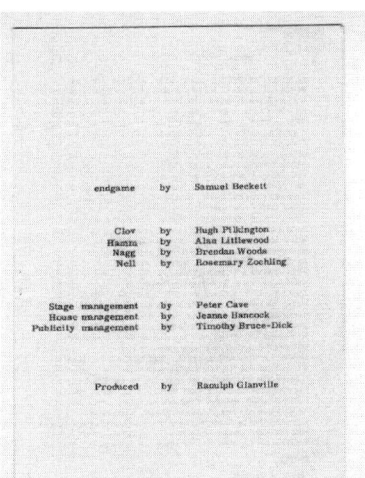

Flyer for Samuel Beckett's End Game, Produced by Ranulph Glanville, AA. Images courtesy of the Glanville Archive and AA, University of Vienna.

Bill: Who was the architect who meant the most to you that you met?
Ranulph: I can't remember an architect who has had the astonishing effect on me, as for instance Peter Maxwell Davies had as a composer. I had a lot of remarkable architecture teachers: Richard Rogers, Norman Foster. All the Archigram guys. I had a very funny meeting with Peter Cook. I was probably the only student of any note at the AA who Peter never talked to. When I was a professor at the Bartlett, he was the senior professor—he was on the point of retiring and he was finding out about life and had used to sort

of grab hold of me and say, "Can you come and talk to me?" A lot of questions would be sort of "Did I ever teach you?" and I would say no, and he would say "Why didn't I teach you?" Me: "I have no idea, Peter, you though I was rather bad."
Peter: "Did I?"
Me: "Well, you banned me from the end of year exhibition."
Peter: "Did I?"
Me: "I was the only student ever banned."
Peter: "Well you're rather good now."
I gave Peter a tutorial on getting the better things out of life when traveling. Cedric Price is one architect I met as a student who had an enormous impact. Cedric gave wonderful lectures. He was not part of Archigram but they all looked to him for guidance. Leon van Schaik refers to him as one of the most intelligent architects he ever met. Cedric had X-ray eyes and he could just see through bullshit. He designed a project called The InterAction Center for a man called Ed Berman. It was a building between some railway lines. And Ed ran out of steam and was leaving the building, and of course all the conservationists were there. Now, Cedric believed that when a building is useless you should get rid of it—so he designed buildings that collapsed and what not, and he led the campaign to have this building knocked down! But it was with an enormous sadness because he just wanted to build and no one would give him the job because they all thought he was much too conceptual. There was a house in Cornwall, with an enormous art collection—and Cedric designed the garden for this. Amongst the things which he designed were two 22 foot high wicker elephants and for me this is the sort of, curiousness of Cedric.
Bill: Was it as a joke, like Jeff Koons kind of art?
Ranulph: No! He just thought the garden needed a couple of life-sized wicker elephants. I used to love that. Somewhere I have a copy of Cedric's slides. If you're interested, his archive is in the Canadian center for contemporary architecture. There you can have a look at his books. Cedric didn't mean for them to be conceptual but many were never developed beyond that and for me it was this ability to create this sort of concept—to cut through the bullshit and the sentiment.
Bill: Was Pask a student of his?
Ranulph: No never. Later Pask and I and John Wells started working on a cybernetic theater. Cedric was also part of that group. I remember it and I remember meetings, and thinking that it just didn't work. When it came to it, I saw Gordon doing what he usually did—leaving everything difficult for

others to figure out, just scooting around doing the easy stuff. And the easy stuff didn't work so the difficult stuff didn't either

Bill: Usually when you meet an architect, the first thing they would do is pull out their book of all their work—what are the most important things, architectural drawings, projects, what should go in here?

Ranulph: I think the most important drawing that an architect does is an axonometric one. Well the thing about an axonometric is that it's a drawing that shows distortion, which perspectives do not show. They tell you a lot, and sections tell you more than plans. The architecture of a building is in the section, not the plan. The plan is the organizer—the generator of the architecture is the section.

Bill: And we should absolutely have drawings from the Internet shopping, pre-Internet shopping? Can we have those? [See earlier section for images.]

Ranulph: I don't know, possibly. I don't know what state they're in.

Bill: Yeah, but we should just say "Yes Bill, we should have those because they're so important to my oeuvre." They've got to be important if we're talking about them!

Ranulph: I remember a final year project I had forgotten. Here is the last final year project: In 1968 or 9, Texas Instruments produced a tiny television screen box, made of LEDs. I looked at this and I reverse engineered it. I said what you could do was etch a pattern of photosensitive material into squares and you could scan the squares, measuring the voltage that each produced and you would have a television camera. Interestingly enough five years later they started doing it. I developed this new type of tube, and a new type of scanner. And I proposed that everyone in the world should be given a camera made like this—a pinhole camera, very simple. You would have three settings, Public – Private – Off. Public meant anyone could look through your camera, so they could "borrow you" to see something. Private basically meant a conversation and then you could just have Off. This was propelled by a large number of radio towers around the world that would pick it up as you moved.

Bill: Yeah, I would love to include these kinds of things.

Ranulph: The point of this was actually, the public viewer. I had already done a little project in which I'd codified the sentences used in news broadcasts about the Vietnam War, and there weren't very many. All you'd have to do was say "Number 37, Haiphong, 251." And that would be "earlier today Americans attacked the port in Haiphong with 251 casualties." My idea was that the problem with the news was that it was all spun and all controlled by the US military. You never knew what was going on. But if you could drop in on some tunnel in the jungle in

Vietnam—through a camera someone was wearing—you would be able to see what was going on there. You would be better informed and you would be able to object much more forcefully and accurately to a thing. The Vietnam War continued on much longer than it needed to so that Dow Chemical could test their new weapons. It was there that the first drones were being flown, all controlled by two big ILLIAC V computers. Agent Orange was being used, the sensors were being used. That was my other project, in an attempt prevent war. But of course, this was what the cellphone is now.

Bill: I also had a project for a poly-sensing environment with a number of distributed sensors. They would end up becoming a video camera. You can have a distributed network of these sensors, and the whole architecture would become the eye. It would have its own kind of vision.

Mental Maps of Living

Bill: What are Mental Maps of Living?
Ranulph: Yeah, they appeared in both of my PhDs and a couple of papers. When I was brought in to teach at the AA, first year—Urban Studies—we had a group of students that explored three main things… Ours was run (in part) with an old friend of Alvin Boyarsky's who ran the AA. I was selected to teach with Leon van Schaik and we were in the same two final years together. He had come from Newcastle, and we had never met each other. We did all this urban stuff and one of the things that I was interested in was ways of showing the way that we thought. I would ask people to draw maps of the structure of London: "Is it concentric? Is it radial? Is it a grid?" and then I would ask them to then draw a concentric and radial structure on a conceptual map of London so the center would be AA, the north would be Camden Eye tube station. Then I asked them to say what would lie underneath the axis at these—all mental maps here—they would name these places and then go find them on a map and trace from a cartographic map which would give you all the distortion. You'd have bits that were longer and you'd ask, "Well, why is that longer?" and they say, "Well, I don't know I've never been there." It was a way of showing differences in how different people understood London. And to provide a certain metric which by the distortion of the circle they thought they were doing showed them where there were inequalities.

Bill: It really relates to the *Object* idea… "What is the individuality of your impression of London?" [Ranulph wrote about Objects the following: "I am

less interested in what we have in common than what we have that supports us and keeps us different" (Glanville, 2012, vol. 3, p. 209).]
Ranulph: Yeah, because people think that everyone sees the same. I wanted to show them that they didn't. But yes, this was long before I'd invented Objects.
Bill: Then when you put them all up at the end, they all realized just how differently they'd viewed the city?
Ranulph: Yup. I got people to draw spaces only being in blindfolds too.
Bill: You're very interested in this idea of—
Ranulph: I was under the impression that we should understand that people see the world differently. For me, it's a basis on which I work.

Famous Dictums

Bill: I was thinking that as there are these famous dictums from the history of cybernetics, if you were to begin to—based on your own ideas—would you make a set of new statements…?
Ranulph: Well it's only recently that I've been able to come up with a characterization for cybernetics that's my own. I've said things like "Cybernetics is a subject which is seriously interested in circularity."
Bill: If you made up a new statement that embodied your approach, which differentiated you from the rest…?
Ranulph: One little adage which I came up with many years ago, which has been misquoted by George Klir is, "A whole is a part in a role." *A part is a whole, in a role* [emphasis Seaman]. You take a whole and you say I'm going to treat you as a part, you'll have the role of being a part in this but it is still a whole.'[81]

Emptiness

Ranulph: You know that I wrote a book of secret pieces and I printed one copy of it, in white, on white paper. I have a paper called "Emptiness."[82]
Bill: Yeah, I was going to ask you about that.
Ranulph: I think all of those conceptual architectural papers are important—philosophers hated them, and of course Robert Trappl hated them.
Bill: Could you talk about him for a minute?

---

81. See Klir (1985)
82. Glanville (1984).

Ranulph: There's not much to say. Robert is an extremely good politician and his work is interesting. He's in medical AI. His big achievement has been keeping a series of conferences going for a long time. But he hated "Emptiness." He said people will laugh at me.

Bill: But emptiness ties back to Cage, this anechoic chamber, *Silence*.[83] I think it's an important part of your teachings…

Delight in Exploring the Senses and Their Temporary Removal

Aartje: I think one of the other things you did was explore the senses by either taking them away, like seeing, or one time you put me in that room with no sound…just to have the experience—just to see what would happen. You've done that a lot I think.

Postscript: Room for Delight

Room for Delight from the outside. Courtesy of Aartje Hulstein and the Glanville Archive, University of Vienna.

---

83. Cage (1967).

Albert Müller in the Room For Delight (detail). Courtesy of Aartje Hulstein and the Glanville Archive, University of Vienna.

Albert Müller in the Room For Delight (Window). Courtesy of Aartje Hulstein and the Glanville Archive, University of Vienna.

Shelves and Hallway. Courtesy of Aartje Hulstein
and the Glanville Archive, University of Vienna.

Bookshelves. Courtesy of Aartje Hulstein and the Glanville Archive, University of Vienna.

Kitchen. Courtesy of Aartje Hulstein and the Glanville Archive, University of Vienna.

Buddha. Courtesy of Aartje Hulstein and
the Glanville Archive, University of Vienna.

Backyard Project. Courtesy of Aartje Hulstein and
the Glanville Archive, University of Vienna.

Ranulph: The little room we're making comes from years of thinking about it but also it comes from doing it. One of the main things is the opportunity.
Bill: Well it's going to be exquisite.
Aartje: It is a combination of real world objects, but it's more about how does it take form...

These conversations took place on September 14th by Phone, and on October 1, 2, and 3, 2014 in person.

Ranulph died soon after this long conversation on December 20, 2014. Most people had imagined he would live much longer, and that they would get a chance to see him again. I was very moved by our meeting, and I feel that his spirit is very much alive in this text. I hope this text has in some way been meaningful to you, the reader (composer).

References for The Long Conversation

Alexander, C. (1977). *A pattern language*. New York: Oxford University Press.
Alexander, C. (1979). *The timeless way of building*. New York: Oxford University Press.
Ashby, R. (1956). Design for an intelligence amplifier. *Automata Studies: Annals of Mathematics Studies* (34), 215–234.
Ashby, R. (1964). *Introduction to cybernetics*. London: Methuen.
Bachelard, G. (1964). *The poetics of space*. Boston: Beacon Press.
Beer, S. (1994). *Beyond dispute. The invention of team syntegrity*. Chichester, UK: Wiley.
Cage, J. (1967). *Silence*. Middletown, OH: Wesleyan University Press.
Carroll, L. (1871). *Through the looking glass and what Alice found there*. London: MacMillian.
Casti, J. (1994). *Complexification: Explaining an illogical world through the science of surprise*. New York: Harper Collins.
Churchman, W. (1967). Wicked problems. *Management Science, 14*(4), B141–B142. Retrieved June 6, 2017 from: https://punkrockor.files.wordpress.com/2014/10/wicked-problems-churchman-1967.pdf
Ertl, M., Korn, W., & Müller, A. (2016). *Ranulph Glanville, Architecture | Art | Cybernetics | Design*. Vienna: echoraum.
Fox, T. (1977). *The labyrinth scored for the purrs of 11 cats*. Retrieved June 8, 2016 from: http://www.kunstradio.at/2012A/01_01_12en.html
Gaugusch, A. & Seaman, B. (2004). (Re)sensing the observer, offering an open order cybernetics. Retrieved June 9, 2016 from: http://www.bcchang.com/transfer/articles/2/14490740.pdf
Glanville, R. (1966). Klee and Messiaen. *Arena, The AA Journal, 81*(898).
Glanville R. (1978) The nature of fundamentals, applied to the fundamentals of nature. In G. J. Klir (Ed.), *Applied general systems research* (NATO Conference Series; vol. 5, pp. 401–409). Boston: Springer.
Glanville, R. (1984). Emptiness. In R. Trappl (Ed.), *Cybernetics & Systems Research 2*. New York: Elsevier North Holland.
Glanville, R. (1997). *The value of being unmanageable: Variety and creativity in cyberspace*. Retrieved June 17, 2016 from: https://www.univie.ac.at/constructivism/papers/glanville/glanville97-unmanageable.pdf
Glanville, R. (2012). Black B∞x, Vol. 1: Cybernetic circles. Vienna: echoraum.
Glanville, R. (Ed.). (2013). *Trojan horses: A rattle bag from the "Cybernetics, Art, Design, Mathematics—A Meta-Disciplinary Conversation" post-conference workshop*. Vienna: echoraum.
Glanville, R. (2015). Thesis on the theory of Objects and the invention of second-order cybernetics. *Cybernetics & Human Knowing, 22*(2-3), 21–25.
Glanville, R., & Varela, F. (2012). Your inside is out and your outside is in. In Black B∞x, Vol. 1: Cybernetic circles (pp.479–482). Vienna: edition echoraum.

Glasersfeld. E. von. (1974). Piaget and the radical constructivist epistemology. In C. D. Smock & E. von Glasersfeld (Eds.) *Epistemology and education* (pp. 1–24). Athens, GA: Follow Through Publications.

Huysmans, J. K. (2009). *Against nature*. Oxford, UK: Oxford University Press. (Originally published in 1884, Charpentier Press)

Irtem, A. (1971). Happiness, amplified cybernetically. In J. Reichardt (Ed.), *Cybernetics, art and ideas* (pp. 71–75). Greenwich, CT: New York Graphic Society.

Jarry, A. (1965). *Selected works of Alfred Jarry*. New York: Random House

Klir, J. (1985). Architecture of Systems Problem Solving, Binghamton, New York: Springer Science and Business Media, LLC. Retrieved January 2, 2020 from https://books.google.com/books?id=tAr3BwAAQBAJ&pg=PA177&lpg=PA177&dq=ranulph+glanville+George+Klir+is,+"A+part+is+a+whole+in+a+role&source=bl&ots=YF_mKN2kZF&sig=ACfU3U05MD8dYQQZjZ8YZr9hqWrUtuMkuw&hl=en&sa=X&ved=2ahUKEwjtg7bhxOXmAhUMhAKHUApCwsQ6AEwAHoECAoQAQ#v=onepage&q=ranulph%20glanville%20George%20Klir%20is%2C%20"A%20part%20is%20a%20whole%20in%20a%20role&f=true

Mead, M. (1968). Cybernetics of cybernetics. In H. Von Foerster, J. D. White, L. J. Peterson, & J. K. Russell (Eds.), Purposive systems (pp. 1–11). New York: Spartan Books.

Monk, R. (1991). *Ludwig Wittgenstein: The duty of genius*. London: Vintage.

Müller, A. (2015). Ranulph Glanville's thesis on the theory of Objects and the invention of second-order cybernetics. *Cybernetics & Human Knowing, 22*/(2-3), 21–25.

Nichols, M. (2011). The Lost Art of Listening (Digest by S. Mclean). Retrieved June 9, 2016 from: http://www.asc-cybernetics.org/2011/wp-content/uploads/2011/01/The_Lost_Art_of_Listening__Digest_Sharon_McLean.pdf

O'Doherty, B., Pellizi, F. (2011). *Vertical thoughts: Morton Feldman and the visual arts*. Dublin: Irish Museum of Modern Art.

Panofsky, E. (2005). *Gothic architecture and scholasticism*. Latrobe, PA: Archabbey Publications.

Pask, G. (1961). *An approach to cybernetics*. Cambridge, MA: The MIT Press.

Peirce, C. S. (1931–1935, & 1958). *The collected papers of Charles Sanders Peirce*. Vols. I–VI [C. Hartshorne & P. Weiss, Eds., 1931–1935], Vols. VII–VIII [A. W. Burks, Ed., 1958]. Cambridge, MA: Harvard University Press. (Citations use the common form: CP vol.paragraph).

Pickering, A. (2010). *The cybernetic brain*. Chicago: University of Chicago Press.

Rittel, H., & Webber, M., (1984). Planning problems are wicked problems. In. N. Cross (Ed.), *Developments in design methology.* (pp. 135–144). Chichester, UK: John Wiley & Sons. Retrieved September 28, 2019 from: http://www.sympoetic.net/Managing_Complexity/complexity_files/1973%20Rittel%20and%20Webber%20Wicked%20Problems.pdf in *Policy Science, 4* [1973], 155–169)

Rössler, O. E. (1998). Endophysics: The world as interface. Singapore: World Scientific Publishing Co.

Roussel, R. (1988). *Impressions of Africa*. London: Calder Publications. (Originally published in 1910)

Seaman W. (2013). Neosentience and the abstraction of abstraction. *Systema: Connecting Matter, Life, Culture and Technology, 1*(3), pp. 50–67.

Seaman, W. (2014). A multi-perspective approach to knowledge production. *Kybernetes, 43*(9/10), 1412–1424, available at: https://doi.org/10.1108/K-07-2014-0145

Seaman, B., & Supko, J. (2014). s_traits (Record album). Manchester, UK: Cotton Goods. See https://www.johnsupko.com/s_traits

Spencer-Brown, G. (1979). *Laws of form*. New York: Dutton. (Originally published in 1969 by George Allen and Unwin)

Steier, F. (1991). *Research and reflexivity*. London: Sage.

Thomas, L., & Harri-Augstein, S. (1985). *Self-organised learning: Foundations of a conversational science for psychology*. London: Routledge & Kegan Paul.

Varela, F. (1976). Not one, not two. *CoEvolution Quarterly, 12,* 62–67. Retrieved September 27, 2019 from: http://cepa.info/2055

Varela, J., Thompson, E., & Rosch, E. (1993). *The embodied mind: Cognitive science and human experience*. Cambridge, MA: The MIT Press.

Von Foerster, H. (Ed.) (1974). *Cybernetics of cybernetics. The control of control and the communication of communication.* Urbana, IL: University of Illinois.

Von Foerster, H. (Ed.). (1995). *Cybernetics of cybernetics. The control of control and the communication of communication.* Minneapolis, MN: Future Systems (Reprint of 1974 ed.)

Von Foerster. H. (2003a). *Understanding understanding: Essays on cybernetics and cognition*. New York: Springer.

Von Foerster, H. (2003b). Objects: tokens for (eigen) behaviors. In *Understanding understanding* (pp. 261–271). New York: Springer. (Originally presented as a paper at the University of Geneva on June 29, 1976 on the occasion of Jean Piaget's 80th birthday.)
Von Foerster H. (2003c). What is memory that it may have hindsight and foresight as well? In *Understanding understanding* (pp. 101–131). New York: Springer. (Originally published in 1969)
Wiener, N. (1950). *The human use of human beings: Cybernetics and society.* Boston: Houghton Mifflin
Wiener, N. (1961*). Cybernetics, control and communication in the man and the machine* (2nd ed.). Cambridge, MA: The MIT Press.
Wilhelm, R. & Baynes, C. F. (Trans.) (1968). *The I Ching, or book of changes* (Foreword by C. G. Jung). London: Routledge & Keagan Paul.
Wittgenstein, L. (1953). *Philosophical investigations.* New York: MacMillan.
Wittgenstein, L., (1956). Remarks on the foundations of mathematics. New York: MacMillan.
Wittgenstein, L. (1967). Zettel. Berkeley, CA: University of California Press.
Yates, F. A. (1966). *The art of memory.* London: Routledge and Kegan Paul.
Zumthor, P., & Bourgeois, L. (2012, January 3). *Steilneset memorial* Retrieved July 25, 2018 from: http://www.dezeen.com/2012/01/03/steilneset-memorial-by-peter-zumthor-and-louise-bourgeois/

Ranulph Walking. Courtesy of Aartje Hulstein
and the Glanville Archive, University of Vienna.

Ranulph Glanville:
Cybernetician of Ignorance (Expanded)[1]

Søren Brier[2]

Recently our columnist Ranulph Glanville received the rare distinction, the degree of Doctor of Science (DSc) from Brunel University, in Uxbridge, Middlesex, Great Britain with this citation:

> Doctor Ranulph Glanville has led a long, varied and distinguished career spanning the fields of architecture, cybernetics, design theory, and philosophy. Having first qualified from the Architectural Association School, London, he studied for two periods at this University and was awarded the degree of Doctor of Philosophy there on two occasions. The first in 1975 was on the subject of "A Cybernetic Development of Epistemology and Observation Applied to Objects in Space and Time as seen in Architecture"; the second, in 1988, was entitled "Architecture and Space for Thought." Over his career, Dr Glanville has worked as a consultant, researcher, teacher, scholar, and adviser within higher education and more widely. His activities have ranged across no fewer than five continents, making him a truly international figure. He has published more than 300 papers, books and other writings, been a member of several editorial boards and received numerous awards. He is also a commissioned musical composer, with his roots in the 1960s Avant Garde London music scene. As a student, he was seen to have been ahead of his time as an architectural thinker. He is now regarded an outstanding and authoritative figure in the field of cybernetic epistemology, a subject at the very core of modern cognitive sciences, design research, and architecture studies.

I find that this is a unique opportunity to give a short overall description of Glanville's work and present his production in an orderly manner for those who want to study it further. Part of the procedure by which such a degree is awarded is a thorough peer evaluation by three senior experts in the area of the whole produced corpus of the candidate's work. Fortunately, the evaluators have given me their texts to use as base material from which this short paper is written. Thanks to Mary Catherine Bateson, Dirk Baecker, and to Stephen Gage, who are—so to say—my co-authors.[3]

---

1. This is an expanded version in the form of a manuscript. The earlier, related document is published as: Brier, S. (2008). Ranulph Glanville: The cybernetician of ignorance. *Cybernetics & Human Knowing*, 15(1), 81–89.
2. Email: sorenbrier@gmail.com
3. I do not reference these scholars' individual contributions, when I use formulations from them, as their is no public access to their evaluation papers. I have added my own contributions from the perspective of philosophy and theory of science as well as from semiotics and cybersemiotics here and there, and they should be easy to identify.

The publications that Glanville submitted for scrutiny—see bibliography at the end of this article—represent about a third of his corpus and included all the columns he has written for *Cybernetics & Human Knowing*, which is another reason for this review article. The selection of texts is due for publication as a 3-volume set entitled (provisionally) *The Black B∞x* in 2009/2010 by edition echoraum in Vienna, as part of the series *Complexity / Design / Society*.

What is a DSc?
The degree of Doctor of Science (DSc) is not common and needs some explanation. In the first place, it is essentially a British degree found in Britain and those parts of the world (e.g., Australia) where Britain has been influential in the institution of University Education. It is a degree that is earned through examination, rather than (as, for instance, in the USA) an honorary degree. Here is the official description of the degree, taken from the regulations of Brunel University:

> The degree of Doctor of Science (DSc) shall be awarded on the basis of distinguished original work which has established a candidate's position as an authority in his or her field of study. The title of the higher doctorate shall appropriately reflect the candidate's field of study. The original work can be demonstrated in the form of either
>
> (a) published material or
> (b) scientific or technological innovation, development, or achievement.
>
> The submission may contain material certified as refereed and accepted for publication by a reputable learned journal.

Thus, there are other degrees of similar status appropriate to non-scientific fields, such as a Doctor of Literature. It is the highest degree awardable and it is rare: Brunel University awards on average one higher doctorate a year. The process of examination is firstly an informal internal submission to the university, using external consultants when necessary to determine that the submission is likely to represent a sufficient standard, and to help select appropriate examiners. The full submission (often consisting of all publications, but, in this case, because Glanville's list of publications tops 300 items, made up of a selection of approximately a third) is then sent to the examiners for comment, and the award is made if all examiners agree the proper standard has been reached. When asked why he submitted work for this degree, Glanville says that one reason is to increase the acceptability and recognition of cybernetics as a field. This of course is also one of the purposes of this review article, which again ties in to the next question one can ask.

How Did Glanville Come to Write a Column for CHK?

After having seen Ranulph Glanville perform at a couple of the American Society's conferences from as early as 1990 in Virginia Beach, and read some of his work, I met him again between August 81, 1994, at the second Orwellian symposium in Karlovy Vary, Czech Republic. This was part of the International Conference on Interdisciplinary Research, arranged by The International Institute for Advanced Studies in Systems Research and Cybernetics who hold their yearly conferences at Baden Baden. We had a long talk about the history of cybernetics and second-order cybernetics and I asked him if he would be able to help with the then fledgling journal, *Cybernetics & Human Knowing*, because he was a central figure in the field of second-order cybernetics with a commitment to keep the field alive and also had a commitment to explain and celebrate the work of the (other) founders in the field, where he maintained a good connection to those who were alive.

He said that he needed to think about it because of his existing editorial commitments, and returned with the suggestion that he could write a column for each issue. We discussed this and since he is one of the still active cyberneticians with the longest and oldest roots in the cybernetic movements I agreed. I suggested that he should write a series of scholarly, problem-oriented, but personally reflected columns on the understanding and use of cybernetic thinking, which is still a very misunderstood area. These should not be formal articles with formal peer review, but the result of a dialogue between the writer and me, where I would defend the "short form," the readability by asking stupid questions when I did not understand the text, and be the devil's advocate around the argumentation. Originally his columns appeared 4 times a year, but with the addition of Louis Kauffman—the father of knot logic and one of the leading authorities on Spencer-Brown's laws of form—as a second columnist and later the American Society for Cybernetics column, we agreed Glanville and Kauffman would alternate, each writing two columns per year.

How Was His Work Judged to Qualify for a DSc?

Ranulph Glanville started out in architecture and design. He has for many years taught in schools of architecture. But Glanville's work in cybernetics has continuously engaged him with fundamental epistemological questions. It has led him to an understanding of the processes of thought and knowledge creation that is basically conversational, recursive, and constructivist. Cybernetics, in its self-understanding as applied epistemology, has had a recurrent identity crisis because its tools and concepts, often in only half-understood forms, have been taken from it to be applied in other disciplines. The first generation of the discipline—the founders—has now died out. Their works are being archived for further use and monographs are written in order to interpret their contributions. In the paradigm shift from first-order to second-order cybernetics led by Glanville's first mentor, Heinz von Foerster, Glanville played a major role. Now standing among the second generation, Ranulph Glanville has emerged to play a key, unifying role within the discipline. Much of his work has been focused on clarifying the connections between cybernetics and other disciplines. He has further emphasized the need for further exploration of these same tools, such as: What really is a black box? What is a deeper understanding of fundamental concepts like distinction and observer? He has also demonstrated the extent to which earlier work, particularly by social scientists, actually fits the new paradigm.

But the prime example is the way Glanville has explored and enriched the relationship between cybernetics and the fields of architecture and design. In these fields, many associate cybernetics with merely using computers as swift and skilful replacements for draftsmen. In contrast, Glanville focuses on the process whereby a design emerges and is developed and refined through multiple iterations of discovery. He points out that this is done in a reflective process of self-correction, whether the designer is using a pencil or a computer. This process is not accelerated and enriched by the computer's capacity to deal with complex wholes. Where architects have called for making the design process more scientific, he has shown that scientific research is itself a self-reflexive design process. The ways the scientist is included in the process, is especially describable in terms of second-order cybernetics. Glanville both refines his thinking and tailors his presentation to different professional audiences, particularly in addresses to different scientific societies. It has been critical to Glanville's professional role that his presentations be persuasive. In practice this means they are frequently playful, using humour and metaphor and phrased in personal terms. In this way many of Glanville's texts deviate stylistically from typical scholarly papers, designed to display erudition and academic authority. Glanville does not display knowledge but rather elicits knowing, exemplifying an educational practice directly reflective of his psychological and epistemological premises. This can especially be seen in Glanville's papers on education, which focus on learning rather than on instruction.

These papers not only reflect internal conversation, they invite it, respecting the autonomy of the reader's processes of interpretation and assimilation. This is a rare quality in scientific writing and teaching. But that is not to deny that many of his papers use complex formal arguments that conform to the standard styles of scholarly presentation. Glanville's scholarship is also carefully situated, through historical discussion and detailed bibliography, and focused on individual scholars who have been important to him, such as von Foerster and Pask, as well as Varela and Froebel.

I am proud that the evaluators point out that the collection of columns written for the journal *Cybernetics & Human Knowing* provides a particularly strong picture of Glanville's range and role in cybernetics. They demonstrate that the originality and quality of Ranulph Glanville's work consists in getting to the very core of the conceptual foundations of cybernetics, architecture, design, and education. It is pointed out that nowhere as in Glanville's work is one able to find such rigorous reflections on the core notions of black boxes, distinction, knowing, and Objects (a technical term in Glanville's work), which points to the success of our personalized short form.

The evaluators also point out how Glanville's work plays a considerable role in foundational studies of social theory, as pursued in German academies, especially by the late Prof Dr Niklas Luhmann. Luhmann was able to identify a highly original line of thought, leading from German idealism and romanticism (Fichte, Hegel, Novalis, Schlegel) via English romanticism (William Blake, Thomas Carlyle) to the epistemology developed in second-order cybernetics. Here part of Glanville's work has been highlighting a general notion of self-reference (or self-referential constitution of subjects) applying not only to human beings, but possibly also to organisms, social systems, and even things or Objects. These Objects need to be complex enough to have to rely on some kind of a self-modeling device in order to be able to play the role they do play in a society. Glanville has been able to show that all kinds of Objects can only work if they are embedded within a formalism of second-order observation. This does not assume identity, unambiguous determination and causality, but distinction and recursion plus synthetically determined, yet analytically undeterminable black boxes. Close to the work of Bruno Latour and Callon's actor networks theory, which, however, is semiologically based, Glanville was able to show in a number of his papers that all kinds of Objects are only possible if embedded within loops of circularity, which both separate and entangle first-order operations (connections) and second-order observations (references). Any kind of knowledge or rather operation of knowing, is of an emergent quality, based on interaction, the only level able to constitute self-transparency of relations (or whiteness). It is one of the basic claims of Glanville's work that any attempt to determine from which substance, or essence, that whiteness comes, ends up with the discovery of two black boxes interacting. Thus from his arguments about the black boxes comes his main philosophical idea that all the knowledge we build from them is based on a profound ignorance. Hence the title of this review essay. His (consequent) argument, developed in, for example, architecture and design, that meaning is a personal construct, rather than existing independently as such in a visual language, still causes great unease, among many designers.

This forces the results of his research into the very core ideas of more recent cognitive sciences, which are *emergence, interaction,* and *boundary objects,* where he draws the attention to processes constituting reality via operations, rather than having to rely on elements being already "out there" and awaiting their combination for various systems' uses. One of his leading metaphors became that of von Foerster's non-trivial machine, whose actual operations produce an emergent reality. Any one Object becomes a boundary Object since it is addressed by operations and observations in a variety of ways. Objects are central to our conceptualization of the space that must exist, in which we recognize those things we take to populate our world (such as buildings) as things with complex attributes, which go beyond immediate function. This is an important argument in Jakob von Uexküll's cybernetically founded biosemiotic idea of organisms' Umwelts and it is carried on in C. S. Peirce's semiotic concept of the object in the triadic semiosis.

In his work Glanville showed what kind of notions, concepts, and models are appropriate in helping the student of architecture and design to construct his way in (a) posing a problem, (b) evaluating possible solutions, and (c) looking at him- or herself being one among the relevant restrictions of the posing of problems and the evaluation of solutions. This is a good philosophy of science reflection that Bourdieu also made. He included a description of the cognitive background of the researcher in his methodology in order for readers of the results to be able to evaluate the character and framework of the researcher who had done the research by using his personality as a tool in the social research. The cybernetic circle is evident here in having to know about yourself first, in order to be able to watch, and accordingly vary, the way you look at things and how things (Objects) look at you as both an epistemologically necessary and an interactively charming way to self-correct how one looks at the world. Bruno Latour's work also tried to take into account different kinds of observers (organs and cells of human bodies, animals, plants, technical artefacts, social networks) observing us while we try to look at them.

Learning to know in both education and research means learning to look at yourself doing the knowing and doing the learning how to do it. Glanville's work consists in founding a kind of ecological research, which undercuts and subverts the Cartesian distinction between subject and object and looks at our scientific research as being just one way by which human beings may inhabit their niche and change, less and less inadvertently, the world surrounding that niche by the way we inhabit it. I guess that is why our—sometimes heated—discussions about the contents and form of the columns have been so productive. Glanville goes past the personal and straight for the knowledge gain of disagreement every time. This is in line with the careful way he looks at the distinction between epistemological self-reference, on one hand, and on the other, ethical consequences drawn for a reflection on our ecological responsibility. His work on black boxes allows us to understand how the observer must "create" the environment on his or her own terms, and how this process develops the need for concepts of conversation in order to discuss the sharing of experience. This is where I see one of the important links to semiotics, as well as the necessity for a cybersemiotic framework, as not even second-order cybernetics includes concepts of understanding and signification. Anyway because of Glanville's modesty, such conversations with him most often turn out to be mutually informative resulting in new knowledge, which is one of the reasons I felt the wish to write this review. His papers on education summarize some of the aspects of this, especially the extent to which conversation theory can be developed in teaching practice.

I think it says a lot about what universities have become in the so called knowledge society that Glanville felt it necessary—in order to carry out his scholarly vision—to leave his secure and tenured post in the University of Portsmouth in frustration in order to become an independent scholar. This has lead him, for the past 13 years, to travel between universities in the USA, Australia, Hong Kong, Amsterdam and Brussels—as well as in the UK—as an expert teacher and consultant in his fields, leaving the support of his wife Aartje (who he counts the best critic of his work) at home too much of the time: He says their marriage is sustained by extensive e-mailing! Busy as ever, always writing and editing even while he travels, he deals with continuing family obligations for his elderly parents, without complaint. Yet I have always been able to reach him within a couple of days when we were discussing his column. How he can stand this pressure of work and travel at his age without becoming ill shows me the strength of a mind dedicated to producing knowledge, and to helping others to do so, too, in a responsible and ethical way. His international network in itself is of an overwhelming size and I do not know how he manages to keep up with it. But in accordance with his theory he is somehow the non-trivial product of the recursive differences and distinctions of that network, of which I am proud to be a part. Glanville has made a major contribution to the network behind CHK in upholding our ethics of keeping disagreements fruitful and in looking at them as circulating differences that we encourage to emerge as differences that really make a difference out of our mutual cybernetic ignorance!

Bibliography to the D.Sc.

0 Black Box (and between)

0.1 (1979). The form of cybernetics: Whitening the black box. In J. Miller (Ed.), *Proceedings of 24 Society for General Systems Research* (pp. 35–42). Louisville, KY: Society for General Systems Research.

0.2 (1982). Inside every white box there are two black boxes trying to get out. *Behavioural Science, 12*(1), 1–11.

0.3 (1992). CAD abusing computing. *Proceedings eCAADe 1992*. Barcelona: Polytechnic University of Catalonia.

0.4 (1995). Architecture and computing: A medium approach. *Procs. 15th Meeting of Association for Computing in Architectural Design in America* (pp. 5–20). Seattle: University of Washington.

0.5 (1997). Behind the curtain. In R. Ascott (Ed.), *Procs. First Conference on Consciousness Reframed*. Newport: UCWN.

0.6 (2001). An intelligent architecture. *Convergence: Journal of Research into New Media Technologies, 7*(2), 12–250.

1 Cybernetics (and second-order cybernetics)

1.1 (1987). The question of cybernetics. *Cybernetics, an International Journal, 18*, 99–112.

1.2 (1990). Sed Quis Custodient Ipsos Custodes. In F. Heylighen, E. Rosseel, & F. Demeyere, (Eds.), *Self-Steering and Cognition in Complex Systems* (pp. 107–112). London: Gordon and Breach.

1.3 (1994). As if (radical objectivism). In R. Trappl (Ed.), *Cybernetics and Systems Research '94: The proceedings of the European meeting on cybernetics and systems research* (pp. 613–620). Singapore: World Scientific.

1.4 (1995). Chasing the blame. In G. Lasker (Ed.), *Research on Progress and Advances in Interdisciplinary Studies on Systems Research and Cybernetics* (vol. 11). Windsor, Ontario: IIASSRC.

1.5 (1997). A ship without a rudder. In R. Glanville, & G. de Zeeuw (Eds.), Problems of excavating cybernetics and systems (pp. 131–142). Southsea, UK: BKS+.
1.6 (2001). An observing science. *Foundations of Science, 6*(1-3), 45–78.
1.7 (2001). Listen! In G. de Zeeuw, M. Vahl, & E. Mennuti (Eds.), Problems of participation and connection (pp. 425–432). Lincoln, UK: Lincoln Research Centre.
1.8 (2002). Second order cybernetics. In Encyclopaedia of life support systems. Oxford: EoLSS Publishers. (Web publication at URL hyperlink http://www.eolss.net)
1.9 (2003). Behaving well. In I. Smid, W. Wallach, & G. Lasker, et al. (Eds.), Cognitive, emotive and ethical aspects of decision making in humans and in AI (Vol. II, pp. 61–70). Windsor, Ontario: IIASSRC.

2 Design

2.1 (1977). Amazing space: For the architectural stimulus-response rat? *AAQ, I 9*(2/3), 40–57.
2.2 (1980). The architecture of the computable. *Design Studies, 7*(4), 217–225.
2.3 (1981). Why design research? In R. Jacques & J. Powell (Eds.), Design/Method/Science (pp. 86–94). Guildford: Westbury House.
2.4 (1999). Researching design and designing research. *Design Issues, 15*(2), 80–9.1
2.5 (2003). An irregular dodekahedron and a lemon yellow citroen. In L. van Schaik (Ed.), *Practice of practice* (pp. 258–265). Melbourne: RMIT Press.
2.6 (2005). Appropriate theory. *Proceedings of FutureGround Conference of the Design Research Society* (unnumbered on CD). Melbourne: Monash University.

3 Distinction

3.1 (1979). Beyond the boundaries. In Ericson, R (Ed.), *Proceedings Society for General Systems Research Silver Jubilee Conference* (pp. 70–75). London: Springer Verlag.
3.2 [with F. Varela] (1981). Your inside is out and your outside is in. In G. Lasker (Ed.), Applied systems & cybernetics (vol. II; pp. 638–641). Oxford: Pergamon.
3.3 (1990). The self and the other: The purpose of distinction. In R. Trappl (Ed.), *Cybernetics and Systems '90: The proceedings of the European Meeting on Cybernetics and Systems Research* (pp. 349–356). Singapore: World Scientific.
3.4 (2000). Living in lines. In R. McLeod (Ed.), *Interior cities* (pp. 178–183). Melbourne: RMIT Press.

3.5 (2002). Francisco Varela: A working memory. *Cybernetics & Human Knowing,* 9(2), 67–76.

4 Education

4.1 (1980). Construct heterarchies (rev. ed.). In M. Shaw (Ed.), Recent advances in personal construct technology (pp. 135–145). London: Academic Press.

4.2 [with R. McKinnon-Wood] (1996). NOAH: The ark of knowing in a learning environment. In R. Trappl (Ed.), *Procs. 13 EMCSR* (pp. 449–454). Vienna: University of Vienna and Austrian Society for Cybernetic Studies.

4.3 (2002). A (cybernetic) musing: Cybernetics & Human Knowing. *Cybernetics & Human Knowing,* 9(1), 75–82.

4.4 (2002). A (cybernetic) musing: Some examples of cybernetically informed educational practice. *Cybernetics & Human Knowing,* 9(3), 117–126.

4.5 [with L. van Schaik] (2003). Designing reflections: Reflections on design. In D. Durling & K. Sugiyama (Eds.), Proceedings of the third conference, Doctoral Education in Design (pp. 35–42). Chiba, Japan: Chiba University.

5 Knowing (and knowledge)

5.1 (1984). The one armed bandit. In J. Powell, I. Cooper, & S. Lera (Eds.), Designing for building utilisation (pp. 124–132). London: Spon.

5.2 (1983). Introduction: Behind the screen. In R. Glanville, & G. de Zeeuw (Eds.), Interactive interfaces and human networks (pp. 7–11). Amsterdam: Thesis Publishers.

5.3 (2003). Ein Wort iiber Wissen—A Note on Knowing (N. Ort, Trans.). In O. Jahraus & N. Ort (Eds.), *Theorie—Prozess—Selbstreferenz* (pp. 187–197). Konstanz, Germany: UVK Verlagsgesellschaft.

5.4 (2003). Architecture and the embodiment of knowledge. In E. Griin & E. del Carlo (Eds.), *Ensayos Sobre Sistemica y Cibernetica.* Buenos Aires, Editorial Dunken.

5.5 (2005). A (cybernetic) musing: Certain propositions concerning prepositions. Cybernetics & Human Knowing, 12 no (1-2), 87–95

5.6 (2006). Design prepositions: Keynote lecture at the Conference on The Unthinkable Doctorate. Brussels, April 2005. (written version, to be published in the proceedings)

6 Objects

6.1 (1976). What is memory, that it can remember what it is? In R. Trappl et al. (Eds.), *Recent progress in cybernetics & systems research: Proceedings 3 European Meeting on Cybernetics and Systems Research* (vol. 7, pp. 27–37.) Washington, DC: Hemisphere Press.

6.2 (1977). The nature of fundamentals, applied to the fundamentals of nature. In G. Klir (Ed.), *Proceedings 1 International Conference on Applied General Systems: Recent developments & trends* (pp. 401–409). New York: Plenum.

6.3 (1980). All thoughts of things. In R. Trappl et al. (Eds.), Progress in cybernetics & system research (vol. 9, pp. 437–446). Washington DC: Hemisphere Press.

6.4 (1980). Consciousness, and so on. *Journal of Cybernetics, 10*, 301–312.

7 Others

7.1 (1996). Heinz von Foerster: The form and the content. In Heinz von Foerster—a Festschrift. *Systems Research, 13*(3), 271–278.

7.2 (2001). And he was magic. Special issue: Gordon Pask—a Commemoration. *Kybernetes, 30*(5/6), 652–672.

7.3 (2002). Doing the right thing: The problems of Gerard de Zeeuw, academic guerilla. Special issue: In G. de Zeeuw—a Festschrift. *Systems Research and Behavioural Science, 19*(2), 107–113.

7.4 (2003). Machines of wonder and elephants that float through air. In Heinz von Foerster—a Commemoration. *Cybernetics & Human Knowing, 10*(3-4), 91–105.

7.5 (2005). Lerner ist Interaktion: Gordon Pask's *An approach to cybernetics*. In D. Baecker (Ed.), Schlsselwerke der Systemtheorie (pp. 75–94).Wiesbaden, Germany: Verlag fur Sozialwissenschaften. (unpublished English original text)

7.6 (2006). Learning from Locker. *Kybernetes, 35*(1-2), 223–227.

8 (Re-)presentation

8.1 (1977). Finnish vernacular farmhouses. *AAQ London, 9*(1), 38–52.

8.2 (1980). Mapping realities. *AAQ London, 12*(4), 20–31.

8.3 [with A. Pedretti] (1980). The domain of language. In R. Trappl et al. (Eds.), *Progress in cybernetics & systems research* (vol 11; pp. 235–242). Washington DC: Hemisphere Press.

8.4 (1984). Distinguished and exact lies. In R. Trappl (Ed.), *Cybernetics & systems research 2* (pp. 655–662). New York: Elsevier North Holland.

8.5 (1984). Emptiness. In R. Trappl (Ed.), *Cybernetics & systems research 2* (pp. 663–664). New York: Elsevier North Holland.

8.6 (1996). Communication without coding: Cybernetics, meaning and language (How language, becoming a system, betrays itself). *Modern Language Notes, 111*(3), 441–462.

9 Variety

9.1 (1994). Variety in design. *Systems Research, 11*(3), 95–103.

9.2 (1998). A (cybernetic) musing: Varieties of variety. *Cybernetics & Human Knowing, 5*(1), 57–62.

9.3 (2000). The value of being unmanageable: Variety and creativity in CyberSpace. In H. Eichmann, J. Hochgerner, & F. Nahrada (Eds.), Netzwerke (pp. 27–38). Vienna: Falter Verlag.

9.4 (2004). A (cybernetic) musing: Control, variety and addiction. *Cybernetics & Human Knowing, 11*(4), 95–103.

Books

[with G. Pask, G., & M. Robinson] (1980). *Calculator Saturnalia*. London: Wildwood House.

(1988). *Objekte* (selected papers translated by D. Baecker). Berlin: Merve Verlag.

Ranulph Glanville's *Objekte*[1]

Bernard Scott[2]

In this review of Ranulph Glanville's book *Objekte*, I begin by covering the following topics: my own relationship with Glanville, his relationships with peers and mentors and how *Objekte* came into being. I then briefly discuss some of *Objekte*'s major themes. I end by positioning Glanville's book as a valuable contribution to cybernetics.
Key words: Ranulph Glanville, cybernetics, Objects, observer, self-reference

Introduction

My discussion of Ranulph Glanville's book *Objekte*[3] takes the following form. First I say something about Ranulph Glanville, as I know him, as my friend and fellow cybernetician. I then, in the section entitled "On peers and mentors," comment on where I see Glanville's contributions sitting in the midst of the work of other cybernetic luminaries. I then say something about how *Objekte* came into being. I have had an opportunity to ask Ranulph about this and found what he had to say quite illuminating both as a description of a personal odyssey of ideation and creativity and also as an interesting case study of how cyberneticians may fail or succeed at doing cybernetics of cybernetics, that is, of fostering the discipline that is fostering them. I then look briefly at the major themes to be found in *Objekte*. This is of course a personal selection. I hope I say enough about *Objekte* in this chapter as a whole to inspire the reader to explore not just *Objekte* but also Glanville's many other writings on cybernetics and related topics for herself.

Finally, I make some concluding comments. These are intended to be inspirational with respect to what cybernetics still has to contribute to the mainstreams of philosophical and scientific thought. As part of my polemic I position Glanville's oeuvre as a central, conceptually clear and comprehensive contribution to the cybernetics enterprise.

---

1. Scott, B. (2005). "Ranulph Glanville's *Objekte*: An appreciation." Invited chapter for D. Baecker (Ed.), Schluesselwerke der Systemtheories, Wiesbaden:VS Verlag für Sozialwissenschaften. (With added abstract and key words.)
2. Email: bernces1@gmail.com
3. Glanville (1988).

About Ranulph Glanville

Ranulph Glanville is my peer and my friend, a friendship that has matured beautifully thanks to Glanville's enthusiasm and generosity. As contemporaries doing PhDs in cybernetics with Gordon Pask at Brunel University, West London, UK, in the late 1960s and 70s, we have much in common—shared interests, shared experiences and have also affected each other in many ways as role models, counter-role models, as sources of ideas and as appreciators of ideas. However, I can also say that after more than 30 years there are blind spots. There are things we do not know about each other, partly because our sense of intimacy, our sense that we do know each other, serves to make us overlook what we do not know. One does not go looking for something one does not know exists.

What an exciting life Glanville has led in his intellectual journeys and his geographic journeys! His enthusiasm for and commitment to cybernetics is indicated by the personal sacrifices often involved, as he is a largely self-financed, independent academic. He makes a point of attending conferences and being an envoy for cybernetics and does so as best he can without the support of an academic institution or other source. He is also prolific as a writer about cybernetics and about a wide range of topics that he enlightens from a cybernetic perspective, including education, language and communication, design and innovation. (See, e.g., his regular column, "Cybernetic Musings," in the journal *Cybernetics & Human Knowing*.)

Having read or reread the papers in the *Objekte* collection, I find much light has been shone into areas I was not familiar with. What I see revealed in his writings is even more breadth and depth than I had understood existed in his thought. I believe Glanville's work stands quite appropriately alongside that of Pask, von Foerster, Maturana and Varela, all of whom are generally better known and with all of whom Glanville has had close personal and professional relationships. Glanville has also had a close connection and involvement with the evolving ideas of Ernst von Glasersfeld on radical constructivism and Niklas Luhmann on social systems.

In his many commentaries on others' work, Glanville illumines their separate positions and brings out similarities and differences. He has also forged his own view. Perhaps it is his acceptance of the uniqueness of any particular self-referential system (a Glanville *Object*, as expounded in *Objekte*) that makes him particularly tolerant of the multiplicity of personal views and to recognise that they are not necessarily in competition. (Note: Glanville uses upper case *o* for his Objects to distinguish them from our accustomed use of the term *object*.) It may seem that the metaphysical edifices of Maturana, Luhmann or Pask are in competition with each other. Gotthard Gunther[4] has commented that "Cybernetics is in search of a theme that is hidden." For Glanville, the many variations on that hidden theme are reflections of the inevitable fact of our differences one from another. As we shall see (below), his own metaphysical edifice, rather than attempting to reveal the hidden theme, sets out the form of our searching for any theme, hidden or not. An understanding of this form may then be reflexively and pragmatically taken as the solution that we seek. As Warren McCulloch liked to say (and as many wise persons have said before him), "Do not look at my finger, look to see where it is pointing."

Although Glanville promotes cybernetics and especially the work of his mentors he does not promote himself, at least not to the same extent. He is astonishingly modest, perhaps because he has failed to be recognised by a narrow-minded, ossified mainstream. His innate sense of personal dignity and integrity forbid him from contesting for attention. When peers fail to recognise one's gifts, passion or commitment, one may become disheartened and give up or one may stoically, perseveringly and energetically soldier on. This latter has been Glanville's preferred choice. As noted, Glanville has been prolific in his commitment to cybernetics with more than 250 papers in print. He has also found time to be an artist, musician and creative educator.

Glanville is Anglo-Irish but he is also unashamedly a cybernetician. Given the general parlous state and lack of appreciation of cybernetics in the UK and most of the English speaking world, it is perhaps no surprise to find he is far more celebrated in Europe and especially Germany and Austria than elsewhere. In those latter countries there is much more interest in self-referential and autopoietic processes than elsewhere. Glanville's carefully argued logical poems have been recognised for the incisive gems that they are. My own observation is that he has also won over hearts and minds by his passion and commitment to living out the cybernetic ethical ideals that he espouses. He does take responsibility for what he says and does as they affect the world around him. He does aspire and strive to act with the generosity and courtesy that he believes one self-referential Object should afford to another. However, he would also be the first to admit that he is not a saint. My observation is that he is intolerant of cant and impatient of others' blatant and not so blatant attempts to push their personal agendas, preferring argument to assertion and intellectual bullying, preferring clarity and simplicity to inflation and obfuscation. My observation is that for

---

4. Gunther (1972, p. 33)

Glanville life and work are inseparable. Life is committed to doing dutiful work, diligently. In that sense Glanville competes only with himself. His life is also a work, an adventure, a creation, a living out of the art of living. He travels, he narrates, he does art, music and design, he fosters a network of relationships—not just professional and collegiate but also an extended network of family and friends, those whom he loves and with whom he has fun.

On Peers and Mentors

At this point, I would like to say something about the way that mentors and students and peers interact within a field. I say this with some awareness that Pask's conversation theory,[5] which has been a great influence on both myself and Glanville (indeed, we both consider ourselves to have contributed to its development), has much to say about how professional relationships and institutions evolve.

There is an explicit discourse between professors and their students and between peers in the field that is documented in lecture notes, published papers and so on. There is also an implicit and largely non-conscious but potentially conscious discourse at work. To know two or three persons fairly intimately through their writings over a period and to know that they have interacted during that time because of their common interests, permits one, at least hypothetically, to construct the conversations by which one actor on reading the work of another has absorbed the other's ideas into his own work. These absorptions, these connections, whether in agreement or disagreement are, I believe, often made without direct reference not because of some desire to deceive but simply because we do absorb ideas from each other as we go along without necessarily realising that this is happening. Thus Glanville in many turns, in many recursions, can be seen in his writings to be rewriting and developing the ideas of his mentors. But it is also the case that in these interactions over time one can observe the mentors' rewriting of Glanville's writings. As this process unfolds ownership of concepts becomes a moot if not an irrelevant point.

Pask mainly drew on and referred to work that could readily be incorporated in or demonstrated by his own theoretical edifice, conversation theory. Glanville's abstraction did not fit this bill. Neither von Foerster nor (especially) Maturana were wont to refer to the work of colleagues and peers. It is as if they wished their ideas to be presented in a pristine, pure form, untarnished by the requirements of scholarly niceties.

---

5. See Pask, Scott, and Kallikourdis (1973) and Pask (1975).

In earlier drafts of this chapter I set out to give examples of the interplay between Glanville and von Foerster and between Glanville and Pask. In the end I decided I did not have sufficient valid data to do this with confidence. Elsewhere I have given an overview of the development of second order cybernetics with some account of the collegiate interplay within the cybernetics community, as I perceived it at the time.[6]

How Objekte Came Into Being

Glanville was trained as an architect. The initial motivation for his PhD was an interest in artificially constructed environments containing interactive objects. Recall, this was in the 1960s and early 1970s when many visionaries, not least Pask, were anticipating a world of interactive cybernetic artefacts.

In architecture, this ranged from the mundane idea of homes that are self-maintaining to autonomous robot servants that interpret and anticipate human needs. Glanville's investigations took him down the artefact designing and building route. However, he was eventually driven to write the thesis that he did by first of all having to respond to the need to formulate a question that was nagging at him to be answered and then to find an answer for that question. I have recently asked him about this period of his work. He tells me he "wrote to find both question and answer," following a daily writing regime for several months generating many drafts. I see this as a particularly fascinating case study of creativity. The writing was not quite automatic writing but it was given a freedom to flow. I would characterise this as giving permission for multiple viewpoints to emerge as multiple voices engaged in conversation. Glanville would in a separate phase of the creative process read over and attempt to understand and critique what he had written. The reader should appreciate Glanville was struggling to expose in language that which had not yet been said, indeed, that which language as a tool or medium keeps hidden from our awareness.

This creative period came to a close one particular day when the question and answer were fully crystallized. In a fifteen hour stint of writing, Glanville wrote the heart, the core of his thesis.

And what was the question that had to be formulated and then answered?

"How is it that observers, all of whom observe and know differently from each other, come to believe, and behave as if they are observing the same thing?" And the answer? He had recognised that any entity, to have

---

6. Scott (2004).

the stability that constitutes it as an entity, has to remember itself, to reconstruct itself as itself. Memory is the process that has itself as a product. With one stroke Glanville places circularity and self-reference at the very heart of an abstract cybernetic algebra. Expressed in later terminology, Glanville develops a second order cybernetics and at the same time establishes that all first order forms (observed systems) intrinsically carry with them second order considerations (all observed systems are observing systems; all systems are, mutually, self-observing). By this stage in Glanville's thought he had discovered that the word *object* had once upon a time had the connotation that we now give to the word *subject*. As noted above, as a quirky twist on this historical change of meaning, Glanville refers to his self-observing entities that are both subject and object for themselves as *Objects*.

On completion of the oral examination of his thesis, he was asked by his external examiner, Heinz von Foerster, what he proposed to do next. Glanville had not thought that far. Von Foerster suggested he could either render the thesis in book form or he could write a series of shorter papers, in which the core ideas of the thesis were elaborated and applied. Glanville chose this latter option. *Objekte* is a collection of some of the first in this series of papers.

Objekte came into being thanks to the great German sociologist and social systems theoriser Niklas Luhmann. Luhmann read several of Glanville's papers and suggested they be published as a book. He helped find a publisher and enrolled Dirk Baecker, one of his students, to serve as translator.

I find it particularly apt that of all the great cybernetic theorisers, it was Luhmann who most strongly acknowledged the importance of Glanville's work, as Luhmann himself has a most sophisticated and clear grasp of the abstract, philosophical issues surrounding concepts of self-reference.[7]

Glanville's Theory of Objects

Lest my efforts at summarising Glanville's work should lack the cogency and clarity of the Master himself, at this point I shall quote extensively from a paper written some twenty years after the publication of *Objekte* in which, as part of an extended essay on the concept and development of second order cybernetics, Glanville gives an account of his own contribution:

---

7. See especially Luhmann (1995, chapter 12).

> My work might be thought of as a generalization of the work of the others. My major initial concern was to develop a set of concepts that might explain how, while we all observe and know differently, we behave as if we were observing the same thing. What structure might support this?[8]

Notice that Glanville is concerned to acknowledge the contributions of peers and mentors. He later lists the key players as Pask, von Foerster, Maturana and Varela. He continues:

> My contribution was a structure developed to accommodate observation and difference. This was achieved by arguing mutualism, here glossed as "the reciprocal arrangement by which what may be of one may be of the other". When drawing a distinction that which can be assumed for one side must in principle at least be possible for the other. This I have called the "Principle of Mutual Reciprocity."[9]

Here, I believe, is the key idea that does indeed generalise over the work of others. The reader should appreciate that the principle of mutual reciprocity is pre-ontological. As yet, there is no commitment to a particular shared reality. If an *I* distinguishes a *Thou*, what is allowed of the one is allowed of the other. If an *I* distinguishes a *me* and a *he, she* or *it*, what is allowed of the one is allowed of the other. Glanville goes on:

> In a universe of discourse determined by individuality and difference in observation, observing entities are taken to observe themselves: they are self-referential. Thus they attain identity and autonomy. (Observation should not be confused with seeing: observation as used here is a formal quality.) Therefore, observed entities must be assumed to have the possibility that they observe themselves.[10]

Here, we are still pre-ontological but we are obliged by the principle of mutual reciprocity to admit of the observed that which we have admitted of the observer: the property of self-observation, giving identity and autonomy. Glanville refers to this class of self-observing entities as Objects, as in the following extract:

> It is considered inconceivable that such entities (called "Objects") are simultaneously both self-observing and self-observed. They are therefore taken to switch roles. This generates time (making time a central and integral concept in second order Cybernetics), allows observation by another Object, and sets up observational time as a way of relating observations of other Objects, giving a relational logic. Objects are seen as oscillating between the two roles, and this oscillation allows the continuity of the observation of self; and the observation of others in time, giving rise to relationships.

---

8. Glanville (2002, n.p.)
9. Glanville (2002, n.p.)
10. Glanville (2002, n.p.)

> Objects generate process, just as they are generated by process: another cybernetic circularity. Since observation can thus take place, it is assumed other activities can also occur.[11]

Glanville has here posited another key pre-ontological principle, which could be named as, for example, the Principle of the Exclusion of Observing and Being Observed. This is first established for the *I* and the associated *me* of one Object and then extended to the case of Objects observing each other. The exclusion principle brings forth the concept of time as the difference between observing and being observed and also brings forth the concept of process, something happening. An Object is seen to be an entity that as a process constitutes itself as itself. It is the memory of itself.

This is the beginning of Glanville's account. Elaborations go on to set out the full structure for the formation of higher orders of self-reflection and for the formation of temporally synchronised social groups and coalitions. It should be emphasised once more that this structure is pre-ontological. As yet there are no *laws of physics*, there are no forms designated living or non-living, there is no beginning or ending. Glanville sums up:

> To use a metaphor: my work is the creation of games fields: others create the games to play in these fields and still others play them. Finally, some are spectators. The point of an account that admits others is not that it is right, but that it is general (and generous). Cybernetics is often considered a meta-field. The Cybernetics of Cybernetics is, thus, a meta-meta-field. My work is, therefore, a meta-meta-meta-field.[12]

What then has Glanville achieved? His structure allows us to play, to observe and be observed, to construct maps and models and to ontologise to our hearts' contents. It also reminds us that this form, the form of playing, is indeed pre-ontological and it behooves us to act accordingly, one to another.

In his *Philosophical Investigations*, Wittgenstein[13] (1953) characterises philosophy as a ground clearing exercise. By an exercise in abstract cybernetic algebra, Glanville has cleared the primordial ground.

Glanville's Points of Reference and Departure in Developing the Theory of Objects
What were the particular roots, the sources, the voices that came together in Glanville's period of gestation? I asked him about this recently. He recalls them as follows.

---

11. Glanville (2002, n.p.).
12. Glanville (2002, n.p.).
13. Wittgenstein (1953).

His first mention was of a long term study of and admiration for the work of Ross Ashby, notably, his *Introduction to Cybernetics*.[14] The reader may not readily appreciate this as a seminal source for leading ideas about second order cybernetics but let me remind her that careful cross-referencing within Ashby reveals that he defines cybernetics as "the domain of 'all possible machines'"[15]; he notes that for him the terms *machine* and *system* are synonyms and that "persistence is ... the most rudimentary property of a machine"[16]; he further notes that cybernetics is primarily interested in systems that are "open to energy but closed to information and control"[17]—they are "information tight."[18] Glanville particularly admired Ashby for his clarity, terseness and precision.

In this period, Glanville's PhD supervisor, Gordon Pask, was lecturing and writing about his emerging cybernetic theory of conversations. Pask was particularly enthusiastic about and inspired by Lars Loefgren's paper, "An axiomatic explanation of complete self-reproduction."[19] In brief, Loefgren shows that one can construct a formally consistent set theory where the existence of self-reproducing entities is accepted a priori. Pask saw that development as giving permission for scientific theorising to include accounts of and models of such self-reproducing entities. In Pask's own theorising he found it useful to distinguish two types of such entity: mechanical (or m-) individuals and psychological (or p-) individuals. In brief the latter are coherent, self-reproducing conceptual systems or systems of belief. The former are the biological or other extant mechanical systems that embody particular p- individuals. Pask's motivations for making the p-, m- distinction are several but he was centrally interested in the processes whereby observers in conversation and interaction come to know and understand each other.

In the same (later 1960s early 1970s) period, Humberto Maturana, produced his seminal prose poem, "The Neurophysiology of Cognition," that provides the central core of his thesis about autopoietic systems (the term *autopoiesis*, meaning self-creating was coined somewhat later). Glanville had access to that essay.[20] He also heard Maturana give a university lecture at Brunel University in 1972). The heart of Maturana's thesis is that what is peculiar to living systems is that they are

---

14. Ashby (1956).
15. Ashby (1956, p. 2).
16. Ashby (1956, p. 108).
17. Ashby (1956, p. 4).
18. Ashby (1956, p. 4).
19. Lofgren (1968).
20. Maturana (1970).

organisationally closed systems, that, whatever else they do, they, instant by instant, reproduce themselves as structures that embody the means to reproduce themselves as structures. This neat formation of a self-reproducing system as a closed system of operations was an immediate conceptual advance over von Neumann's earlier formulations of automata that build replicas of themselves, of Ashby's notion of a system being information tight, of Beer's concept of a viable system and of von Foerster's, Pask's and others' accounts of self-organisation.

Von Foerster himself, as a member of the collegiate that he fostered at the Biological Computer Laboratory, University of Illinois, was a central figure in the development of cybernetic thought. (Indeed, it was he who, in 1974, formally articulated the distinction between a first and second order cybernetics.) A particular source for Glanville's developing thought about *self-observing objects* is von Foerster's "Notes for an Epistemology of Living Things."[21] In it we find the aphorism, "An observer is his own ultimate object."[22]

Interestingly, Glanville tells me that it was only after completing his thesis that he gave particular attention to Spencer-Brown's *Laws of Form*,[23] although he was aware if its currency and influence at the time, not least in the work of Maturana's colleague Francisco Varela. Glanville also tells me that he did not read Wittgenstein's *Tractatus*[24] (Wittgenstein, 1922) until later, either. By his own admission, it is only in hindsight that he himself can appreciate that he was indeed writing an abstract philosophical thesis.

Themes in Objekte

The contents page of Glanville's *Objekte* lists eleven papers organised into four sections: "Objects," "Black Boxes," "Distinctions," and "Cybernetics."

There are four papers in Section I. The first, entitled "What is memory that it can remember what it is?" is a terse recapitulation of the theory of Objects as first presented in Glanville's PhD thesis.[25] The second is entitled "The nature of fundamentals applied to the fundamentals of nature." In it, Glanville builds on Gregory Chaitin's idea that only the random is fundamental, as only the random cannot be further reduced.[26] Glanville

---

21. Von Foerster (1974).
22. Can be found in the English language version of the 1974 article published in Von Foerster (2003, p. 256).
23. Spencer-Brown (1969).
24. Wittgenstein (1922).
25. Glanville (1975).
26. See, e.g., Chaitin (1975).

applies this concept to his own Objects. Objects as unique entities cannot be reduced to a description or rule. Glanville then reflects on Jacques Monod's discussion of the roles of chance and necessity in biological evolution[27] and generalises them to apply to his Objects. Chance produces a variety of Objects each of which is unique. (Pask in later writings captures the same idea in the aphorism "There are no doppelgangers.") Necessity refers to the mandatory requirement that Objects must indeed, whatever else they may do, remember (reproduce) themselves.

The third paper is entitled "The Same Is Different." It makes the point that in a world where the fundamental entities are unique Objects, any notion of sameness is a property of observation, not of the Objects observed. As observers, we set observed Objects in synchrony with one another as members of classes and class hierarchies. We impose order, hierarchy, predictability and connection. We distinguish an ontology, a universe of lawfully behaving entities.

The fourth paper in Section I is entitled "Consciousness, and so on." In it Glanville generalises from the theme of the previous paper to the case where, as observers (Objects), we come to distinguish a world of Objects in which we can come to know about each other and the universes we inhabit. The first step in this construction is to set up a triadic relation in which for an observer one Object serves as a sign for another Object. Observers in interaction may then establish systems of agreement in which signs have shared meanings. Glanville notes that it is this same construction of sharing meanings with another that enables an Object to share meanings with itself and to be conscious with itself of its self. Glanville is following Warren McCulloch, one of the founders of cybernetics, in his insistence that consciousness implies far more than an undifferentiated state of awareness. It implies, as in the Latin *conscire*, to know with another.

The importance of triadic relations as being the minimum required in setting up a theory of signs was first emphasised by C. S. Peirce. G. H. Mead, a follower of Peirce, coined the term *significant symbol* for the idea that a particular sign may call forth similar responses in both sender and recipient. Glanville's own sources for concepts of signification were Ferdinand de Saussure and Gottlieb Frege. Of interest here is that Glanville sets out these forms from the fundamental starting point of his Objects and does so with an elegant, terse simplicity.

Section II of Objekte, "Black Boxes," has just two papers. The first paper, "The Form of Cybernetics: Whitening the Black Box," discusses the relationship between first and second order cybernetics. Glanville notes that,

---

27. Monod (1972).

"Cybernetics is a study not just of control and communication but, since control and communication interpret all our experience, it is a study of form itself, hence the cybernetics of form (and, of course, the form of cybernetics)."[28] The form of cybernetics is that of the observer and the observed. Whilst first order cybernetic focuses on the observed as in classic black box theory, where the observed is a black box that the observer attempts to *whiten* by manipulating it and deducing the rules that govern its behaviour, second order cybernetics teaches us that the observer-observed relation is itself a system, a black box, to be observed and understood. This second order study is reflexive, as in the phrase, *cybernetics of cybernetics*. Considered thus it is clear that the notion of the observer controlling the observed is about allocation of roles to the two parts of the system. From a different perspective the roles may seem to be reversed: The laboratory rat may be seen as a controller of the behavioural psychologist; the heat source may be seen as the controller of the thermal switch.

The second paper in this section develops a complex argument in which the proposition that "inside every white box there are two black boxes trying to get out" (this is the title of the paper) is established as a fundamental form that can be recursively drawn as nested hierarchies. The base of the recursion is a Glanville Object with its two inseparable roles of self-observer and self-observed. Higher order forms correspond to socially constructed bodies of knowledge and norms of behaviour where observations of observer-observed relations become established as matters of fact or belief. In the case of the natural sciences this (to use a term from Suppe[29]) is a system of *justified true beliefs*. Glanville makes clear that this construction is fundamentally a social endeavour that is premised upon the existence of a stable shared reality and which, in its operation establishes just such a stable shared reality. Any such construction is of course subject to that stability being called into question and to logical critiques that reveal inconsistencies or lack of coherence in the conceptual structure so established. Glanville further notes that it is an attribute of human, as distinct from machine, intelligence that we can transcend levels. We can contemplate formal structures and, to use Peirce's term, *abduce* new forms.

Section III, "Distinctions," has four contributions. The first is entitled "Beyond the Boundaries." The reader is encouraged to read this contribution for herself. Its significance is best appreciated by those already familiar with Spencer-Brown's *Laws of Form*.[30] Here I can only give a terse

---

28. This quotation is from the original English language manuscript that Dirk Baecker translated into German.
29. Suppe (1977a, 1977b).

summing up of Glanville's argument. Brown begins with the form taken out of the form, for example, a piece of paper, and makes distinctions in this form as examples of *perfect continence*, for example, the shape of a circle. The laws of form are developed from this starting point. Glanville offers a critique, suggesting that we should really begin with the form itself. This is the realm of Glanville's Objects. Glanville argues that his Objects are the form of indication; they are self-indicators. He goes on to suggest they are representable by a Moebius strip, that is, by a boundary that does not distinguish an inside and an outside. (To construct a Moebius strip, take a strip of paper, give it a twist and then join the ends together; the strip now has just one edge and one side.)

The second paper of Section III is entitled "Your Inside Is Out and Your Outside Is In." It was jointly written with Francisco Varela. At the time, Varela was being lionised as the coming next generation cybernetician of note. Glanville invited Varela to work with him on a joint paper as a way of getting to know each other and each other's work. The theme of a boundary that does not distinguish an inside and an outside is again explored. This time the construct is used to question both the concept of elementary entities (as in quantum physics) and notions of the universal, the whole, everything (as in cosmology). Glanville and Varela show that neither can be indicated in the Spencer-Brown sense of lawfully constructed forms, since a Spencer-Brown distinction always implies at least one more distinction: the distinction between a form (the mark of distinction) and its content (the value assigned to the form distinguished. They go on to show that both concepts, the elementary and the universal, collapse into each other as the form of a self-indication, a Glanville Object, imaged (as earlier) by a Moebius strip. The paper is short and my account is shorter. Again, I urge the reader to read this for herself. I particularly like the penultimate paragraph:

> Things are brought into being out of no thing by distinctions being drawn which insist on boundaries. That these, in as far as we external observers can tell, are illusions does not make them any less real or necessary. Such illusionness —ephemerality—ungraspably shimmering quality we often forget, especially in laying down the laws of the bounds of the imagination.[31]

The third paper in Section III is entitled "Distinguished and Exact Lies." The argument here is that in a world where that which is unique cannot be

---

30. Spencer-Brown (1969).
31. Glanville and Varela (1981, p. 241).

described and communicated all we can do is try to provoke shared understandings by exchanging the distinguished and exact lies of the title.

The fourth paper in Section III is entitled "Emptiness." The third paper introduces it by saying, "We live in a world of interaction in which we tell, and can only tell, distinguished and exact lies. Just what we can do if we do not accept this argument I explore in ... (the next paper)." "Emptiness" consists in a blank sheet.

It may be helpful to the reader to appreciate the significance of the papers in Section III if she recalls that Glanville's Objects are pre-ontological. A universe of objects has not yet been brought forth. The full emptiness of the void has not yet given way to the empty fullness of a world of distinctions. This prefatory no-place cannot be described but it can be experienced. Glanville and Varela[32] refer to "the ungraspably shimmering quality". For a recent extended journey in and out of the no-place, see Lou Kaufmann's poem "Void Selecta."[33] For a classic text where the Moebius strip is used as a metaphysical metaphor, see Alan Watts Zen Buddhist oriented commentary (Watts, 1967).[34] For a recent essay on "cybernetics as universal knowledge" using the same metaphysical metaphor, see Geoffrey Bowker's *Synchronization 4: Hermes, angels and the narrative of the archive*.[35]

Section IV of *Objekte* contains just one paper entitled "The Question of Cybernetics." In it Glanville briefly sums up the argument that shows that with second order cybernetics we have problematised the status of knowledge and what it is to know. He argues that, "We need to radically change our conception of knowing, knowledge, laws of nature, reality, causality and so on." He then raises the question of cybernetics: "How to effect this change, how to be confident that the change will be productive, and what will the change mean to us as cognitive entities?"

I think this paper may be usefully seen as signalling a plateau and a stepping off point in Glanville's work. Although he has continued to write about epistemology and communication, he has also quite explicitly addressed ethical concerns, such as responsibility, trust, nurture, and generosity, a move that he has shared in their different ways with Pask, von Foerster and Maturana. As I think is quite clearly and fairly brought out in his own account of its development,[36] with his peers and mentors, Glanville has given second order cybernetics depth, maturity and authority.

---

32. Glanville and Varela (1981).
33. Kauffman (2004).
34. Watts (1967).
35. Bowker (2004).
36. Glanville (2002).

Concluding Comments

The collection of papers in *Objekte* were of their time but also, as very early meditations on second order aspects of cybernetics, systems theory and the natural sciences, they were also ahead of their time. The pre-ontological formal truths that they assert are, of course, timeless. They stand alongside the other great, formal truths of cybernetics and carry a validity and relevance for all generations.

In the 1970s many within cybernetics and the systems movement heralded the *new cybernetics* and the new second order paradigm but this heralding went rather unnoticed in other circles. Cognitive science, chiefly with the concept of the physical symbol system, established and maintained a stranglehold on cognitive psychology and artificial intelligence research. The so-called connectionist, sub-symbolic revolution merely emphasised how conceptually poverty stricken cognitive science had begun. Simplistic notions of self-organising networks were reinvented to give some account of learning, albeit in the form of trivial adaptation. Accounts of how symbolic processing emerged from sub-symbolic processing were as confused as the ideas about symbols that they were trying to replace. It was even the case (thanks to Harnad and others) that the so-called symbol grounding problem was posed. Roughly, this is the problem of how symbol systems are connected to or grounded in some ontology.[37]

The two great omissions in the cognitive science paradigm were the related concepts of (i) organisational closure and (ii) the social basis of communication that gives rise to the illusion that language has the form of a symbolic code.

The cybernetics community in particular has systematically drawn attention to these great omissions over several decades. It is a rather wonderful irony that the concept of organisational closure obligingly offers an explanation of why it, itself, as a central concept has been ignored by the mainstreams of physics, biology, cognitive science and sociology. Scientific research programmes, like other conceptual systems, have founding predications that, whatever else, cannot be challenged lest the system itself fails. The systems are organisationally closed. They exhibit immune reactions against ideas that would undermine them. Alternative systems are ignored or vilified.

Thus has the mainstream dealt with cybernetics and second order cybernetics. I believe this state of affairs is changing. Some voices in the mainstream are noting the existence of the alternative paradigm. It may be a

---

37. For a second order cybernetic approach to this problem, see Scott and Shurville (2011).

little while yet before the mainstream recognises that the alternative is not just an antithesis to it as thesis. Rather it offers a synthesis, a higher order perspective in which the problem of symbol and meaning is properly seen as a pseudoproblem within the larger context of the pragmatics of communication and interaction amongst organisationally closed systems.

Glanville's self-referential objects in their coming together and their evolution of ways to share understandings offer a comprehensive, terse and elegant account of this higher ground. They stand comparison with other attempts at higher order syntheses but are philosophically purer in intent.

Maturana's grand synthesis built around the concept of autopoiesis has an ontological orientation towards biology.

Von Foerster's grand synthesis built round the concept of what it is for an observer to compute objects has an ontological orientation towards the physical world, albeit a physical world that, prior to observation, "contains no information but is as it is."[38]

Pask by human admission was a "philosopher mechanic." His theories were embodied in and inspired by artefacts. His conceptual systems (p-individuals) require a medium for their embodiment (his m-individuals).[39]

Luhmann unashamedly wishes to ontologise social systems as entities alongside the biological and the psychic.

Glanville presents us with abstract forms and abjures us to recall that we are the forms that bring forth forms. We are the form of bringing forth.

References

Ashby, W. R. (1956). *An introduction to cybernetics*. New York: Wiley.
Bowker, G. (2004). *Synchronization 4: Hermes, angels and the narrative of the archive.* Retrieved May 5, 2012 from: http://www.peterasaro.org/writing/bowker2.pdf.
Chaitin, G. (1975). Randomness and mathematical proof. *Scientific American, 232*(5), 47–52.
Glanville, R. (1975). The object of Objects, the point of points—Or, something about things. Ph.D. thesis, Brunel University.
Glanville, R. (1988). *Objekte*. (D. Baecker, Trans.). Berlin: Merve Verlag.
Glanville, R. (2002). Second order cybernetics. In *Encyclopaedia of life support systems*. Oxford, UK: EoLSS Publishers. (Online at http://greenplanet.eolss.net/MSS/default.htm)

---

38. Von Foerster (1970, p. 48).
39. See Scott (1980, 1982).

Glanville, R., & Varela, F. (1981). Your inside is out and your outside is in. Glanville, R., & Varela, F. (1981). Your inside is out and your outside is in. In G. Lasker (Ed.), *Applied Systems and Cybernetics, Vol. II* (pp. 638–641). Oxford, UK: Pergamon.

Gunther, G. (1972). Cognition and volition: A contribution to the theory of subjectivity. In *Collected Works of the Biological Computer Laboratory*. Urbana, Illinois: University of Illinois.

Kauffman, L. (2004). Virtual logic: Fragments of the void—selecta. *Cybernetics & Human Knowing, 11*(1), 99–107.

Loefgren, L. (1968). An axiomatic explanation of complete self-reproduction. *Bull, Math. Biophysics, 30*(3), 415–425.

Luhmann, N. (1995). *Social systems.* Stanford, CA: Stanford University Press.

Maturana, H. (1970). Neurophysiology of cognition. In P. L. Garvin (Ed.), *Cognition: A multiple view* (pp. 3–24). New York: Spartan Books.

Monod, J. (1972). *Chance and necessity*. London: Collins.

Pask G., Scott, B. C. E., & Kallikourdis, D. (1973). A theory of conversations and individuals, exemplified by the learning process in CASTE. *Int. J. Man-Machine Studies, 5*, 443–566.

Pask, G. (1975). *Conversation, cognition and learning*. Amsterdam: Elsevier.

Scott, B. (1980). The Cybernetics of Gordon Pask, part 1. *Int. Cyb. Newsletter, 17*, 327–336.

Scott, B. (1982). The Cybernetics of Gordon Pask, part 2. *Int. Cyb. Newsletter, 24*, 479–491.

Scott, B. (2004). Second order cybernetics: An historical introduction. *Kybernetes, 33*(9/10), 1365–1378.

Scott, B., & Shurville, S. (2011). What is a symbol? *Kybernetes, 48*(1/2), 12–22.

Spencer-Brown, G. (1969). *Laws of form*. London: George Allen and Unwin.

Suppe, F. (1977a). The search for philosophic understanding of scientific theories. In F. Suppe (Ed.), The Structure of Scientific Theories (2nd.ed.; pp. 3–230). Urbana, IL: University of Illinois Press.

Suppe, F. (1977b). Afterword 1977. In F. Suppe (Ed.), The Structure of Scientific Theories (2nd ed., pp. 617–729). Urbana, IL: University of Illinois Press.

Von Foerster, H. (1970). Thoughts and notes on cognition. In P. L. Garvin (Ed.), *Cognition: A multiple view* (pp. 25–48). New York: Spartan Books.

Von Foerster, H. (1974). Notes pour un epistemologie des objets vivants. In E. Morin & M. Piatelli-Palmerini (Eds.), *L'unité de l'homme* (pp. 401–417). Paris: Editions du Seuil.

Watts, A. (1967). *The book: On the taboo against knowing who you are*. New York: Collier-Macmillan.

Wittgenstein, L. L. (1922). *Tractatus logico-philosophicus*. London: Routledge and Kegan Paul.

Wittgenstein, L. L. (1953). *Philosophical investigations*. Oxford: Basil Blackwell.

On the Contrary

Ted Krueger[1]

Bill asked me to write about Ranulph, but I won't. Instead, I'll write about something both more general and specific than that—the contrarian's approach.

Pedagogical Practices

On several occasions, I attended doctoral presentations in which Ranulph countered one-for-one most every assertion made by a presenter. And it seemed, the more confident the presenter, the more likely this response. Anyone witnessing, or directly experiencing, these moments already has a working definition of a contrarian. But witnesses are also likely to agree, that the counter-points were, while firm to emphatic, never belligerent. These were moments when the objective was to shift habits of thought, perhaps ones firmly held and successful in the past, into an altered and more open pattern.

I have seen university painting studios that contain a large mobile full-length mirror enabling the canvas in-progress to be examined in mirror-image. Anomalies in form or errors in composition to which the painter has become accustomed are reversed and thereby doubled in the reflection. They become more apparent when seen in a new way. The work is evaluated from a second vantage point through the mirror and this shift in perspective allows a familiar work to be seen anew.

Design critics in an architectural school might use a related tactic, to take the physical model from a beginning student and turn it upside down explaining why it is a better solution than the one proposed. To the student, it may seem that everything has changed, but that's not so. The basic organizational logic and spatial relationships remain intact, only the gravitational vector has shifted. One fundamental and invisible assumption altered in an effort to encourage a more flexible and creative re-reading of the work.

These pedagogical tactics are fundamentally related. They encourage seeing differently—opening the possibilities that one's personal, cultural, or disciplinary predispositions had closed. It is to challenge assumptions, and

---

1. Email: tedkrueger3@gmail.com

thereby, to make them visible. By asserting the contrary, it is sometimes possible to prompt a reflection on how these predispositions are constructed and so to foster the ability to construct them differently and multiply.

The proliferation of possibilities was important to Ranulph. He often recommended the von Foerster paper on Ethics that appeared in the Stanford Electronic Humanities Review.[2] This is one where von Foerster asserts that it is only the undecidable things that we can decide—a proposition itself conspicuously within the contrarian domain. This paper also contains the core of von Foerster's ethics—to always act in ways that increase the number of available choices. This is the essence of Glanvillian contrariness. Its not concerned with opposition, but with generosity. The effort is to open up that which is already closed. It was an opening offered to the presenter. Although perhaps difficult to see at first, it was an invitation to generosity.

Embrace Undecidability

It is in the context of undecidability that the true contrarian appears. This is evident in discussions about mind-independent reality and constructivism. If one position is asserted in preference to the other, the contrarian will not assert its opposite and so be controlled by the first. Instead, the contrarian maintains the undecidability, as Ranulph always did, to emphatically elect not to choose rather than to have one's options circumscribed by an oppositional pair. In this case, one is not driven away from the stated positions, but has already been drawn to a meta-position; one that does not come into being in relation to a statement, but instead, through an understanding of the consequences of that kind of statement making.

A contrarian, then, might look beyond the content of a proposition and reject it based on its structure alone. It is frequently the case that oppositional pairs are offered to the exclusion of a wide variety of other possibilities. The pair seems to charge the field of potential, and to nucleate this potential around itself obscuring what might otherwise be a rich variety of potential responses. In *Ontogeny of Information*,[3] Susan Oyama has an especially clear exposition of how this operates within the field of biology in the chapter "The Ghosts in the Ghosts in the Machine Machine." In it, she traces the structural similarities across a wide variety of theories showing that their differences hide a common underlying argumentative strategy that finds it roots in the dualism of Descartes. Her point is that by examining a range of competing theories of biological development what one sees from

---

2. Von Foerster (1995).
3. Oyama (2000).

the group is less concerned with an inherent logic of the biology and more a characteristic thought pattern through which biological phenomena are examined and explained.

The dualist trope exists within culture, not within the processes examined by biology. This way of organizing the world has very deep cultural roots that are continually being reproduced and reinforced. I have often observed parents who consider themselves enlightened and empowering offering their children "choices." Whether apple or banana, MacDonald's or Chuck E. Cheese's, behave or take a time-out, this manner of interacting prepares the child instead to be comfortable making decisions within the constraints imposed by the more powerful—Mac or PC, Donald or Hilary, or any of 500 channels of cable television. This is trivial choice, the illusion of choice.

And, the possible choices are limited only by the imagination. This statement is not an assertion about how large the potential is, but how small. We might consider the human imagination to be boundless, but that is far from the case. The choices available by means of the imagination are very constrained indeed. Knowledge is an interconnected and mutually supporting system of understandings. Imagination, with respect to conceiving viable alternatives choices, operates from this base of understandings and is constrained by them. It is intellectually and emotionally expensive to make alterations to this structuring and so the possible is circumscribed by the already understood. Richard Feynman is unable to explain magnetism to a journalist in any detail because the journalist does not have in place foundational understandings that would enable extension into the new territory of Feynman's descriptions of magnetism.[4] The journalist does not have the background from which to imagine how magnetism operates. The creation of the new, at the level of the individual, requires a linking to that already known. When the supporting antecedents are not in place, a proposition is indistinguishable from magic[5] or simply a crazy idea, usually also designating its proposer in the same way.

If we return to von Foerster with Glanville, it is only the undecidables for which we can choose not to decide, because, as noted, the decidable has already been decided.[6] By maintaining undecidability, both positions have viability and so the question remains an open one. A contrarian asserts

---

4. Richard Feynman interview about magnetism. Video retrieved July 22, 2018 from https://youtu.be/487Dolja2vg
5. "Any sufficient advanced technology is indistinguishable from magic,"—Arthur C Clarke's third law.
6. Von Foerster (1995).

undecidability as a distinct position rather than as something that is able to, hopefully someday, be decided with additional information, better arguments, or larger explosives. It also keeps open the possibility that there are other, potentially viable propositions not contained within the opposition that might be, or become, available. An essential aspect of the open question is its acknowledgement of the future. Some things may move from undecidable to decided. And so, it allows for and expects what cannot now be seen, understood, anticipated, or imagined. It is not only generous, but also optimistic, and humble.

A Lost Cause

Once, a student asked me what I found interesting about cybernetics, what part it still played in contemporary culture. My response was that it rejects the simple and linear causality to which most are still accustomed, indeed addicted. It is evident, especially during election years that simplistic causality remains in vogue, together with oppositional pairs. Cybernetics replaces this with circular and reticulated causalities that, I might hope, are better able to account for the things that I experience.

But of course, there is no causality beyond me. Causality is simply a human cultural product—an interpretive aid in efforts to account for the flow of experience. Causality exists within the domain of explanation, not in that of phenomena. Circular causality cuts a section through my model of reality and examines the cut plane. Circularity falls nicely onto this two-dimensional surface. On it, I can see a limited set of relationships between entities that my prior experience, perspective, and language have segmented into objects or agencies that I interconnect into a loop. Like an architectural section, this way of looking might have analytic or conceptual value, but it is neither the model, nor the architecture itself.

Circularity is a geometric preference and an aesthetic choice. It is an analytic tool that allows a simplified model to be constructed. It has the significant advantage over other geometries in that it can accommodate an arbitrary number of elements along it and so can be used for *A causes A* on up. One value that circular causality has is in providing an opportunity to show how differences in framing relationships can invert previously simple causalities and so to question the way these simple causal agencies were constructed. Ranulph recounted the setting of a room's thermostat to illustrate the difference between first and second-order cybernetics. The thermostat is a commonly available and understood process governor; the classic first-order feedback loop that is easily expanded to include the human observer within it. Ranulph was fond of pointing out how the mechanical system controlled the human causing the thermostat to be set to a different level. Of course, the desire is not to replace one causal vector with another, but rather to show that to speak about causality as unidirectional is suspect and shows that the standard interpretation is only a human-centric perspective masquerading as a neutral and comprehensive view.

This way of shifting conventional perspectives of causality away from the human-centric was also undertaken in the 19th century. Samuel Butler, in the *Books of the Machines* in the novel *Erewhon*,[7] posits an evolutionary trajectory for machines explicitly noting that they have domesticated humans. He suggests that machines act through the agency of humans to supply their own needs, and by supplying at the same time the needs of humans in great abundance, create the conditions in which humans cannot easily exist without them, and therefore, willingly continue to serve them.

Clearly this is a radical inversion of the perspective that would have been held in Victorian England at the time that the book was written. Today, similar descriptions can be written about the domestication of humans by various plants and animals which with we closely co-habit—whether pets, or a broad range of agricultural organisms. While we generally see humans through time as being much the same as ourselves only adding these plants or animals to their lives, domestications, none-the-less have altered humanity, and settled it, perhaps civilized it, so that it can care for these other species. Those that have succeeded in domesticating humans are now wildly successful both in population and in geographical distribution.

These shifts are especially important for us in the 21st century. There are deep cultural traditions that put humans in a privileged and dominant position with respect to all other species. These biases are reflected in the organization of species into a pyramid where humans occupy the apex. Those higher up the pyramid are thought to prey upon and control those that are lower. They are considered more powerful, capable, and intelligent. The anthropocene is another cultural expression of this same perspective.[8]

But, to prey upon is to be fed by. Predator and prey are roles that have particular qualities that we have assigned. Though, I admit that I prefer eating to being eaten, on an individual basis. Studies of population dynamics, however, show that the prey does not benefit from a lack of predators, but rather overpopulates and succumbs to disease and eventual starvation. Predators face much the same fate in time from an over

---

7. Butler (1974).
8. End of anthropocene.

abundance of prey. A balance is necessary and desirable as they are mutually dependent species. The position at the top of the pyramid is dependent on all those that are below and so it is not more powerful or capable, but only stands in relation. While these can seem but amusing re-readings, they are simple shifts in perspective that circular causality allows. There is value in seeing relations reversed both in a painter's mirror and in other pictures that we paint for ourselves. Both von Foerster and Glanville were enamored of such inversions.

Circular causality has a significant limitation, however. The ring is composed by perhaps an unlimited chain of simple cause and effect relationships. Many, probably most, situations cannot be so simply described. Multiple simultaneous factors with varying strength would seem the rule. Reticulated causality has a closer identity with my own model of reality. But for its part, my construction of reality is most often an accurate and comprehensive portrait of its builder rather than a description of the dynamics of the "world." This is to suggest that the way that one explains things sheds more light on the experiences, capacities, and culture of the explainer that it does on the phenomenon in question [also a von Foersterism]. *Reticulated causality's* network metaphors became popular at the end of the 20th century. Like the mechanistic or computational locutions that preceded it and drove earlier imaginations, its rise and fall will have nothing to do with the things it purports to describe, but is simply a product of its moment within the human cultural conversation. I expect that some form of biological metaphor, perhaps the ecological or metabolomic, will soon join and then supersede the network in popularity. By noting the cultural dimension of these metaphors I don't intend to diminish their importance but rather to acknowledge it.

We often see the contrarian in a social context as above, a relation between people or positions on a particular topic. But contrarianism also operates, and perhaps most effectively, as a component of an internal dialogue, as a way of thinking. In this case, it serves to check otherwise unexamined predispositions. And might often serve to open up the possibilities.

The Last Shall Be First, and the First Shall Be Last[9]

I'll close with a list of paper titles each of which has two parts in uncommon dialogue. I've drawn them from a list of Ranulph's papers. The sample is convenient, but perhaps not comprehensive. Many instantly

---

9. Matthew 20:16.

recognize these titles as Glanvillian. But while somehow similar, there is no easy formula here, nor simple way of capturing the relation between the first and second part of these titles, rather there is a (varieties of) variety[10] This evidences a practice of thought, not a habit of thought. When reading these as a list, I sense something impish, too. I see the flicker of a smile. Because that too is a quality of the contrarian; well, at least that particular contrarian.

1978.	The Concept of an Object and the Object of a Concept.
1981.	Your Inside Is Out and Your Outside Is In. [w/Varela]
1986.	The Cybernetics of Ethics & the Ethics of Cybernetics.
1995.	The Cybernetics of Value and the Value of Cybernetics: The Art of Invariance and the Invariance of Art.
1998.	Varieties of Variety.
1999.	Acts Between and Between Acts.
1999.	Researching Design and Designing Research.
2003.	Designing Reflections: Reflections on Design. [w/van Schaik]
2003.	Inter View: Designing Interfaces and Inter-facing Design.
2006.	What Makes the Difference—Reflecting on Reflecting (on Reflecting on Reflecting).
2009.	Reflecting and Acting: Reflecting on Acting and Acting on Reflecting.
2010.	Architecture of Distinction and the Distinction of Architecture.
2011.	The Boundaries of Distinction? The Distinction of Boundaries?

References

Butler, S. (1974). *Erewhon*. London: Penguin.
von Foerster, H. (1995). Ethics and second order cybernetics. In Constructions of the mind: Artificial intelligence and the humanities. *Stanford Electronic Humanities Review, 4*(2), 308–327.
Glanville, R. (1998). A (cybernetic) musing: Varieties of variety. *Cybernetics & Human Knowing, 5*(1), 57–62.
Oyama, S. (2000). The Ghosts in the Ghosts in the Machine Machine. In *The ontogeny of information* (pp. 84–128). Durham, NC: Duke University Press.

---

10. Glanville (1998).

Monument
The Writings of Glanville

Albert Müller

There are three volumes of Ranulph Glanville's collected writings that are known as the *Black B∞x Series*. Glanville's selection of works is monumental and also indicative of his many-sidedness and his great eloquence as an author. Such monumentality and ingenious composition lets one assume the presence of an expansive intellectual edifice, a coherent movement of thought, a systematic, planned approach and this all in keeping with the author's theory-architecture. However, the monument of the *Black B∞x* Series is constructed out of many building blocks, which, to my mind, certainly reflect the author's intellectual biography and an interesting development. Of course, this is neither remarkable nor surprising. After all, the articles evolved over a period of more than forty years. In addition to the wide range of subject matter, the main point here is to document the chronology of this oeuvre. Such a perspective can contribute to making Glanville's better known and more appreciated. When Glanville began writing it was not so much science that influenced him but rather art, architecture, literature, and the related theoretical writings and bodies of knowledge. The study of architecture at the Architectural School of the AA (Architectural Association) in London brought Ranulph Glanville into contact with what was happening at the cultural hubs of the 1960s. Looking back at the Ranulph Glanville's early career and biography it quickly becomes clear that he was not predestined from the very start to become a scholar and academic teacher. The early phase still left open the option of young Ranulph becoming a writer, avant-garde artist or avant-garde musician. And in spite of his conflict-laden period while studying at the AA, a career in architecture was also in the cards. In fact, Ranulph Glanville completed a mandatory internship, working at an architectural firm, as part of his studies. It should also be noted that in the second part of his internship year Glanville worked as a teacher at the school he himself had attended. It is certainly plausible to assume that this contributed in large part to his becoming aware of his own exceptional teaching skills long before he began to theoretically address the difficult link between teaching and learning. How strong Ranulph Glanville's links to, and his interest in, art, literature and music were—along the related theoretical developments and innovations can be seen most strikingly in the works that he presented on completing his studies at the AA. Gordon Pask proved to be doubly inspiring for R.G.'s future path as cyberneticist.

Ranulph would time and again describe how his first meeting and conversation with Pask left an indelible mark on him. What really laid the groundwork was the fact that following R.G.'s graduation from the AA, Pask suggested that he pursue his Ph.D. and even offered to secure him a scholarship. The completion of this dissertation was beset with great difficulties. His thesis advisors were Gordon Pask and Heinz von Foerster. Ranulph's transition to a cyberneticist was shaped by this experience. At the same time, it should be pointed out that he had selected an extremely difficult subject for which he developed a separate formal language so as to be able to deal with it. This epistemological work also shows a clear link to the field of architecture and design and even to the social sciences. Elsewhere I myself have examined the content of the dissertation and the connection to the theme of parallel developments within cybernetics leading up to a second-order cybernetics. Even though this unpublished dissertation marked a new direction, there were still other themes that stood out in R.G.'s publications. These included architecture and the history of architecture. This phase of his work found only little or indirect reflection in the *monument* of the Black B∞x Series. However, several important examples have been reprinted in the volume of R.G.'s architecture work. In a certain sense this volume constitutes an addendum to the three volumes. Among these works, he devoted relatively large space to Finnish architecture. In addition to work on theories, like those relating to Alvar Aalto, a rather structuralist work on Finnish amateur architecture and the Finnish language is particularly salient. Even if he was not able to have his dissertation of 1975 published, R.G. began addressing aspects of his theory of Objects at congresses.

At the same time Glanville proceeded to explore some of the basic categories of cybernetics of the 1950s. With extraordinary, fundamental contributions to the theory of the black box, which was also substantially expanded by them, he was able to draw international attention. But in doing so he also acquired the reputation of being the enfant terrible of cybernetics. At the same time Glanville was working on ways to put application examples to the service of his theoretical work. These examples existed of course, since he had taught—first at AA, then at the University of Portsmouth. R.G. was one of the early protagonists in the field of CAD—and at the same time he was also one of the early, and staunchest, critics in this field. R.G. had an almost intuitive technical understanding of design and architecture. This by no means clashed with his openness toward any kind of innovation, which could be seen in all phases of his work. Only a small part of this phase of his publication work found entry into the *Black B∞x Series*. However, several of the early works were documented in the volume edited by Ertl/Korn/Müller—a volume that also provides good access to the early R.G. It seems as if Ranulph Glanville needed some time to make up his mind as to where to present his works. Ultimately, he opted for all spheres that were accessible to him. Much more than other discipline-oriented individuals, he was able to see, understand and analyze existing problems. A chronology of the works included in the Monument of Black B∞x shows, as has already been stated, a highly fascinating spectrum. The early works reflect hardly any knowledge of, and only little interest in, cybernetics. This may well have been due to the general research situation at the time. And one should not forget the suppression of cybernetic research in favor of Artificial Intelligence around 1970. [Note from Seaman: This refers to a discussion in Müller, A. and Müller, K. (2007) *An Unfinished Revolution, Heinz von Foerster and the Biological Computer Laboratory* | BCL 1958-1976, edition echoraum, Vienna in the chapter entitled "Interview on Heinz von Foerster, The BCL, Second Order Cybernetics and the American Society for Cybernetics" by Stuart A. Umpleby.]

Organisation

Volumes 1 and 2 are structured in the same way, while vol. 3 follows a separate logic of linearity (*39 Steps*). Glanville structured 1 and 2 as a triptych following the pattern of "Before – During – After." The format of a triptych once again highlights the monumental dimension. The distinction between "Before – During – After" introduces a temporal dimension that, however, has little, if anything, to do with the genesis of the individual chapters. The sections "Before" and "After" also contain—in addition to significant material on Glanville's way of thinking and working, including interviews—tributes from others, as for instance, Bernard Scott or Karl Müller. Especially Scott's assessment of Glanville offers surprising, interesting insights. This is also true of the two outside-perspective articles in vol. 3, written by Søren Brier and Richard Jung. In closing the table of contents of the three volumes reflects the clear descriptions.

Black B∞x volume 1

Table of Contents
Before
>Acknowledgements
>Foreword: Ranulph Glanville and His Magic Cybernetic Circles
>*Karl H. Müller*
>Introduction: The Black B∞x Introduction Vol. 1
>Abstracts
>Ranulph Glanville's Objekte
>*Bernard Scott*

During

1   Cybernetics
1.1   Abbreviation 85
1.2   The Question of Cybernetics Ranulph Glanville
1.3   Sed Quis Custodient Ipsos Custodes?
1.4   as if (Radical Objectivism)
1.5   Chasing the Blame
1.6   A Ship without a Rudder
1.7   A (Cybernetic) Musing: the Gestation of second-order cybernetics, 1968–1975—A Personal Account
1.8   A (Cybernetic) Musing: the State of Cybernetics
1.9   An Observing Science
1.10 Listen! A Tribute to the Listen Inn, Shaftesbury, Dorset, UK
1.11 Second-order cybernetics
1.12 Behaving Well
1.13 A (Cybernetic) Musing: Desirable Ethics
1.14 A (Cybernetic) Musing: The IFSR, Diagrams and Inclusive Logic
1.15 A (Cybernetic) Musing: Five Friends

2   Objects
2.1   The Object of Objects, the Point of Points, or—Something about Things
2.2   What is memory, that it can remember what it is?
2.3   The Nature of Fundamentals, Applied to the Fundamentals of Nature
2.4   All thoughts of things
2.5   Consciousness, and so on
2.6   The Same Is Different
2.7   The Cybernetics of Value and the Value of Cybernetics.
2.8   The Art of Invariance and the Invariance of Art. Et Cetera
2.9   Acts Between and Between Acts
2.10 Comparison

3   Black Box
3.1   The Form of Cybernetics: Whitening the Black Box
3.2   Inside Every White Box there are Two Black Boxes Trying to Get Out

3.3 Behind the Curtain
3.4 A (Cybernetic) Musing: Ashby and the Black Box
3.5 A (Cybernetic) Musing: Black Boxes

4   Distinction
4.1 Beyond the Boundaries
4.2 Your Inside is Out and your Outside is In
4.3 *Ranulph Glanville | Francisco Varela*
4.4 The Self and the Other: The Purpose of Distinction
4.5 Living in Lines
4.6 Francisco Varela (1946 to 2001): A Working Memory

5   Variety
5.1 Variety in Design
5.2 A (Cybernetic) Musing: Varieties of Variety?
5.3 A (Cybernetic) Musing: Variety and Creativity
5.4 The Value of being Unmanageable: Variety and Creativity in CyberSpace
5.5 A (Cybernetic) Musing: Control, Variety and Addiction

After
   Between Now and Then: The Auto-Interview of a Lapsed Musician
   Grounding Difference
   Bibliographic References of the Chapters in this Volume
   Bibliography
   Index

Black B∞x volume 2

Table of Contents
   Before
      Introduction to volume 2 Freedom and the Machine
      Inter-view: Designing the Interface and Inter-facing Design—A
         Conversation
      Abstracts
      Freedom and the Machine

   During
6   Design
6.1 The Architecture of the Computable
6.2 Why Design Research?
6.3 CAD Abusing Computing
6.4 Architecture and Computing: A Medium Approach
6.5 Researching Design and Designing Research
6.6 An Intelligent Architecture
6.7 Not Aping the Past: Mirror Men

6.8   An Irregular Dodecahedron and a Lemon Yellow Citroën
6.9   Appropriate Theory
6.10   Construction and Design
6.11   Design and Mentation: Piaget's Constant Objects
6.12   Design Prepositions
6.13   Try again. Fail again. Fail better—
       the Cybernetics in Design and the Design in Cybernetics
6.14   Designing Complexity
6.15   Conversation and Design
6.16   A (Cybernetic) Musing: Design and Cybernetics

7   Representation
7.1   The Domain of Language
7.2   The Model's Dimensions: A Form for Argument
7.3   Distinguished and Exact Lies
7.4   Emptiness
7.5   Communication without Coding: Cybernetics, Meaning and Language
       (How Language, becoming a System, Betrays itself)
7.6   A (Cybernetic) Musing:
       Language and Science in the Language of Science
7.7   A (Cybernetic) Musing:
       Encyclopaedias and the Form of Knowing International Encyclopaedia of
       Systems and Cybernetics second edition
7.8   Visual Logic

8   Knowing
8.1   The One Armed Bandit
8.2   Introduction: Behind the Screen
8.3   A (Cybernetic) Musing:
       Encyclopaedias and the Form of Knowing
8.4   A Note on Knowing
8.5   Architecture and the Embodiment of Knowledge
8.6   A (Cybernetic) Musing:
       Certain Propositions Concerning Prepositions
8.7   Knowledge Creation and Research in Design and Architecture

9   Education
9.1   Construct Heterarchies
9.2   NOAH: the Ark of Knowing in a Learning Environment
9.3   A (Cybernetic) Musing: Cybernetics & Human Knowing
9.4   A (Cybernetic) Musing:
       Some Examples of Cybernetically Informed Educational Practice
9.5   Designing Reflections: Reflections on Design
9.6   Reflecting and Acting:
       Reflecting on Acting and Acting on Reflecting.

10 Others
10.1 A (Cybernetic) Musing: Robin McKinnon-Wood and Gordon Pask: A Lifelong Conversation
10.2 Heinz von Foerster: The Form and the Content
10.3 And He Was Magic
10.4 Doing the Right Thing: The Problems of ...Gerard de Zeeuw, Academic Guerrilla.
10.5 A (Cybernetic) Musing: Machines of Wonder and Elephants that Float through Air
10.6 A Cybernetic Serendipity
10.7 Gordon Pask's "An Approach to Cybernetics"
10.8 Learning with Locker
10.9 The Importance of Being Ernst
10.10 Pask at the Centre
10.11 Richard Jung

After
Ranulph Glanville: A Conversation
Bibliography
Index

Black B∞x volume 3

Table of Contents
Acknowledgements Introduction Abstracts
Ranulph Glanville: The Cybernetician of Ignorance
Søren Brier

1 A (Cybernetic) Musing
2 A (Cybernetic) Musing: Control 1
3 A (Cybernetic) Musing: Control 2
4 A (Cybernetic) Musing: Communication 1: Coding
5 A (Cybernetic) Musing: Robin McKinnon-Wood and Gordon Pask: A Lifelong Conversation
6 A (Cybernetic) Musing: Communication: Conversation 1
7 A (Cybernetic) Musing: Communication: Conversation 2
8 A (Cybernetic) Musing: In the Animal and the Machine
9 A (Cybernetic) Musing: Varieties of Variety?
10 A (Cybernetic) Musing: The Gestation of Second-Order Cybernetics, 1968–1975—A Personal Account
11 A (Cybernetic) Musing: Variety and Creativity
12 A (Cybernetic) Musing: Language and Science in the Language of Science
13 A (Cybernetic) Musing: Encyclopaedias and the Form of Knowing
14 A (Cybernetic) Musing: The Millennium Bug
15 1A (Cybernetic) Musing: Thinking the New Millennium
16 A (Cybernetic) Musing: The State of Cybernetics
17 A (Cybernetic) Musing: Constructing my Cybernetic World

18	A (Cybernetic) Musing: Cybernetics & Human Knowing
19	Francisco Varela (1946 to 2001): A Working Memory
20	A (Cybernetic) Musing: Some Examples of Cybernetically Informed Educational Practice
21	Heinz von Foerster: A Personal Farewell
22	A (Cybernetic) Musing: In Praise of Buffers
23	Foreword: The Ouroboros and the Glass Bead Game
24	A (Cybernetic) Musing: Machines of Wonder and Elephants That Float Through Air
25	Understanding Systems: Conversations on Epistemology and Ethics: A Review Article
26	A (Cybernetic) Musing: Desirable Ethics
27	A (Cybernetic) Musing: Control, Variety and Addiction
28	International Encyclopaedia of Systems and Cybernetics, Second Edition: A Review
29	A (Cybernetic) Musing: Certain Propositions Concerning Prepositions
30	A (Cybernetic) Musing: Invisibility and Silence
31	Dark Hero of the Information Age: A Review Article
32	A (Cybernetic) Musing: The IFSR, Diagrams and Inclusive Logic or—[both [either [either/or] or [both/and]], and [both [both/and] and [either/or]]]
33	Life is a Verb: A Review Article
34	A (Cybernetic) Musing: Ashby and the Black Box
35	A Cybernetic Musing: Design, the User, and Klaus Krippendorff 's The Semantic Turn: A Review Article
36	A (Cybernetic) Musing: All The 8's
37	A (Cybernetic) Musing: Five Friends
38	A (Cybernetic) Musing: Black Boxes
39	A (Cybernetic) Musing: Design and Cybernetics

Ranulph
Richard Jung
Bibliography
Index

The Mentoring Process
Texts From Students and Others Ranulph Has Supervised in Some Capacity

Ranulph in Full Regalia. Courtesy Aartje Hulstein
and The Glanville Archives, University of Vienna.

Difficult Student

Ted Krueger

Best I could tell, it was an aberrant form of psychotherapy.

Of course, from the beginning I understood the need for designers to have a body of knowledge about the nature of design activity and it was best if designers built it themselves. I taught architectural design for years, long after having been subjected to the same treatment. I worked in professional offices as well as doing my own more experimental work. Design was something that you learned in the doing of it, initially in the company of critics and a cohort of other learners and afterwards in the day-to-day. That's why they call it a practice. But when someone asks "How do you design?" ("How do you walk?") It is easier to say where you go than how you get there. There is little substantive material to offer beyond the usual iteration. There are so many different approaches.

What was mine? I had no idea. Except I was certain that it was not of sufficient general interest. To believe that I had something to offer the design community would take more than the available arrogance and I felt that I had already accumulated a substantial reserve. Like most designers, my interests were in the products of design activity not in the activity itself. Initially, I thought that the work of the doctorate would be focused on the designs, not on the designing.

But over time, the shift happened. In retrospect, this must have taken considerable patience for Ranulph. I'm inclined to think that I was his most difficult student, but I know some of the others. Many of them might think the same with justification, and we were probably all individual challenges.

Shortly after I began working on the PhD, I found myself with a new slate of responsibilities, one of which was directing a new PhD program in the Architectural Sciences. Doctorates have many traditions, the one I was directing was built along very different lines from the one that I was undertaking. Its traditions were clear, so you understood what you're supposed to do. I found that a reflective understanding of my own practice was far more difficult and involved than Schön would have us believe. He thinks it's something we do all the time. Not at this level.

Ranulph was a master at reading your work and process, of seeing and then suggesting what you did as a designer. Over the course of what would become a maximum-time-limited doctorate, he offered me a score of suggestions on how to tie the whole thing together. All perfectly reasonable, but none seemed right. I tried several on for size, but they didn't fit very well. All seemed imposed from the outside rather then derived from the design process itself. But I expect that Ranulph was never concerned with solving the puzzle. What he offered was not the solution, but the opportunity to see that there were many solutions that could be made. He understood that I wouldn't accept anything that was not my own so he stayed well away from those I might eventually see for myself, and instead, showed me how to do it, rather than what to do.

I met with Ranulph three times a year, once at the Graduate Research Conferences in Melbourne. Of course, he had many other meetings and so the time together was limited although it always managed to include Aussie-rules football. Another time we would meet in the States. Now, I had the full schedule as he inevitably came while teaching and administration were in full swing. But once each year we met in South Sea, and these were by far the most therapeutic. Ranulph and Aartje provided a place of calm and conversation, good food, trips in the countryside—to an ancient ring, chapel, or Bletchley Park—and time to think between the conversations. It was here that the best ideas came to light and the weaker ones perished. Ranulph was firm and his standards were very high, but he was extremely generous and always confident that I could succeed, especially when I was not.

This examination of ones designing, though difficult, is productive work, most especially for design educators. When you can finally see your own, you have a better chance to help others. Coming to a deep understanding of ones own practice is a great deal like psychoanalysis, except that it has an upside. Reflective practice is something you learn in the doing of it initially in the company of critics and a cohort of other learners and afterwards in the day-to-day. That's why they call it a practice.

On one of his many visits, Ranulph observed my activities at the university commenting that I was working too hard in part because I was concerned with teaching students when instead I should be concerned with their learning. Their education did not depend on my activity, but rather was something that they needed to accomplish. Ranulph had many students who were full or part-time faculty members and I expect others may have heard similar comments. Like most others in the design professions, I had no education in education. But, Ranulph did.

Ranulph was an educator of teachers and a teacher of educators. He was a strong proponent of learning rather than teaching. Ironically, the difference between teaching and learning is one of the most important things he taught me. He taught me many other things, of course, some of which I've learned. And, some of which I haven't. Yet.

We Had Never Met Before

Craig Bremner

In 1999, only a few hours after the Queen had officially opened the Lighthouse, a newly restored Mackintosh building in Glasgow, I met Ranulph Glanville as arranged on the opening night of this new museum for architecture & design. We had never met before. I was nearing completion of the fieldwork for my PhD in Glasgow and my supervisor in Melbourne had asked Ranulph to come and check on my progress. We had an enjoyable evening wandering around the newly amplified building, looking at the exhibitions inaugurating the museum and chatting to each other, and to anyone else Ranulph spontaneously engaged in conversation.

Ranulph had supervised my supervisor, and together they were developing a typically atypical PhD program for RMIT in Melbourne, from which I was one of the first to graduate. But when we met all I knew was someone was coming to check on me.

Thinking about it now our meeting in Glasgow was timely because at that moment what I thought I understood was not what I came to understand after thinking about Ranulph's questions, and I stress thinking about his questions because I couldn't answer them at the time. Later, when I explained my new understanding to Glasgow City Council, who had commissioned the research, it inspired them to request more research adding another two years to my candidature, and made Ranulph my co-supervisor and then more importantly my friend. Among the many things I might cite why I have come to love and treasure Ranulph's friendship, without meeting in Glasgow and his influence on my thinking I would not have spent more time in that city where I eventually cemented friendships with people that I still hold dear today.

Normally it is Ranulph weaving connections between people who he thinks might enjoy meeting, but only by being connected remotely in Glasgow did Ranulph enable me to make enduring connections with people with whom he had no connection. From that meeting over the years I observed that Ranulph is an active observer of the possibilities of loops and linkages—connecting dots that no-one else can see—which is of course no surprise to anyone who knows him.

Another reflection about meeting Ranulph in a place where we had no history, nor history of each other, was there appeared to be nothing to loose in the ground we covered, and it was obvious to me instantly that Ranulph had covered a lot more ground than I. So the meeting in Glasgow was timely. I was immersed in what I was doing and hadn't given thought to how to call an end to my PhD. Ranulph guided me to the end and that opened a new beginning for me.

When I returned to Australia I realized I was in an unusual position in the field of design—someone with a PhD at the very moment when universities began to demand exactly that (previously they had expected a higher degree but would compromise on some measure of equivalence—usually practice). As a result I began to supervise other academics under increasing pressure to upgrade their qualifications. At the same time my university seemed ambivalent about giving me a promotion while another offered me a professorship if I also became head of a school of design and architecture, which, with the guarantee of Ranulph's support, I did. After one of the typical structural changes brought about by new Vice-Chancellors, the School became a Faculty and I became a Dean, so for several years I was an administrator and supervisor. Before I anchored myself to any of these roles I made sure I could secure Ranulph's services, so he would come twice a year to listen to the staff in the school and offer guidance (if they were willing to listen), and then tell me what he had heard, listen to my response and then offer me guidance—and we did this for many years.

If I had to pin it to one thing, it was through listening.

We cemented our relationship. Ranulph often says it is the listener who constitutes the conversation. I had no choice but to learn to listen when I had studied design in Italy (in Italian) many years before I met Ranulph in Glasgow, and I am grateful for their lesson every day. But as always with Ranulph there is a trick to his notion of listening. It is possible to form the opinion that he doesn't listen because he is a prolific communicator of his own ideas and he likes to tell you about these ideas and anything else that prompts him. However he has an amazing skill to listen and hear what was said and then to demonstrate in his reply that he has listened and understood what was said, to explain why what was said is possibly more interesting than the speaker may have thought initially, and to demonstrate how what he has just explained about what was said might be even more knowledgeable than the speaker could have known. In a world geared for speaking Ranulph's skill is a gift because, as I have experienced, we sometimes get to hear our own ideas for the first time.

While Ranulph listens to us we get to listen to him and when he talks about research there is little anyone could argue with. Only by working with Ranulph for the past decade have I observed that too much research activity has been premised on only one of the two basic questions every researcher in every field must face, and that is explaining to oneself what one is already doing in the hope to be able to do it better. This preoccupation has meant the other question has more or less been forgotten, and that question is what needs to be done? This has slowly dawned on me, but when I think back to my first meeting with Ranulph I was looking at what needed to be done without making use of the utility of what I was already doing. Almost instantly Ranulph connected these two fundamental questions at which point I could articulate the questions I had thought I was answering and could now see even the utility of what I had thought was useless.

Ranulph and I have looked at research, examined research, attended exhibitions and concerts, and socialised for many years, all springing from meeting in that new museum in Glasgow. Forgive me if I keep coming back to our first meeting but while museums have always functioned to archive time I classify my friendship with Ranulph as timeless.

Listening

Ben Sweeting

Ranulph often emphasised the importance of listening as part of conversation.[1] Where we are overly focused on what we are saying, we miss the significance of what is said by others. This is partly an ethical matter, concerning our respect for others, and partly a practical one, as without listening conversation cannot occur. Indeed, as Ranulph has noted, ethical qualities such as respectfulness are required in order to sustain a conversation, something which unites practical and ethical considerations.[2]

---

1. Fantini van Ditmar & Glanville (2013) and Glanville (2001).
2. Glanville (2004); see also Sweeting (2018).

Listening is a key part of what makes conversation creative. This is made especially clear in the analogy Ranulph developed between Gordon Pask's conversation theory[3] in cybernetics and characteristic design activities such as sketching.[4] Part of the distinctive quality of sketching is that it is driven not just by the way we draw but also by the way we look at what we draw. When we return to a drawing, we often see something not previously intended in it and to which we can respond through more drawing (and so on). When sketching, designers quickly move between these two modes, alternating between looking (listening) and drawing (speaking). This process takes a circular form and is often thought of as a conversation that designers hold with themselves.[5] While we tend to focus on the making of the marks that compose a drawing, it is the way we reinterpret the marks that we have already made that helps us to create new possibilities and reframe the questions we are exploring. Through this, design, much like conversation, leads us to places we did not expect or foresee. It is important, therefore, for design students to learn not only how to produce drawings to present their ideas but also how to develop creative ways in which to look at (listen to) what they make. Similarly, in face-to-face conversation, listening is not only a matter of respectfulness but also what sustains a conversation and moves it on. We often think of speaking as the creative part of conversation. Yet, we generally know what we are going to say when we speak, whereas when we listen to what others say (and also when we listen to ourselves) we are always constructing a new understanding. We can even see speaking (and drawing) as part of listening (and looking). Just as designers make models with which they can explore new ideas not just to represent existing ones,[6] so too we can learn to speak in ways that help us to listen rather than only to express our ideas.

Listening was a key part of Ranulph's manner as my PhD supervisor.[7] Many of our meetings would take place sitting outside at a café at the back of the Royal Festival Hall, where we would meet after Ranulph had arrived from Portsmouth to nearby Waterloo station and before he headed to the RCA or elsewhere. I would often find myself talking and talking and, whenever I talked myself out, Ranulph would somehow set me going again

---

3. Pask (1976).
4. Glanville (2007, 2009).
5. This understanding of design in terms of conversation is relatively common. See for instance Schön (1983/1991), who characterizes design as a "reflective conversation with the situation" (p. 76).
6. See the distinction between "models of" and "models for" (Glanville, 2005).
7. My PhD research was funded by an AHRC doctoral scholarship and supervised by Ranulph and Neil Spiller at UCL (Sweeting, 2014a).

with some simple comment or by putting something I had said into question for me to respond to. This took on a more intense form on the few times I went to Portsmouth to meet Ranulph at his home. We would have tea or lunch with Ranulph's partner Aartje who would ask me the sort of simple, direct questions that are the hardest to answer (such as "what is your PhD about?"). I would then speak about my research to Aartje while Ranulph would quietly listen. In this careful listening Ranulph was primarily doing something that is important to all teaching: taking care to understand what I was doing and what I understood of this so he could better advise and challenge me in my work. There was also something more to this in the specific context of design research. Ranulph's listening to me also encouraged me to listen to what I was saying and doing. In this way I came to a deeper understanding of my own project and its possibilities. Similarly to the way that the importance of the conversational setting of design tutorials is as a performance of design's conversational structure[8] (Sweeting, 2014b), Ranulph's listening demonstrated the kind of reflective conversational thinking that is so important in design and in practice based design research.

References

Fantini van Ditmar, D., & Glanville, R. (Eds.). (2013). Listening: Proceedings of ASC conference 2011. Special double issue of *Cybernetics & Human Knowing, 20*(1-2).

Glanville, R. (2001). Listen! In G. de Zeeuw, M. Vahl & E. Mennuti (Eds.), *Problems of Participation and Connection*. Lincoln, UK: Lincoln Research Centre. Retrieved 5 November, 2019 from http://www.asc-cybernetics.org/2011/wp-content/uploads/2011/01/Ranulph_Glanville_on_Listening.pdf

Glanville, R. (2004). Desirable Ethics. *Cybernetics & Human Knowing, 11*(2), 77–88.

Glanville, R. (2005). A (cybernetic) musing: Certain propositions concerning prepositions. *Cybernetics & Human Knowing, 12*(3), 87–95.

Glanville, R. (2007). Try again. Fail again. Fail better: The cybernetics in design and the design in cybernetics. *Kybernetes, 36*(9/10), 1173–1206. doi: 10.1108/03684920710827238

Glanville, R. (2009). A (cybernetic) musing: Design and cybernetics. *Cybernetics & Human Knowing, 16*(3/4), 175–186.

---

8. Sweeting (2014b).

Pask, G. (1976). *Conversation theory: Applications in education and epistemology.* Amsterdam: Elsevier.
Schön, D. A. (1991). *The reflective practitioner: How professionals think in action.* Farnham, UK: Arena. (Original work published in 1983.)
Sweeting, B. (2014a). *Architecture and undecidability: Explorations in there being no right answer—Some intersections between epistemology, ethics and designing architecture, understood in terms of second-order cybernetics and radical constructivism.* PhD Thesis, UCL, London. Retrieved from https://iris.ucl.ac.uk/iris/publication/972511/1
Sweeting, B. (2014b). Not all conversations are conversational: A reflection on the constructivist aspects of design studio education [Open peer commentary on the article "Radical Constructivist Structural Design Education for Large Cohorts of Chinese Learners" by Christiane M. Herr]. *Constructivist Foundations, 9,* 405–406.
Sweeting, B. (2018). Wicked problems in design and ethics. In P. H. Jones & K. Kijima (Eds.), *Systemic design: Theory, methods, and practice* (pp. 119–143). Tokyo: Springer Japan.

Generosity, Bombast, Love, and the Third Eye

Caroline Vains

I had never really seen Ranulph as a mentor until I was invited to write this piece.
He'd been a friend.
A dear friend.
With whom I could discuss anything, academic, professional, personal and even spiritual.
In retrospect, however, of course he was a mentor.
But one who so easily and readily shared ideas, insights, provocations, and counsel that any sense of being tutored receded into a background of lively conversation and shared laughter.
Such was his generosity.

He enjoyed people, all sorts of people, and engaged with everyone in the spirit of a shared humanity.
Though never an elitist, he didn't suffer fools.
Or, more precisely, foolish behaviour
When I was foolish he called me out.
When he lapsed into bombast I called him out.
So we had some arguments.
And sometimes didn't speak for months.
Then he would reach out.
A bigger human than me.
Bridge-building.
In these reconciliations we would each defend our respective positions.
Then agree to disagree and recommence our friendship.

I first met him the year I commenced my Masters, in 1998.
He had two PhD's.
He invited me to accompany him to a dinner with some university heavyweights.
The first of many such dinners.
Seating me beside people of interest or influence, he chaperoned my way-finding in this world.
He did this for many young academics and others interested in ideas.
Instigating introductions and facilitating creative relationships.
Stitching us all into his extended, global network.

I also invited him to dinner at my house.
Many times.
He pointed out that every time he came to Australia I had a new live-in boyfriend.
He was often blunt like that.
Most of the time I appreciated this bluntness.
Sometimes not quite so much.

One night we stayed up talking ideas till the early hours of the morning.
It was time for me to go to sleep, I explained, as I was teaching my first class in the morning.
He said he was usually good with 4 hours sleep and, regarding teaching, suggested I prioritise improvisation over preparation.
Some years later, he had recanted on the sleep and I had learned how to improvise.

The Mentoring Process

We went on long conversational road trips.
He took me places I hadn't previously been—an Englishman showing me my own country.
Sometimes these places were forgivably far from Melbourne—like Canberra.
Other times they were embarrassingly local—Point Addis, Apollo Bay, Yarra Valley and King Lake.
He said this was common amongst Melbournites.
They didn't get outside Melbourne much.

Ranulph was the third eye of supervision during my Masters degree.
But this was an informal role and, therefore, not officially recognised.
I have always regretted not thanking him on the day of my examination.
So nervous, I forgot to publically acknowledge his kindness and commitment.
I would like to do that here, honour his important role by detailing three of his many crucial contributions.
Knowing I was interested in the phenomenology of light, he referred me to two of the most remarkable books I have ever read; *Silence* by John Cage and *The Dancing Wu Li Masters: An Overview of the New Physics* by Gary Zukav.
Knowing I was interested in the Australian desert and sea, he suggested I design two remote pavilions and use live satellite feeds to video project the light conditions of each into the other.
And knowing (more than me) how underprepared I was for examination, he rehearsed my verbal presentation with me for eleven hours straight just before the big day.
With regards to the references, he knew exactly what would inspire me, and they have remained north points in the sea of readings I have since absorbed.
With regards to the pavilions, he foresaw the role of new technology in design and design research (well ahead of many others) and, with characteristic generosity, was trying to coax me in these new directions.
Finally, with regards to the examination rehearsal, because of his huge effort I passed with commendation. Without it, the outcome would not have been so successful.

Ranulph remained my generous, insightful, forgiving, blunt, bombastic, big-hearted friend and mentor to the end.

His lessons in life, scholarship, and what it is to be human will never be forgotten.

Lodged in heart and mind, they are his legacy to me.

My Experience of Ranulph

By Michael Trudgeon

I have been thinking carefully about how to write down my experience of Ranulph. I'm finding it very hard to succinctly frame the impact of Ranulph's friendship and his knowledge. Let me know if the following is at all useful.

Ranulph was introduced to me by Leon van Schaik in 1992. Leon had recommended that I meet Ranulph in London, as someone who was very interested in experimental architecture and design, as challenging and difficult territories. Leon had arranged the meeting and given me very simple instructions. They were: go to the Architecture Association in London. Go to the café/bar and say to the barman: "I am here to meet Ranulph Glanville." He will direct you to Ranulph. I loved the idea that you could travel 17,000 kms with such a minimal set of instructions and meet someone so effortlessly. With his characteristic generosity, Ranulph took me to meet all of the most interesting and technologically innovative architects and designers in London. I delivered a lecture at the AA and the University of South Hampton, reviewed his students' work and had articles published in World Architecture, all as a result of his introductions.

Later in the early 2000s, doing my PhD with Leon and Ranulph as my supervisors, Ranulph would always come to the point about designing as a messy, chaotic process. I began by not understanding, at all, what this meant. Over many years, as I tried to present my PhD work as a clean organised journey, Ranulph would return to this idea of mess and chaos. The significance of this observation took a long time to sink in for me and Ranulph was always very patient about allowing new ideas to seed in the minds of his students. Now many years later, many of the strange, and at the time hard to grasp, ideas have become central to the way I think about the design process and how to communicate it to others.

Tracks Left by a Walk Walked

by Thomas Fischer

Ranulph Glanville was my semi-official advisor during my time as a PhD student at the Spatial Information Architecture Laboratory at the Royal Melbourne Institute of Technology between 2002 and 2007. As I approached the completion of my studies, Ranulph invited me to join the American Society for Cybernetics where I was soon elected as an executive board member, alongside Ranulph who served as the society's president from 2009 until his untimely death in late 2014. In this way, I had the fortune of having Ranulph as a mentor, as a close collaborator, and as a good friend for about a decade. With the profound impressions he made on me in these roles, it is safe for me to say that I could not have wished for a kinder, brighter, or more inspiring teacher.

In this piece I discuss three particular ways of acting which set Ranulph apart as an educator. While Ranulph also had many other ways of acting that distinguish him as an educator and otherwise, I focus on three which I experienced as particularly impressive, and which I hope capture and convey his style. These three ways of acting are his hands-off teaching from behind, his creating of "games fields" for others, and the example he set by having great demands towards himself. As aspects of Ranulph's living in cybernetic circles[9] these ways of acting embodied his theories: He talked his cybernetic talk, and he walked his cybernetic walk.

At the heart of Ranulph's living in cybernetic circles lies a loop between the subjective self and the other, who may be imagined or not, a person, or anything "listened" to carefully enough to allow it to talk (or teach) back. This cycle consists of two arcs, namely an arc of perception and understanding, which connects the other to the self, and an arc of expression and acting which connects the self to the other. Acting consistently with cybernetic theory, Ranulph argued, is neglected in today's cybernetics. He showed how to act—not by instruction but by example; and he made explicit the tenets of his cybernetic ethics in his writing: Generosity, honesty, learning, mutuality, open-mindedness, respect, responsibility, selflessness, sharing, and trusting.[10] To describe the relationship between the acting-understanding cycle and its various manifestations in the self and

---

9. Glanville (2014a)
10. Glanville (2009, pp. 293–303).

in the other (ideas, music, buildings, etc.) Ranulph[11] used the metaphor of a turning wheel leaving tracks in the sand: As the wheel of acting and understanding turns, it produces in its wake what we accomplish.

Ranulph described the first of my selection of three ways of his acting as "teaching from behind."[12] In keeping with the navigational imagery that is often used in cybernetics, Ranulph used a car journey as a metaphor for graduate research and for his influence on it. He refused to take the metaphorical driver's seat and insisted in the student taking the wheel. Ranulph would sit in the back of the car as a passenger, and, in his own words, for most of the time "do absolutely nothing at all." Just every once in a while, he said, he would use a very long arm to reach ahead of the car to rearrange some traffic signs, or maybe the road. It was up to his students to find the way.

Ranulph described my second pick of his ways of acting as the creation of games fields for others to play in.[13] His insight, his clarity and his general way of being gave Ranulph a natural gravitational pull towards the centre of attention in many settings. He disliked being at the centre of academic attention and tended to leave the room when he saw recognition or praise coming his way. Instead of shining light onto himself, he preferred others, including his students, to engage in their cycles of acting and understanding for their own delight and benefit. For that to happen, he explained, he preferred to create games fields to enable others to play, or for them to create rules for yet others to play by.

The third of Ranulph's ways of acting discussed here was his directing reciprocally high demands towards his students as he did towards himself. I perceived a large gap between the challenges Ranulph posed to his students, including me, and the challenges I dare to pose to my own students. One evening during a conference Ranulph organised at the Experimental Media and Performing Arts Center at Rensselaer Polytechnic Institute in 2010, my wife Candy and I met Ranulph and his wife Aartje in their hostel room. After planning this and that for the coming conference days we talked about teaching, and I asked him what made him so sure that his students and reviewees were up to his demands. Ranulph refused to accept that it was him who posed the demands others around him perceived in his presence. After all, he was teaching from behind. Then he explained that he could not help having great expectations towards others because he could not help but being "incredibly demanding" towards himself.

---

11. Glanville (1997, p. 4).
12. Glanville (2009, p. 251).
13. Glanville (2002, section 4.1).

To me, and surely to other graduate students of his, these ways of acting (which Ranulph summarised more comprehensively[14]) created contexts of learning where the source of rigour lies within the student. This is a rare exception within the broader field of higher learning. With his ways of acting Ranulph inspired his students to create new questions, new answers, new ways to investigate these, and our own value systems from which to derive standards and internal guidance along the way. Ranulph neither instructed, nor did he prescribe formal methods, nor did he peddle his own writings. Those who expected simple answers and instructions from him could take Ranulph's ways of acting as ways of sidestepping commonplace educational responsibilities. Others, especially those of us who work as educators, recognise these ways of acting as expressions of a valuable and rarely practiced ethics. Enabling exceptional learning, this ethics requires trust towards students and defiance towards commonplace institutional frameworks that are not easy to muster. Ranulph was able to offer these at the expense of a "cushy" full-time academic appointment, working as an "itinerant, peripatetic, vagrant professor of odd jobs" [15] that allowed him the requisite independence to not only talk his talk but also to walk his walk, leaving beautiful tracks in the minds of others.

References

Glanville, R. (1997). A ship without a rudder. In R. Glanville & H. de Zeeuw (Eds.), *Problems of excavating cybernetics and systems* (pp. 131–142). Southsea, UK: BKS+.

Glanville, R. (2002). Second order cybernetics. In *Encyclopedia of life support systems, systems science and cybernetics, Vol. III*. Oxford, UK: EoLSS Publishers. Retrieved October 31, 2019 from https://www.pangaro.com/glanville/Glanville-SECOND_ORDER_CYBERNETICS.pdf

Glanville, R. (2009). *The Black B∞x, Vol. III: 39 Steps*. Vienna: edition echoraum.

Glanville, R. (2014a). *The Black B∞x, Vol. II: Living in cybernetic circles*. Vienna: edition echoraum.

Glanville, R. (2014b). *RSD3 Keynote Speakers*. Retrieved October 23, 2019 from: http://systemic-design.net/rsd3-keynotes (Keynote presenters' biographies on the *Relating Systems Thinking and Design 2014 Symposium* website.

---

14. Glanville (2014a, pp. 250–251).
15. Glanville (2014b).

My PhD with Ranulph

Tim Jachna

Pre-

"I hear you are thinking about doing a PhD." These are the first words I can remember Ranulph ever having said to me. This was when I showed up at a screening, in Hong Kong, of a short film co-authored, as he described it, by himself, his son Severi, and the auto-focus mechanism on his video camera. I suppose that's when I started thinking about doing a PhD.

He and I had both been told that we had likely met before, since we were both around the Architectural Association in London at the same time in the early '90s. It was good to have this background of previous acquaintance and shared experience as a foundation for our eventual PhD teacher-student relationship, even if neither of us could ever definitively say whether or not we had actually crossed paths prior to that day in Hong Kong.

I believe Ranulph would balk at this characterization of our relationship—not the probably-but-maybe-not-fictional prehistory part (he'd relish that), but the part about teachers and students, known as he is for his self-deprecating stance that (the pretense of) teaching is a scam—that there are no teachers, only learners.

Peri-

During my PhD process, my conversations with Ranulph typically led to the dismantling of whatever clever turns of phrase, trendy theory or well-honed taxonomy I was currently using as scaffolding to prop up my work, in order to try to get at what, if anything, I was proposing that was of value.

Working on producing knowledge? No. You're working on producing *you*. So get to a reasonable, honest and unpretentious way to say what you mean—first of all to yourself.

Ranulph was the one with whom I could have conversations that drew out of me a way of thinking, of talking and, after considerably more struggles, a way of writing about what I was doing that made sense to me. I suppose this is what one calls discovering one's cybernetics. We seldom used that word in our conversations, though, and it was when reading what was intended to be a penultimate draft of my thesis that Ranulph asked something along the lines of "Where are all the cybernetic references?"

So my thesis did not develop by arguing *from* cybernetics (as an a priori world view that structured and validated my methods, theories and conclusions), but it did gradually evolve into an argument *for* cybernetics as the only way of seeing, describing and discussing that allowed me to give coherent expression to the value and meaning that I perceived in what I was discovering/constructing.

Post-

A few years after the completion of my PhD, I found myself among an entourage being treated to an impromptu guided tour, by Ranulph, of the 1986 reconstruction of Mies van der Rohe's German Pavilion for the 1929 International Exposition in Barcelona.

Demonstrating the endlessly multiplying spatial and optical discoveries that the building revealed with each new angle, Ranulph claimed that this experiential depth and richness was not the result of some consciously-aimed-at design intention or goal, but rather was the emergent and residual evidence of a designing mind intensively engaged with an idea, a place, a time, and with itself.

It would be disingenuous to present the thesis document that is the residue of my PhD process as the end result of a structured and intentional application of a "methodology" or pursuit of intentions (even if convention requires that it be presented in that way). It is an artifact that emerged from a valued conversation, with which I continue to converse, and in which I am constantly discovering new angles each time I return to it.

Each of us may see and experience for ourselves, and construct our own meaning from these experiences, but we, as social animals, rarely experience alone. It matters profoundly with whom one experiences, and who one engages in conversations about these experiences. If I feel that I am the one learning more from this process of shared experiencing and conversing, and if my companion and conversation partner takes exception to being called a teacher, then I'll call him a mentor.

Thank you, Ranulph.

Ranulph Working. Courtesy of Aartje Hulstein
and the Glanville Archive, University of Vienna.

Ranulph. Courtesy of Aartje Hulstein
and the Glanville Archive, University of Vienna.

Composing Composing

Bill Seaman

Young Ranulph. Date Unknown. Courtesy of Aartje Hulstein
and the Glanville Archive, University of Vienna.

Introduction

This text is a focused conversation with some of the salient artifacts of thought that Ranulph has provided for us. It is derived in part from a close reading of *The Black B∞x*,[1] a three volume series, and other of Ranulph's texts. Volume 3 brought together many of the columns Ranulph authored for *Cybernetics & Human Knowing*.[2] The following text often contains quite succinct definitions provided by Ranulph for us to converse with, often drawn from the wonder and inspiration of *The Black B∞x*, and Volume 3 in particular.[3]

Composing Composing Composing[4]

> How is "Composing Composing" to be composed?

---

1. Glanville (2012, 2014, 2009). Numbered edition set of the 3 volumes presented in a black box, signed by the author.
2. See also the published single book, *The 39 Steps* by Glanville (2009), published by echoraum editions.
3. Other texts that explore this method include Cage conversing with Satie in John Cage's *Silence* (1967); *Conversing with Cage* by Richard Kostelanetz (2003); and Glanville conversing with Pask in *Gordon Pask, Philosopher Mechanic—An Introduction to the Cybernetician's Cybernetician* (Glanville & Müller, 2007).
4. The above title reminds me of the poetics of Gertrude Stein, A rose is a rose is a rose… I hear a pun—eros, oscillating with a rose.

Let me reflect for a moment on the theme of form and content. In my contribution to the festschrift I edited for Heinz von Foerster, I introduced the argument that one reason certain pieces of von Foerster's were distinctive was because in those pieces, the form reflected the content, the content the form. (3_164)

I point out the unusual quality of the column, that it both says something, and also does what it says. (3_207)

The Black B∞x definitely plays with form and content, I unpack the title/object later in this text. In my own teaching, I point out that we have a circle of form and content, we must have both in order to make meaning—sometimes with more emphasis on the form, and sometimes with more emphasis on the content, but in the end, each must be one with the other to function. I also like this idea to be seen from the perspective of living in cybernetic circles…

Heinz von Foerster at the Biological Computer Laboratory. Copyleft.

Parts and Wholes

Ranulph's last article for *Cybernetics & Human Knowing* is where the term *composing* in my title comes from. In this text Ranulph presents an interesting thought… the composer replaces the observer.

I love the fact that in our long conversation Ranulph revealed to me his idea of augmenting the term *observer* after 40 years or so, with the notion of the composer—a lovely musical/artistic metaphor for how we compose our understandings of the world. He also expressed this more formally in a late text about parts and wholes—his last chapter for his long series of texts, in the *Cybernetics & Human Knowing* journal—"A (Cybernetic) Musing: Wholes and Parts, Chapter 1."[5]

> 1. parts are themselves wholes, to be treated as such;
>
> 2. their "part-ness" comes from what in Second Order Cybernetics has been for more than 40 years called "observing by an observer," but which I shall, in the course of the column, reword as "composing by a composer." Wholes are composed into the roles of parts in relation to other wholes.[6]

In his own footnote to this quote he wrote: "In changing observe to compose, I hope to move away from what some see as the quasi-objectiveness associated with observing." I personally understand this notion of composing to be pointing anew at the concepts inherent to Ernst von Glaserfeld's (and Ranulph's) ideas surrounding constructivism.[7] This might be considered in part to be Ranulph's own introspective conversation with Von Glaserfeld, after the fact. It also points to how creativity might become enfolded with other more historical concepts central to cybernetics: feedback, the black box, variety, the study of circular causal systems, and so forth. Composing becomes an organizational metaphor to help us examine the tenets of cybernetics in a subtle new way,[8, 9] as well as thinking in general.

The One and The Many

There is a kind of permission that one gives to oneself to be many people, to pursue many different kinds of interests despite what society or academia prescribes as being a "success." I believe this in part is what interests me in the architecture of ideas that Ranulph has explored through his huge oeuvre of texts, designs, audio compositions and artworks. I enjoy, in particular, the incredible breadth of his intermingling of theory and practice, his own conversations with notions of the academic status quo, and conversations with himself in the pursuit of creativity; his love of being a playful and pointed contrarian; his deep appreciation for learning; his employment of

---

5. See *The Black B∞x vol 3: 39 Steps* for many of his texts presented in the *Cybernetics & Human Knowing* historically.
6. Glanville (2015).
7. See *An Exposition of Constructivism: Why Some Like It Radical* by Ernst von Glaserfeld (n.d.), Retrieved January 19, 2017 from http://www.oikos.org/constructivism.htm
8. *Definitions of Cybernetics*, Retrieved January 20, 2017 from https://www2.gwu.edu/~asc/cyber_definition.html
9. *Definitions of Cybernetics, Overview* by Stuart Umpleby (2000), Retrieved January 20, 2017 from http://asc-cybernetics.org/definitions/

differing forms of scholarship; as well as his pursuit of variety functioning in the service of creative production.

In the following text, I wish to continue my conversation with him through his textual artifacts but in a more abstract manner, especially in terms of creativity, and his interest in the arts and design and their relation to second-order cybernetics. This is similar to how Ranulph, himself, had a conversation with Pask via his quotes in *An Approach to Cybernetics*,[10] as found in the book *Gordon Pask, Philosopher Mechanic: An Introduction to the Cybernetician's Cybernetician*.[11]

This essay begins each section with a title and then I often present a simple question that the quote(s) address, or I lay out some salient point related to the quote(s). I have performed a close reading of *The Black B∞x* as well as many others of Ranulph's texts. In "Composing Composing" often I use the shorthand of the volume I have taken a quote from (out of the three volumes, 1, 2, or 3) and then present the page number or numbers as a means of reference (e.g., 3_164). I have tried to have these quotes unfold a pointed concatenation of salient ideas and processes that have been central to Ranulph's architecture of ideas.

The 5 Friends

How do these 5 overarching themes become relevant to so many areas of inquiry? Ranulph's writings often circle back to these themes. It somehow made sense for me to start off "Composing Composing" using them as a central set of foci.

> This column presents five concepts associated with cybernetics (the Five Friends of the title), explaining why they are of value and what they are. They are discussed in terms of the delight of thinking with them, and in exploring them, thinking about thinking. The choice of the five is entirely personal.
>
> 1) A Description of a thing is not the thing itself (the description is not the thing described)
> 2) Circularity
> 3) The Turing Test
> 4) The Black Box
> 5) The Principle Law of Mutual Reciprocity (3_393/4)

Ranulph often discusses these concepts and applies them to the understanding of new contexts, in particular in relation to design, art and

---

10. Pask (1961).
11. Glanville and Müller (2007).

music. Perhaps the most import idea is that he used second-order cybernetics as a springboard for design and art theory. Using these concepts, in particular, and others germane to this transdisciplinary field of fields, he advanced the scope of design and art, while simultaneously bringing second-order cybernetics to a new intellectual, pragmatic and creative plateau.

The Black Box

Given, Ranulph has named his *monumental* work of works, the playful, punning and poignant title, *The Black B∞x* (see also Albert Müller's section in this book related to the Monument). It seems clear that perhaps he wished to point, in particular, to the importance of the black box concept in his writings, thought and action.

Information: Ross Ashby operating the Elementary Nontrivial Machine [Black Box] (foreground). The larger machine in the background is probably a project Ashby called "Grandfather Clock," but it's uncertain. Despite appearances, the two machines are entirely separate. Date is unknown, but probably during the 1960s at the University of Illinois Biological Computer Laboratory. Courtesy University of Illinois (special thanks Jamie Hutchinson).

The back box is a concept developed by James Clerk Maxwell and brought into cybernetics by Norbert Wiener and additionally, by W. Ross Ashby. Ashby claimed a great range of differing applications for the black box—suggesting its relevance to the whole of science. Can one provide a simple definition of the black box?

> The Black Box consists of a closed box with an input and output, both of which display behaviors. The task is to try to find a pattern that links input to output. (3_365)

> The Black Box is, I have argued, a far subtler device than we had originally thought. It is truly a Second-Order device, and the outcome of its employment is a description that accounts for observed behaviors that belong neither to the observer, nor to the phenomena supposedly contained within the black box, but to both together, in and through interaction: it is perhaps the archetypical Second-Order Cybernetic investigative device. (3_298)

> Over the years…I have often referred to the Black Box. I believe the black box is such a powerful device, that it is time to explore it seriously, in its own right; for it allows us the most magical of tricks, a way of acting confidently, with/from the unknown/unknowable. From this position and with this device, we can build an understanding of the world regardless of whether we can honestly regard it as "real" or existing knowably independent of our knowledge of it. We can also build descriptions of the world that, ultimately, are not based in presumed knowledge, but in ignorance. (3_366)

Sometimes it is rendered in physical form.

Søren Brier has explored in part the last section of the quote concerning ignorance, in his chapter "Cybernetician of Ignorance" elsewhere in this volume.

Ranulph quoted Ashby at length discussing the uniqueness of the black box problem set:

> Ashby: We imagine that the investigator has before him a black box that, for any reason, cannot be opened … in its original, specifically electrical, form, the problem was to deduce the contents in terms of known elementary components. Our problem is somewhat wider. The questions we are interested in … are such matters as … What general rules of strategy should guide the exploration, when the black box is not limited to electrical but may be of any nature whatever?
>
> When the raw data have been obtained from the outputs, what operations should in *general* be applied to the data if the deductions made are to be logically permissible? What can in principle be deduced from the box's behavior and what is fundamentally not deducible? (Ashby in Klir, 1991, 252; see 3_369)

So from this we ask what relation do the Black B∞x have to the black box?

The Black B∞x ↔ The Black Box

The Black B∞x by Ranulph Glanville. Images © 2020 courtesy of Bill Seaman.

First, the work is a minimal black sculpture with three volumes in it—a beautiful sculptural object in its own right, functioning literally and metaphorically as a black box. The title is a slippery, slightly off pun, oscillating between books and box. It is also a pun built by misusing the infinity sign, or better, repurposing it for the sake of poetics and sign-related polysemy. The sign functions here as the two "o"s of the word *book*, (oh, oh, I say…or uh, oh). So, perhaps it alludes to an infinity of interpretations opened up via the published set—or in a more focused way, to the infinity of differences inherent to each composer's interpretation of Ranulph's texts. I associate infinite time, and also the finite time of a human life—the book is basically a container for a life's work. Also, Ranulph's death was imminent, lurking in the background…a kind of anticipation of blackness. The black box, by Ranulph's own words, cannot be opened. Yet, this major work is designed to be opened and read cover to cover. Was this his ironic "whitening" of the black box (which he writes is an impossibility)? Or perhaps he was suggesting that this is an example of nested black boxes...you open one only to lead to yet another set, and through the "ignorance" of reading and interpreting things, perhaps one can even explore a different variety of knowing based on the contents of the texts themselves. Weaving together this set of sets of living foci, with a name like the Black B∞x, Ashby's spirit must also certainly be looming in the background!

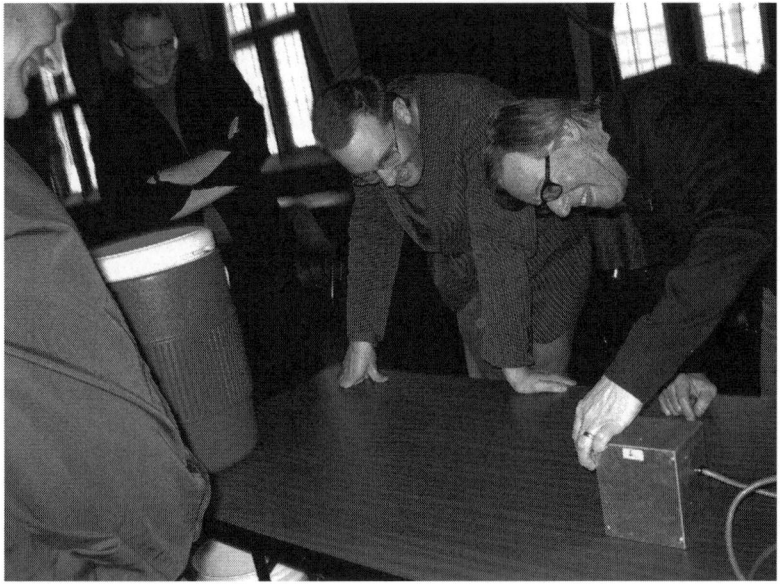

Information: (Left to Right) British author Matthew Fuller, University of Illinois professor Kevin Hamilton, University of Illinois technician Skot Wiedmann, and Texas Tech University professor Bruce Clarke operate Ashby's Elementary Nontrivial Machine at the University of Illinois YMCA, March 2012. Courtesy Jamie Hutchinson, University of Illinois.

Ranulph talks specifically about the black box, perhaps shedding some light on his design for the Black B∞x:

> The Black Box is an invention. We can't open it because it is not there. It is a device, a trick, a deceit, a phantasm. In its absence, we can't open it because it is black by definition, unopenable by dictat. So we build knowledge (knowing) through considering our interaction with it, our behavior together. But this knowing is always based on a profound ignorance. This is the most amazing ability of all, that we can make and know a world based on nothing.
> We can't see inside the black box because there is no inside.
> We can't see the Black Box because there is no Black Box. (3_421)

So, like this text entitled "Composing Composing," we enter into interaction with the artifacts of thought that Ranulph has spun for us—an embodied, sculptural, spun pun. For me, the Black B∞x becomes a kind of object-based Zen koan exploring poetically the concept of the black box, as well as the limits of what and how we can come to know our world via composing and via ongoing interaction with the world, the self and others—living and conversing in cybernetic circles. It must be noted that Ashby's black box (sometimes called the nontrivial machine) was also made into a functional object, somewhat playfully contradicting Ranulph's observation above.

Thinking Companions

Ranulph took delight in finding interesting pathways into the re-understanding of design and of cybernetics, as well as being extremely interested in increasing variety in the service of creative production.

> Why have I referred to these thinking devices as friends? Because, as with human friends, I like to be with them. I enjoy their company. I like doing things with them, especially when I am surprised in our interaction and can consequently increase the range of my ideas. (3_400)

His ongoing interaction with the Five Friends, to my mind, has opened out a series of new conversations for second-order cybernetics (the cybernetics of cybernetics),[12] shifting the overarching flow of ideas and conversations through the constructed interactions of conversational conferences, and in particular by attracting many new art and design participants to the field. Where he has kiddingly called himself a cybernetic necrophiliac (3_335), I see his activities as being reanimating! To me cybernetics is very much alive and being redefined in an ongoing manner by the people that explore it, as well as live it. This, in particular, includes a series of people that have been influenced by how Ranulph lived his life, was generous with his ideas, and was an excellent listener. He excelled in giving feedback to the ideas of others in a form that was somehow appropriate to the content and needs of their manner of thought. He did this using many different strategies that were intuitively applied as needed.

---

12. Von Foerster (1974, 1995).

Five Themes

Additionally, in a similar manner to his ongoing relationship with the Five Friends, Ranulph discussed 5 important themes that he also returned to over and over again in his writing and thought. In one example of his explorations he pragmatically pointed to these themes in discussing the following—Ranulph was enamored of Charles François's *International Encyclopaedia of Systems and Cybernetics*[13] and wrote a review of it in one of his papers. He stated:

> In the Encyclopaedia (and in building and the arts) there is a significant and important link between the content and the form within which the content is integrated, organized and presented. Secondly, that the nature of the link, that is the form of the Encyclopaedia, is not fixed, but presents particular epistemologies associated with particular times and contents. (3_166)[14]

This quote also circles back to the first quote in this text about a deep interest concerning the circle of form and content as it relates to von Foerster's work, and also epistemology (mentioned below). Ranulph would often take a concept and apply it to multiple contexts, bringing out nuanced understandings from each differing use. Can "Composing Composing" avoid being redundant by bringing out new ideas through each re-contextualized instance of circling back?

Ranulph pointed to a set of five overarching themes of integration that he would often return to in his writings:

— General information
— Methodology or model
— Epistemology, ontology and semantics
— Human sciences
— Discipline oriented (approaches) (3_167)

These foci articulate different perspectives for Ranulph's architecture of ideas. It is also interesting to note that Ranulph did not like the "list" or "network" methodology that both Pask (as I understand it) and I, have employed at times. He was much more interested in juxtaposing disciplinary

---

13. François (2004).
14. See also Wittgenstein's notion that "the meaning of a word is its use in language" in his book *Philosophical Investigations* (1958).

pairs and articulating their relationality to define new wholes, as opposed to more rhizomatic[15] positionings.

Transcendence and Release Through Idea Development

One asks what stimulated Ranulph to work through ideas from this central set of intellectual perspectives, many drawn from the core of cybernetic approaches to ideation?

Ranulph elaborates on Pask's discussion of Lofgren's ideas in the text "A (Cybernetic) Musing: The Gestation of Second Order Cybernetics." He points to the fact that Pask came to articulate some thoughts related to system-related meta-concepts. Pask asked, "What if the model of the model was actually the object modelled in the first instance?" (3_129). This quote takes a kind of Mobius strip form, another device Ranulph uses in relation to critique and the making of distinctions, discussed in our Long Conversation. In Pask's case, Ranulph points out the fact that:

> The meta-system may eventually come to simultaneously, completely and consistently model all of the system. (This of course was a calculus of limits and could be translated into a concatenation of metas). (3_129)

Additionally, This ongoing interaction with Pask's ideas proved to open out a salient set of processes for Ranulph in terms of defining reciprocal relationalities,[16] as well as finding pleasure in the playful, language-driven interrogation of ideas:

---

15. Deleuze and Guattari's definition of the Rhizome: Let us summarise the principal characteristics of a rhizome: unlike trees or their roots, the rhizome connects any point to any other point and its traits are not necessarily linked to traits of the same nature; it brings into play very different regimes of signs and even nonsign states. The rhizome is reducible to neither the One or the multiple. It is not the One that becomes Two or even directly three, four, five etc. It is not a multiple derived from the one, or to which one is added (n+1). It is comprised not of units but of dimensions, or rather directions in motion. It has neither beginning nor end, but always a middle (milieu) from which it grows and which it overspills. It constitutes linear multiplicities with n dimensions having neither subject nor object, which can be laid out on a plane of consistency and from which the one is always subtracted (n-1). When a multiplicity of this kind changes dimension, it necessarily changes in nature as well, undergoes a metamorphosis. Unlike a structure, which is defined by a set of points and positions, the rhizome is made only of lines; lines of segmentarity and stratification as its dimensions and the line of flight or deterritorialization as the maximum dimension after which the multiplicity undergoes metamorphosis, changes in nature. These lines, or ligaments, should not be confused with lineages of the aborescent type, which are merely localizable linkages between points and positions... Unlike the graphic arts, drawing or photography, unlike tracings, the rhizome pertains to a map that must be produced, constructed, a map that is always detachable, connectable, reversible, modifiable and has multiple entranceways and exits and its own lines of flight (Deleuze & Guattari, 1987, p. 21).
16. See also the paper O. Perriquet and W. Seaman. "Art ↔ Science Relationalities." *International Symposium on Electronic Art Proceedings* (2011), See also William Seaman, "A Multiperspective Approach to Knowledge Production," *Kybernetes*, Publication date: 3 November 2014 for discussions of the use of the word relationalities.

> I don't know if I can communicate the astonishment and sense of revelation I experienced at this. The process was exciting. The man (Pask) was exciting and brilliant. But the ideas, the ideas were pure magic. Suddenly there was a world sucking me in, where ideas could be played with, developed and turned in on themselves in wondrous ways. For me it was a (long) moment of sudden passionate clarity, and transcendence and release. (3_129)

This architecture of ideas, as it is played out via circular causal accretive activities, can become very moving indeed! Here dialogic thought production opened out through participation in conversation is essential. I will circle back to notions surrounding conversation and Pask's conversation theory later in this text.

The Principle of Mutual Reciprocity

When I explore the breadth of Ranulph's ideas, one of his most interesting is *the principle of mutual reciprocity*. This approach opens out a huge set of ramifications, and can be applied in many different contexts. Perhaps one the most important ramifications deals with how intelligence is articulated. Ranulph defined the principle of mutual reciprocity as follows:

> The principal (or law) of Mutual Reciprocity states that, if through drawing a distinction we are willing to give a certain quality to that (which) we distinguish on one side of the distinction, we must also permit the possibility of the same quality being given to that which we distinguish on the other side of the distinction. If I distinguish myself from you and I consider I am intelligent, I must consider that you (which I distinguish from I) might also be intelligent. (3_399/400)

Ranulph would also say that we attribute intelligence related to people and/or machines (3_300). I will later more fully explore intelligence as a topic in this text. In short, Ranulph's position is thus:

> Intelligence is... a quality attributed by one to the other in an interaction. Intelligence requires interaction and is shared: it is found in the contribution of both participants and it is held between them. (2_175)

What does this say about computational intelligence—intelligent systems? How do current computational systems fit Ranulph's criteria? Needless to say, he also pointed out that making distinctions implies "both responsibility and mutuality" (3_300).

Subject and Metasubject

How might one use second-order cybernetic concepts as well as design concepts to shed light on many different fields? Ranulph spoke of this potential:

> Cybernetics is one of those rare subjects (another being mathematics) which, while being a distinct field worthy of consideration in its own right, is also a subject that casts light upon other subjects. It is an abstract subject which has often been applied to enhance our understanding of other subjects. In its incarnation as Second Order Cybernetics, it is both its own subject and its own metasubject.
>
> Design is another such subject: a subject in its own right, that can cast light onto other subjects, and which ... needs to be studied in the light of its own criteria, as a design equivalent of Second Order Cybernetics: the (recursive) cybernetics of cybernetics and the (recursive) design of design.[17]

Ranulph was thus central in helping to redefine second-order cybernetics in terms of design and design in terms of second-order cybernetics.

Assembling Concepts

How should I go about assembling concepts related to assembling concepts in my unpacking of composing composing? How can I design the organization my ideas and thoughts where the form reflects the content and the content reflects the form? It must be noted that as one reads through Ranulph's works, he greatly expands the reading of the meaning of design. I will return back to this thought often in defining an approach to composing composing:

> There is another aspect of the design of concepts that I must mention, if briefly. It is the assembling together of different concepts such that we can form new concepts or we can organize the different concepts we have designed into heterarchical organisations...This act of construction, whether concerned essentially with the organisation of concepts or with the creation from several concepts, of new more general concepts, or splitting the concepts into new, smaller and more detailed concepts, is also an act of design. (2_228)

I have tried to design this text (and many of my poetic texts) with a series of modular sections. I have laid them out in a thoughtful manner, one after the other, yet I am also interested in how they can be approached in a recombinant heterarchical manner. In my own writing about neosentience, I have discussed recombinant informatics (Seaman & Rössler, 2011, 2012). A

---

17. Glanville (2007).

series of micro-chapters populate that book (not unlike the short sections presented in this text). I discuss the idea that one can build bridges between any two of these micro-chapters to open out new lines of inquiry, taking a clue from the idea of bisociation as drawn from Arthur Koestler's book, *The Act of Creation*.[18] I like the idea that one may use a recombinant-informatic approach to this text as well, and in particular enact the assembly and re-assembly that Ranulph mentions above. I have also become interested in what I call poly-association.

In making titles Ranulph was often interested in presenting coupled sets of relationalities, to open out new interpretations and/or to aid in the unpacking of particular subjects from an alternate, perhaps even poetic juxtapositions of vantage points and/or a surprising set of perspectives. These were also often reciprocal positionings, inverting the word order to present new ideas. Yet this approach is different to what he describes above as the principle of mutual reciprocality.[19] So this might be considered a pun-like take on reciprocality. In puns, often one fork of meaning makes sense, while a second meaning may be playful or nonsensical in its relationality to the first. This relationality may exhibit a sense of humor, or this may take the form of an embodied contrarianism as Krueger points out elsewhere in this volume, in his listing of Glanvilleian titles and his explication of contrarianism in general. This tactic gets us thinking about thinking, and also the subtleties of meaning production, as well as the importance of disruption in the normal ordering of our ongoing approaches to coming to understand a particular topic. Yet, these pointed relations can also be quite telling.

A Way of Thinking About Mind and Matter

In a section titled "Freedom and the Machine," Ranulph states:

> I would like to conclude with a quote from James Clerk Maxwell: "The only laws of matter are those that our minds must fabricate, and the only laws of mind are fabricated for it by matter."
>
> I end with this because of the way Maxwell ties mind and matter together. He doesn't say mind comes from matter or matter comes from mind. He says that you need both, that they work together, each producing the other. (2_81)

I have a very similar approach; I suggest that mind and matter are co-arising… In conversation, Albert Müller pointed out that the term co-arising

---

18. Koestler (1964).
19. See Krueger's text, "On the Contrary," in this volume for a list of some of these titles.

that I use, might be compared to the concept of co-evolution that Bateson employed.[20]

> The observer and the observed are linked. There are circles. The Macy conferences' circular causality drives these modern cybernetic machines that liberate rather than enslave us. These are the things that matter. They give us new insights into what it is to be human and to act in the world now. (2_81)

I often point out that in doing research into AI and into neosentience in particular, ironically one learns more about just what it is to be human. I am trying to understand what is at operation in the body that enables sentience to arise, and abstract that into a new variety of the machine. Ranulph often discusses the cybernetic interest in a relationality between mechanisms and animals, stemming from Wiener's early books, yet taking on a new interpretation as time goes on.

Cybernetics and the Machine

Ranulph, in his text "A (Cybernetic) Musing: In the Animal and the Machine," in part discusses the beginnings of cybernetics and also speaks about concepts surrounding *Control and Communication in the Animal and the Machine*,[21] the seminal text by Norbert Wiener. Ranulph's text discusses the role of the Macy Conferences in articulating cybernetics as a field. In particular, Glanville made some interesting observations about the relationality between the animal and the machine—pointing both at the beginnings of cybernetics, and then articulating how those ideas changed over time compared to more recent writing exploring second-order cybernetics and in particular his own thought on the subject:

> In this column I argue (along with Heims'[22] book about "The Cybernetics Group", where he discusses the role of the Macy Conferences in the development of cybernetics and the social sciences) that the early cyberneticians lived in an era where mechanism reigned supreme. Yet they also insisted that the whole was greater than the sum of its parts. I capture this through two reciprocal metaphors. The early cyberneticians treated the animal through the metaphor of the machine, yet understood that this was reductive of some aspects of human experience. These understandings of mechanism and wholism, can be seen as leading to the development of Second Order Cybernetics, which I propose, we can understand through the idea that the machine can be treated through the metaphor of the animal. (3_107/108)

---

20. See Bateson (1972, 1979).
21. Wiener (1961).
22. Heims (1991).

In a form of graphic shorthand, I here present this reciprocal set of metaphors as follows: animal ↔ machine. I am deeply interested in biomimetics and bio-abstraction as they are applied to new forms of robotics and computation.[23] How can I best design systems enabling the examination of multiple of the above topics to help me explore this relationality? My *Insight Engine* project discussed in our "Long Conversation" has been my latest approach at the creation of a transdisciplinary, science/conceptual art-related, database and search engine exploring the potentials of neosentience. Since writing the neosentience book with Otto Rössler, I have worked on this project, which is currently being coded as version 2.0. Through interaction with the system we enable the exploration of poly-association—a computational linguistic search methodology we have developed. This allows a user to select multiple papers and/or topics and the system retrieves a potentially relevant series of related transdisciplinary entries. Ranulph and I discussed in our long conversation the extension of this system to enable the exploration of differences in how multiple people define particular cybernetic terms. Also, one must remember that in the sciences and the arts a single term might be used in a different context and have quite a different meaning.

Cybernetics—A Brief Pragmatic Definition

What, in Ranulph's eyes, what would a brief pragmatic definition entail?

> Cybernetics is a way of thinking that bridges perception, cognition and living-in-the-stream-of-experience (the involvement of the observer), which gives important value to interaction and what we hold between ourselves and others whether animate or inanimate. It is concerned with circular causality and the wish to control in a beneficial manner. It comes with a mechanical metaphor for the animate, which is now partnered with an animate metaphor for the mechanical. While it can be of great use in its traditional business of modelling control systems (and hence control engineering), for me its interest lies in the significance given to the involved observer and the consequent individuality of and responsibility for his/her actions. My position lies at the radical extreme of Second-Order Cybernetics and is unrecognisable to some others in the field. (2_256)

My own beginnings explored circular causal feedback in the design of interactive computer-based artworks. I later sought out each of the original volumes of the Macy Conference proceedings (see now Claus Pias's book[24] which brings them all together in one volume) and this was my introduction

---

23. See Seaman (2013).
24. Pias (2016).

to cybernetics, along with Wiener's major books on the topic: *Cybernetics or Control and Communication in the Animal and the Machine*[25] and *The Human Use of Human Beings: Cybernetics and Society*[26] and later Ashby's *Design for a Brain*,[27] as well as Pask's *An Approach to Cybernetics*.[28]

In certain of my works, the interactant makes decisions, in particular about navigating, combining and re-combining poetic interactive video, music and text (as well as other media objects), in time-based 2D space and in later artworks, in virtual space. Meaning in these works is emergent and individualized for each participant/observer via their use of a specifically authored/designed circular causal variety producing computational system. Over time I moved from interactive generative video to the design of interactive world generating systems, also exploring interactive image, music and text and other media objects and computational processes. I, in each case, designed the systems so that the observer could become mindfully aware of how meaning was arising and changing via their interactions. I call this meta-meaning.[29] The name I have given to this field of embodied exploration of navigable, combinatory digital media and concomitant examination of varying computational functionalities, is recombinant poetics.[30] I see these works as exemplifying second-order cybernetic processes. I was intuitively doing cybernetics before I knew fully what cybernetics was, as was Ranulph. I also had an early interest in developing an electrochemical computer. I was excited to see this was also an interest of Gordon Pask's. Pointing at thought processes and the relationality of thought to computation was also part of the Biological Computer Laboratory's brief, headed by Heinz von Foerster. [31, 32]

Second-Order Cybernetics—A Philosophical Positioning

Ranulph had favorite topics or guiding principles related to cybernetic research as discussed above, and in particular second-order cybernetics. He

---

25. Wiener, (1961).
26. Wiener (1950).
27. Ashby (1954).
28. Pask (1961)
29. Albert Müller asks, shouldn't this just be meaning and not meta-meaning? I have often used this term (meta-meaning) in relation to authored systems which are designed to point at meaning as it is arising and changing though embodied interaction. For me this suggests the term *meta-meaning*, or a situation that enables the exploration of meaning that is about meaning. Instead of writing about meaning, in this kind of work, one addresses it through dynamic interaction and generative human experience. I believe this is very close in spirit to Ranulph's approach.
30. Seaman, (1999, 2010).
31. See my paper on von Foerster (Seaman, 2017).
32. See Müller, A. & Müller, K. (2007).

championed a re-definition of the field. How can these principles be composed and pragmatically form the dynamic composition of composing composing? Is this a question that we must each answer through our own conversation with the self, and/or with the artifacts of thought that Ranulph left us with? This leads to a new question: How is second-order cybernetics relevant to contemporary philosophy?

> As a philosophical position, Second Order Cybernetics makes sense of certain conditions we could not previously make much sense of, and validates the approach of Cybernetics (circular causality, in a nutshell) by applying it to itself, thus acting with a particular and acute rigor and consistency. It also gives us a way of considering when the observer is not excluded: a situation understood as more common and relevant nowadays. (3_335)

As stated above, central to second-order cybernetics is the inclusion of the observer (composer) in the system. It is certainly a unique operative function of the nature of the human observer that they can define meta-level systems as part of a system. In my discussions with Ranulph, I often pointed to a new idea of what might be called "open-order cybernetics."[33] This might include new varieties of observer like intelligent robots and/or *neosentient*[34] *machinic systems* (machines exhibiting a form of sentience), that would be *of themselves* yet would extend the definition of the observer in new ways. Ranulph clearly stated that these new levels always collapsed back down to the second-order cybernetic level. Here, the observer is not a specific person, but also might be a robotic or machinic observer. To my mind, the ability to easily shift meta-levels and discern context, as well as function creatively, is still central to what it is to be human. Much growth is needed in terms of the coding of machinic observers to exhibit these functionalities. I would suggest that this includes defining new kinds of logic, like those undertaken in the BCL by Goddard Gunther.[35] We can begin to think of biologically inspired algorithms—"bio-algorithms" as I call them—as yet non-existent to the level I am interested in. The goal is the emergent exhibiting of sentient–like behavior in machines. Yet, it will take second-order cybernetic concepts to help elucidate such a field, and new forms of logic to enable its becoming. The thought is that this might be achieved by mapping carefully all functional entailments in the body that contribute to sentience arising (no small task!), and then building a new form of combinatoric bio-algorithm that functions in an analogous manner

---

33. Gaugusch & Seaman (2004).
34. See Seaman & Rössler (2011, 2012).
35. Gunther (2004).

to this mapping. Does this suggest the need for a new form on n-dimensional bio-logic? I still think that open-order cybernetics may come to be... Time is long. Will autonomous machinic sentience enable an exact form of shared knowing between machines? Perhaps this suggests a new level that doesn't collapse to the two as Ranulph suggested in our talks, without an infinite regress?

Characteristics

In this case Ranulph starts out by asking the very question I wanted to ask him...in a text entitled "Characteristics" Ranulph laid out some highly relevant salient characteristics:

> So, what are the characteristics of this version of cybernetics? ... Goedel's problem is not at the heart of it. Goedel's problem was one of the drivers that lead to it, but the cybernetics of cybernetics transcends it and relegates it to a corner because it belongs to the class of phenomena that, while logically impossible, always seems to happen. My position is that we must therefore change logic,[36] not what we find to be actual. The crucial characteristics that have been understood include the following:
>
> The observer is included in the system (Famously von Foerster's aphorism in his amazing source book which gave us the name "The Cybernetics of Cybernetics" is "the cybernetics of observing systems." First order cybernetics is characterized as "the cybernetics of observed systems.") Unless the observer is included there is no observing. All knowing presumes the active participation of an observer.
>
> Systems are seen as circular. Circularity is taken to be the general case. Cutting circularity leads to hierarchy and suchlike, i.e., to chains and cause. A metaphor is a wheel and its trace/track. In other words, the circular form of feedback is taken seriously and not dismissed as a minor physical variant acceptable only because the energy required is insignificant.
>
> Thus, the preoccupation is with process rather than object. Objects become apparently stable embodiments of their own (internal) processes (hence, self-description), i.e. autonomous. And the difference between internal and external (within the system and without it, particularly vis-à-vis the observer) is appreciated, especially in relation to stability.
>
> Herein lies the interest in that most diabolical of local arrangements, self-reference. In a universe of reference, stable entitles (Objects) must be taken as referring to themselves, if they are to be taken to be autonomous objects in this universe and to be based on reference. (3_132)

What role does self-reference and the easy shifting between meta-levels have for the autonomous observer? Can this be authored in terms of new kinds of biologically inspired coding for machinic systems? How do

---

36. I discuss new forms of logic as articulated by von Foerster in conceptual conversation with Goddard Gunther, in my invited Von Foerster talk entitled "Towards a Dynamic Heterarchical Ecology of Conversations" (Seaman, 2018).

shifting meta-levels function within circular causal systems? Is this articulated through the embodiment of nested circular causal systems, or does this all just become one time-based, circular system? I will later return to the discussion of parts and wholes as derived from multiple perspectives of composing in this text as articulated by Ranulph.

Circularity

How can we point to circularity, a time-based phenomenon, as being intrinsic to our interactions with the self and the world?

> Generalisation to the circular characteristics, Cybernetic systems and the Cybernetic way of understanding the world, where interaction, conversation, and other similar actualisations of circularity provide both material for study, and Cybernetic models through which to carry out that study. Understanding circularity-as-form must include the observer (as an actor in the system), and admitting that which is examined as examinable: confirming a separation between form and content that allows us to, for instance, discuss experience-in-general as form, while insisting that each particular experience is unique, its meaning belonging to each occasion-and-(actor-)observer.
> Second order cybernetics particularly insisted on the involved observer making (and accepting responsibility for) his/her observations: a circular process where each observation changes both observer and observed, leading inexorably to new observations. (3_396/397)

Thus time, emergent and accretive experience, as well as learning and change, become the norm via circular causal interaction with environment and self.

Second-Order Cybernetics and Design

How can we employ cybernetic qualities and processes to help us enact design in all of its possible forms? In my teaching of Experimental Interface Design, I use a simple equation to help designers and artists think about their potential interface design: Input + Functionality = Output. I always discuss the fact, when talking about the equation, that the output provides an opportunity for the interactant/observer, to take part in a second-order circular causal system and provide new input. Albert Müller pointed out in our past conversations that I should add a recursive line from the output back to the input (which I have done in the figure below). This is implied by my contextualizing discussion with the students. He compares the equation to von Foerster's trivial machine, yet it could also be a nontrivial machine as defined by von Foerster.

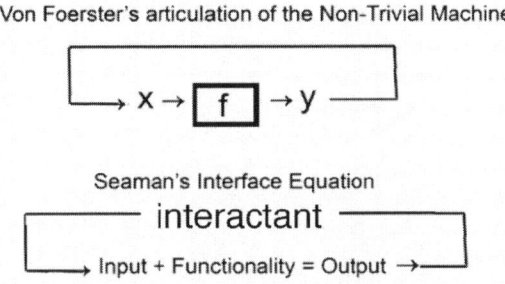

The form of my interface equation allows us, the class and myself, to define a huge set of variables for each possible input system, each possible functionality, and each potential output system for sharing! A small program has been written so one can use chance and/or choice to peruse the variables. In the class, the individual steers this combinatoric system as a brainstorming methodology based on their own interests, some pragmatic and some more on the side of poetics. Some humorous nonsensical solutions can also emerge! This system becomes an interface to help brainstorm possible interface systems. Thus, it is an emergent system.

It must be noted that I was doing cybernetics very much in the Glanvilleian intuitive learning mode, without having initially read his texts. Also, I developed the approach long before knowing about the trivial and nontrivial machines of von Foerster.

So briefly, what is this pragmatic relation of cybernetics to design?

> Design is a basic human activity. And Design is second order cybernetics in action: it is, if you like the application of second order cybernetics, making it useful. Thus, the strong connection I see between cybernetics and design. (3_219)

For second-order cybernetics, thinking is always one with the process.

Designing Design Variety

Why is production of variety so important? Ranulph was enamored with the writings and concepts of Ross Ashby and in our "Long Conversation" he returned to the concept of variety on a number of occasions. In Ranulph's paper "A Cybernetic Musing, Variety and Creativity" he states:

> I argue for the benefits of sharing and co-operation in design (common practice) as a way of increasing available variety, and for us to learn to use computers not so that we control them as tools that carry out our will, but so that they both facilitate sharing and are used to generate surprise, thus also increasing the variety available to the designer.
>
> Behind this a more important concept lurks. It has to do with how we treat an imbalance in variety, when we find that we have less than the system we are interacting with and which we have, traditionally, handled by closing down the variety of the system to be regulated, so that we can control it. (3/145)[37]

Ranulph asserts a unique aspect for second-order cybernetics in his discussion of the exploration of generativity and emergence through enhanced variety, by suggesting exploration and inclusion of the unmanageable as a means of variety generation. One might see this a bit like a Dadaist or contrarian act (see Ted Krueger's text in this volume: "On The Contrary"), where historically anti-art (dada) actually ended up being a fertile ground for creativity and the growth of art. In this case anti-cybernetics enables cybernetics to expand in an interesting and fruitful manner. One should also look at the concepts of Herbert Brün, in particular his notion of anticommunication. Brün's definition sheds light on Ranulph's anti-cybernetics through viewing the relationality that his notion of Anticommunication analogically suggests. In his paper "For Anticommunication" Brün states:

> As this is the point where the arts, including music, come in, let me formulate a useful term. Where a new thought is presented, the speaker's problem is not any longer only a problem in communication, but one of communication. My useful term is introduced thus: A speaker with a new thought has to solve a problem of anticommunication. The syllables "anti" are used here as in antipodes, antiphony, antithesis; not meaning "hostile" or "against" but rather "juxtaposed" or "from the other side". Anticommunication faces communication somewhat as an offspring faces its progenitor. And just as the offspring eventually will in turn become a progenitor, so will anticommunication, in time, become communication. …
>
> Anticommunication is an attempt at saying something, not a refusal of saying it. Communication is achievable by learning from language how to say something. Anticommunication is an attempt at respectfully teaching language to say it. It is not to be confused with either non-communication, where no communication is intended, or with lack of communication, where a message is ignored, has gone astray or simply is not understood. Anticommunication is most easily observed, and often can have an almost entertaining quality, if well-known fragments of a linguistic system are composed into a contextual environment in which they try but fail to mean what they always had meant, and, instead, begin showing traces of integration into another linguistic system, in which, who knows, they might one day mean what they never meant before, and be communicative again. (Brün, 1972, p. 3)

---

37. See also "The Value of being Unmanageable: Variety and Creativity in CyberSpace" (Glanville, 1994).

We substitute the word *anti-cybernetics* with the word *anticommunication* we can begin to see the relationality in operation here.

Anti-Cybernetician | Cybernetician

How can taking the role of an anti-cybernetician actually create new variety in second-order cybernetics? Ranulph articulates the following:

> I position myself as much as an anti-cybernetician as a cybernetician, much in the manner I value ignorance as much as I value knowing (for ignorance is a source of knowing). Is not the unmanageable and the purposeless equally the source of variety and purpose, of cybernetics? If so, telling the story with design as the conceptual source, expands and enhances cybernetics, as our understanding of cybernetics helped us account for design. Remember when I wrote of unmanageability as a way to enhance creativity (Glanville, 1998). Normally, cybernetics is interested in systems which conform to its one universally accepted law, Ashby's law of requisite variety, thus being manageable. In contrast, I propose we should develop an interest in the unmanageable: a form of anti-cybernetics.[38]

Ranulph stated the following in the text "A (Cybernetic) Musing Design and Cybernetics":

> Describing the central act of design through the metaphor of conversing with oneself connects to the idea that different personae don't see the world in the same way, from the same understandings, or know the same things. We can reconstrue this: The variety of any one persona cannot equal the variety of all (other) personae. (3_423)

Unmanageability enables us to fall outside of our comfort zone to increase variety, and to at times take in knowledge derived from other personae, which potentially also falls outside of our comfort zone. This is a circular accretive process where the unknown becomes the known via process and iterative change brought about through our conversations with the self and others.

Unmanageability

What are the processes available to help generate this state of unmanageability in art, architecture, music and design? I believe it is through a shift in self-understanding in terms of second-order cybernetics and a re-understanding of the definition of second-order cybernetics itself. That by seeing these from a slightly different perspective, one can begin to explore unmanageability as a fruitful state. Time and accretive learning are central here. Similar to this re-positioning, one of my own quotes is "displacement illuminates placement":

---

38. Glanville (1994).

> Control is circular, not linear. It exists between the controlling and the controlled systems. If systems are of such variety and such complexity that it is inconceivable that we can satisfy the Law of Requisite Variety and thus properly control them, we must consider them unmanageable. When a system in unmanageable, we have three options: to reduce complexity, to change the organisational structure (how control operates), and to alter our attitude to unmanageability. Unmanageability lives between the (nominally) controlling and controlled systems as interaction. Unmanageability is unavoidable. In unmanageable systems, the communication needed for control cannot be coded: an alternative form for communication is the conversation, with its personal meanings. Unmanageability is not bad. It can give us many options and opens up possibilities for us, if we listen carefully and keep an open mind.[39]

In terms of creativity, Ranulph sought out various processes that would enable the focused exploration of unmanageability to function in the service of variety generation.

Ranulph's Music and Unmanageability

How could this approach be employed in terms of experimental music production? We can see Ranulph applying this notion to his music production in exploring unusual compositional processes, in relation to differing architectures of ideas—chance, the employment of non-trained musicians, the folding together of taped sound sources in a fluid manner enabling rich time-oriented tape-based juxtapositions, experimentation with indeterminancy and the use of unconventional sound generating electronics.[40] Many of these ideas are discussed in the book by Ertl, Muller and Korn.[41] Here John Cage's ideas, and the abstracted influences of others, were made operative in a completely Glanvilleian manner, embodied through Ranulph's own unique approaches.

Unmanageability in Teaching/Learning

How might this approach be explored in the teaching arena? In terms of his design teaching, this meant extensive exercises that pushed notions of observation though eclectic processes related to limiting sensual intake, and/or exploring the senses in new ways, including the use of blindfolds, etc. as well as observing for very short durations, and then unpacking that observation in a focused manner. This also included becoming mindfully aware of this unmanageability by the interactants conceptually observing themselves. Varela, Thompson and Rosch in *The Embodied Mind*, speaking about Buddhist mindfulness/awareness provide this perspective:

---

39. Glanville (1994, p. 10).
40. Ertl, Korn,& Müller (2016).
41. Ertl, Korn,& Müller (2016).

> Its purpose is to become mindful, to experience what one's mind is doing as it does it, to be present with one's mind. What relevance does this have to cognitive science? We believe that if cognitive science is to include human experience, it must have some method of exploring and knowing what human experience is.[42]

Ranulph sought to generate variety via the unmanageable, often exploring experimental research processes that enhanced creativity, and promoted self-conversation, and mindful awareness to the processes at hand.

> To accept unmanageability as wonderful: which gives us endless novelty, a richness beyond our wildest imaginings, magic; and demands from us trust, open mindedness and generosity, the sorts of qualities I assume we would like to claim for ourselves. (3_215)

And these qualities were very much part of Ranulph's deep charm—the output of the embodiment of his own particular magic.

The Unruly Definition of Design

Given the overarching unruly linguistic design of the word design, is design a bit slippery to define?

> Design is (in English) a confusing word, for it has so many different meanings and language roles, from noun to verb, and from the almost criminal "having designs on something" to the pidgin "de sine". For me a design (object) may result from (the activity of) design(-ing). This activity I depict, mentally as a conversation with the self, usually via pencil and paper. (It is possible for some, sometimes, to do it non-materially.) In the process of this conversation, ideas grow and take on their own life, developing and changing in a manner that is essentially unpredictable and which increases variety, richness and depth. (2_122)

In the beginning of our long conversation, speaking about Ranulph's "definitive" final linguistic cybernetics project, he spoke about trying to create a curated, participatory system to reflect the differences of definitions inherent to individual thought and practice related to first order cybernetics as well as second-order cybernetics. Ranulph was deeply interested in how individuals would each define differing terms related to their own individual history, and so forth. Above, Glanville unpacks the word *design* related to its dictionary usage, and also points to the conversation with self, inherent to iterative design processes. Alternately, Ranulph pragmatically re-articulates the broader definition of design in the diversity of his columns; many fragments of these are reproduced in this text.

---

42. Varela, Thompson, & Rosch (1991, p. 23).

Novelty

Does Ranulph suggest that we should intentionally bring error into play?

> Whereas in most models of communication the concern is to reduce error, in design the so-called "error" may be a source of novelty. What is often thought of as error is welcomed as a means of enhancing creativity. (3_431)

Here novelty arises through a circular accretive process, where error becomes a natural part of this process. One gathers over time their errors and steers themselves to new creative spaces which were earlier unknown. In the end, these aspects become manageable.

To Do Design Research is to Design Design

How does designing design become enfolded in design research? Is design research a science?

> Designing is (at least a form of) research. To do Design Research is to design design, and the designer is always in the system. Ways of assessing the outcome of design processes are a completely different type of research which is research on design, rather than in or through design—as I have discussed when writing about prepositions in English (Glanville, 2005). Those who wish to make design scientific (in the conventional sense) miss the point: Design is not science, nor should it be. To try to turn design into science, to try to get from it the certainty we hope for in science, is to try to turn wine into water. There is no design without the designer (just as there is no design without the user): design is inherently a second-order cybernetic activity. (3_379)

> I characterized an activity I hold is at the heart of design, and how this activity—the conversation with oneself via paper and pencil—is so very cybernetic: and I extended it to include the mechanisms of accommodation, affordance and assimilation.
> This is why I claim cybernetics may be thought of as the theoretical arm of design and designing may be thought of as the practical arm of cybernetics. I can summarise the position argued, quoting from a talk I did in 2006 (Glanville, 2006a):
>
> —Design according to Vitruvius, deals with 3 qualities: firmitas, utilitas, and venustas.[43]
> —Conversation is essentially constructivist: each participant constructs his/her own meaning and value (therefore, each is responsible for this).
> —Design is a conversation held primarily with the self (but also others): self-conversation emphasizes the significance of listening/being receptive.
> —Designers develop and amplify ideas, make the new from differences in meanings—when difference in expression is welcomed, not hidden.

---

43. Ranulph (2014, p. 261) states: The earliest, and still arguably the best, definition of architecture was by the Roman architect and writer Vitruvius, who called for "firmness, comodotie, and delight" in translation by Sir Henry Wootton: in today's terms, being well built, functional and delightful. (2_261)

—The process of design is circular, iterative, unknowing (including rejecting and restarting), constructive: explanations are post-rationalised.
—The new is beyond prediction.

Implicit in conversation (and thus design) are many ethical qualities we think of as deeply human and desirable. (3_435)

Cybernetics ↔ Design

How is it that design and cybernetics are mutually supportive?

I believe at the heart of what designers do are acts well explained by these cybernetic understandings which can be seen as a design conversation: the notion of forging pattern, the notion of circularity, the notion of parallel understandings. The structure of a conversation leads to the freedom of individual meaning, just as Objects allow difference in observation nevertheless to be taken as the same Object of attention. (2_80)

Human (and potentially machinic) attention is often shifted via aspects of self-observed conversational processes, forming new patternings.

Definition of Design Writ Large

Ranulph had a very broad set of descriptions of what design was and how it might be approached and understood through cybernetic thought. How does design research fit into this design milieu?

There will be no need for a special area of design research, for all research will be seen as part of design research, with that which we call, now, design research will be the most basic of all. (2_120)

In his broad definition of designing the re-understanding of design, Ranulph suggested that scientists are designers in terms of designing their experiments, and that we all are in fact designers in terms of our thought processes. We are all composers in this light.

Design Process

How does design alter our understandings?

Through designing we come up with new understandings, which we don't have before: and resulting objects, which are new to us, at least. This is how the process of designing is sustained in the metaphor of wandering ... I see this as akin to wandering, with us designers absorbed in where we are, deciding at each instant how to proceed: and then finding we've arrived and the whole process of our wandering suddenly making sense, becoming a progress. (2_22)

This is of course progress arising through process. A form of wandering until the destination becomes found and fully articulated.

Wandering

Can one wonder and wander about wondering, about becoming one with experiencing the wandering? Composing composing accepts self-reflective wandering as being part of a process of finding. Ranulph was interested in this larger metaphor of wandering as part of the creative process both in design and in conversation. I also very much like this notion of going for a walk or a drive and not knowing the destination at the start, being both literal and metaphorical in my case. Here, an emergent outcome is central. Laughingly, I thought this to myself: I wonder about wandering and I wander about wondering. My own term for this is *I Ching Driving* (mentioned above also)—I use this process-oriented found non-path as a means to explore new landscapes and/or subject matter for my photography, music, and video production, examine new approaches for generating poetic texts, unleash new methods for the creation of digital collages and drawings, and empower the opening out of new ideas related to the study of AI. Such wandering also potentially opens up new areas of study as a researcher. This wandering becomes a means to promote new conversations with a variety of environments, natural ↔ computational. How can one elaborate on this creative conversational wandering?

> We can consider a conversation as being like wandering in the country; perhaps in a wood, maybe carrying a hamper for a picnic … As we talk, we follow paths that, to someone else, will almost certainly seem arbitrary. Even talking around a topic we will move away in a manner that is both unpredictable and seemingly without purpose. We will end up somewhere, and will decide that this is a good place to stop…Swap the word "walk" for the word "talk" and the word "topic" for the phrase "feature in the landscape" and the similarity is clear. The place where we end up is the place where we "decide to have our picnic." Arriving at this point, we can make sense of our journey: we can explain the trip and give it purpose. The word we use for this sort of walking is wandering: designing and conversation are both like wandering. (2_281)

Wandering is another form of opening up possibility—of generating variety. This was central to our "Long Conversation." Is "Composing Composing" in part just a form of wandering though Ranulph's textual gems related to his broad definition of design, in that I am just adding in this sentence late in game?

> This is the process of design translated once more. We do not really know where we are going when we design, but when we arrive we know that we have arrived [sometimes we do not know that we have arrived, emphasis Seaman—sometimes we must learn when to stop], and can make sense of the progress. This is not to deny the importance of

those aspects of design that we can treat as specifiable and which we can solve (in the traditional sense), but to recognise and allow the central act that makes (almost stumbles on) the new without quite knowing how or why, and can then explain this, by means of explaining the route taken, as a seemingly sensible (even logical) path. This account is, however, not a purposive problem-solving activity. It is a post-rationalisation, an explanation after the event. (2_282)

I earlier spoke of assembling a variety of thought assemblies. Perhaps this is one goal of composing composing, after the gatherings are informed by our multiple perspectives or frames of conversation. "Composing Composing" went through multiple iterations before ending up in this particular form.

Thinking as a Design Activity

In thinking about thinking, how is it that we can come to compose our thoughts? Ranulph discusses thinking as a design process.

> When we think we often form concepts. These concepts become stable: that is, they can be identified and reproduced: new thoughts and observations can be attached to them. But these concepts are not given: we **compose** them, or put another way, we design them. We also design how we fit them together. They usually don't just fall into place: we have to fit them—and doing so we change them and we change what they are fitted to, and this changes everything we know: in other words, there are knock-on effects. This is how thinking can be characterized as a design activity. I believe we undertake this design process both individually and in groups making up societies. (2_22/23)

Thus thinking is both the subject and the object of the design of composing composing and "Composing Composing."

Cybernetics Applied to Art and Design

Given this process of composing, how can we apply second-order cybernetic concepts to produce insight related to art and design production, given the pointedness of this metaphor?

> There are at least two ways in which cybernetic understandings give insight into aspects of art and design. Firstly, there is a concern with circularity. At the centre of the act of designing is, I argue, a circular process that may be thought of as a conversation held with oneself, in which we alternate the act of mark-making and mark-viewing. This move between roles involves switching what we might think of as personae, or as Pask calls them, p-individuals (Glanville 2007f). Switching between personae gives rise to novelty, just like the exchange with another in regular conversation. Conversation is thus right at the centre of the creative process and is intensely cybernetic. As a version of Paskian conversational structure it is necessarily circular. Thus, design may be seen as a cybernetic practice while cybernetics can, reciprocally, can be seen as design theory. (3_389/390)

I will discuss Pask's conversation theory at length later in the text.

Homo Designans

What is the role of pattern building for design? Ranulph states:

> I asked the professor of Latin, here at UCL, to tell me how she would translate "Man, the pattern maker"? I had not expected her answer: Homo designans. I didn't brief her: let's get the word design in here. Her translating brings making patterns and design together. That is wonderful because (I am convinced) we design how we think: after creating patterns, we **compose** them together. For me, design is the other really fundamental human activity (along with creating constant objects). Designing the way we think and what we think, Humans build patterns. (2_71)

This begins to point at the breadth of the definition of design that Ranulph is articulating.

Pattern Relations

An analogy is a particular variety of pattern relation—the act of comparing two things that are alike in some way. Here we also point to qualities of correspondence. Ranulph did multiple articles where he intentionally pointed out correspondences between relational architectures of ideas. The most famous of which were his study of Messiaen and Klee,[44] and the second was his interest in Finnish architecture as it related to the Finnish language. (Discussed at length in the "Long Conversation.")

We could say that Ranulph was having a conversation with the artifacts of thought that differing groups posited. These conversations took a variety of forms both linguistic and extra-linguistic and/or expanded linguistic (seeing all of these different patternings as modes of differing languages of communication). It is interesting to consider how creativity arises in the individual in that we are not privy to the depth of each other's pattern-related thought processes. Ranulph often pointed out that we each become creative in relation to our own sensibilities and background of learnings. To a certain degree we must infer certain kinds of creative operations as we observe creative practices playing out in others. I believe Ranulph's concept of utilizing a multiplicity of modes of conversation to some degree came out of certain of Pask's approaches to the concept of conversation, although I believe Ranulph found his own voice in this area, in particular in relation to self-oriented conversation. I wrote about Pask's different forms of conversational communication in a paper entitled "Language and

---

44. Glanville (1966).

Computational Creativity, Beyond the Expanded Conversation—Cybernetic Pattern Flows."[45] Differing modes of conversation that Pask employed were discussed there. If we define interactive systems as consensual domains, we can also call certain interactive processes particular forms of conversation. From Gordon Pask's multi-perspective uses of the term conversation drawn from research in differing disciplines, we can hold a cybernetic conversation by exploring natural-language discussion; via analogical processes; as well as through metaphor. Pask's work explores each of these approaches at different times. Here both analogue and digital computers can become conversational vehicles of communication processes and creative thought.

In *Neosentience | The Benevolence Engine* (2011, 2012), the book I wrote with Otto Rössler, we made a list of pattern-related activities. This seems like an appropriate place to quote it. Yet another list… this must be a pattern!

> Pattern of patterns / Meta-patterns / Pattern topologies / Pattern sensing / Pattern orientation / Pattern comparison / Pattern abstraction / Pattern imagination / Pattern recombination / Pattern generation (fragment collages) / Pattern gestalts / Pattern projection (intermingling with environment) / Pattern confluence / Pattern transference (technological production) / Pattern implementation / Pattern re-orientation (categorization) / Pattern strings / Pattern fields / Pattern actions (spatial/conceptual/relational) / Pattern navigation / Pattern recognition / Pattern truncation / Pattern abbreviation / Pattern inversion / Pattern mistreatment / Pattern realignment / Pattern surgery[46]

Piaget's—Constant Objects

How is Piaget's notion of constant objects relevant?

> Piaget asserts this essentially intellectual act (of the construction of Constant Objects via accretive experience, emphasis Seaman) is carried out by the baby, and indeed we humans continue to do it all of our lives. Our method is, in Piaget's account, to compare events and to find in them this sense of the common. From this we propose (to ourselves) objects to which we add other experiences, assimilating them into our objects that, thus remain constant. On occasion, we have to change how we see the object (that is we change the object, or even dispose of the object all together) in order to accommodate the unexpected. (2_233)

I, in my text "Pattern Flows | Hybrid Accretive Processes Informing Identity Construction."[47] proposed a related set of ideas to that of Piaget. I described these as pattern flows of multiple-sense perturbations, enabling us

---

45. See Seaman (2018).
46. Seaman & Rössler (2012, p. 87).
47. Seaman (2005).

to build up an identity over time. This particular identity is being built through subtle ongoing hybridization. Thus, identity is changed slightly with each new instance of patterned observation. Ranulph also discusses the importance of patterning as related to thought processes.

Piaget and Constructivist Design Production

How do constructivist processes fall in relation to the science underpinning the design of composing composing?

> I cannot leave this introductory section without making two more points about design. The first ... is that design (as I use the word) is a necessarily constructivist approach to the world. Designers construct the new, and their new is part of their world view. Accounts that deny a constructivist position are not, in my view, accounts of design. In my extension of this view, I argued that the constructivist design process is the one that Piaget (1955) proposes as the root of the (and hence our) child's construction of reality. This being so, design is the fundamental process of thinking, which returns us to the notion that design is the superset of science ... (2_89)

This concept where design is the superset of science provides a deep broadening to the concept of design that most of us have come to understand. It is very rich in its implications—as is second-order cybernetics—in its inclusion of the composer/observer.

Patterns of Experience and Piaget's Constant Objects

How should we enfold Piaget's concepts into Ranulph's notion of composing?

> In building theory from (and of) theory, we use the same devices we use to build theory: simplification, pattern finding. As well as the objects we have found, we treat relations and patterns they are held to pertain to, also as objects. Using the devices of theory on theory, we act self-referentially: and self-reference is necessarily, circular. We make theory about theory just as we make theory: we find the pattern of patterns. (2_159/160)

At the center of composing composing is finding patterns and responding, functioning through ongoing processes—iteratively, accretively, and in turn creatively. For Ranulph this kind of composing could be constructed from a myriad of foci drawn from differing disciplinary, interdisciplinary and even transdisciplinary contexts, as well as lived experience, and self-oriented conversation as discussed above. It is thinking itself in all of its guises that is being composed.

Teaching as Derived From A Process of Knowing About Knowing

How can we design a situation where the ongoing learning for ourselves becomes of equal importance to that of the learning undertaken by our students? How does knowing become operative in teaching?

> The teacher does not instruct. The teacher does not know subjects: (s)he is not an expert. Or, at least, this is not his/her job. The teacher's role is, rather, to have a more generalised "knowledge" upon which to act ("knowledge" for action, knowing about knowing). The teacher may, of course, be an expert in some subject a learner wishes to learn, but it is not necessary. In contradistinction to the Irish wag's witty comment (He who can does, He who cannot, teaches), the teacher teaches: teaching is what he (she, emphasis Seaman) does. And teaching is a second order activity, not a matter of having and pumping out knowledge, but of supporting the other as (s)he finds his/her understanding, of giving form in their learning, in making their experience information. (3_227)

> If we want to make a subject constituted of knowables more accessible to potential learners, we should allow many different approaches, with first access through any of the knowables that seem familiar, interesting or challenging, to us as the learner. (3_227)

Ranulph intuitively used a series of differing approaches, tailor-made on-the-spot for the differences in each of his students, and their particular needs as he saw them.

> And (s)he can inspire, enthuse, tell stories, engage, encourage, focus, create coalitions, care, dramatise, explain, discuss, even test and bring love: all of the while keeping distinct in his/her mind not only each learner but also the difference between what we think and feel and what we do with it. (3_228)

Earlier in this book a series of Ranulph's students discussed the individual approaches he took with each of them—articulating their particular learning relationality to him as a teacher and mentor.

Cybernetically Informed Learning Tenets

What are Ranulph's central learning tenets?

> Cybernetic Themes:
> Error: autonomy: responsibility: generosity
> Learning is designing is constructing.
> The learner:
> Error and ignorance: difference (in what's known, what's wanted and how to learn)
> Teachback.[48]

---

48. The origin of Pask's interest in conversation is traced to the technique of teachback and the recognition of the individual student difference: the psychology of R. D. Laing. 3_23

The teacher:
Waiting, listening, concern for the other
Conversation—listening: drama (theatre) (3_245)

Teaching and Learning—Some Imperatives

One asks, what are the salient imperatives related to teaching and learning that were important to Glanville?

- Respect the autonomy of each student's understanding
- Place understanding above knowledge: while a subject to be studied has its own structure and autonomy that should be respected (it is a discipline), we learn through the understandings each of us develops.
- Recognise that how we assemble our understandings is the business of each of us. It is at once a matter of our own design, and of respecting the autonomy of each.
- Only the learner can do her/his learning and acquire her/his understanding. You cannot learn for them. When I tell another, if they make an understanding, their understanding is what they think I think. It is not what I think. So, my understanding of your understanding is (precisely) not your understanding. (This is the central point of Pask's Conversation Theory.)
- Teach from behind. Follow the student's lead. You may decide whether to lend the occasional guiding hand, or let them follow their own path unimpeded. Whichever, you are enjoined in a conversation.
- If you want a learner to set the standard, you cannot do it for him/her. Therefore, you must wait. Intervention removes the responsibility from the learner. (I know it is hard to wait.)
- Encourage errors so learners can learn from their errors: no penalty for errors, and no force feeding!
- Treat students as you would be treated yourself: show concern for them and for their vulnerability. Thus, we learn generosity.
- Remember, if we have to present in "preaching" modes, (e.g. lectures), at least to entertain. Lecturing, especially "educational" lecturing at university, is a performance art (that's why it is so hard). Tell a good story well.
- If you wish to be interesting, you must be interested.
- Learn, as a teacher, to listen, and not to determine and control, at least at first. Each learner has their already learnt knowledge and understandings/approaches. To be able to respond, it is necessary to first listen.
- Education is involved in helping others learn. It is concerned with the learner first and foremost. It is not an opportunity for a teacher to display his/her knowledge or authority. We, as teachers, are also learners. There is an endless conversation to be had among us all.
- Above all, practice learning to learn and encourage it in all of your deeds. (3_250/251)

Teaching and Creativity

How would one go about enhancing the creativity of the learner? Ranulph discusses this in his text "A (Cybernetic) Musing: Variety and Creativity":

> I am interested, as a professional matter in my life as a teacher of architecture, art and design, to see if it is conceivable that, through some teaching arrangement, I might in principle be able to enhance the creativity of students. Actually, I'm not quite that ambitious (or foolhardy). Creativity is a scientific can of worms. How on earth can it be shown that creativity, of all things, has been enhanced? What does creativity mean, anyhow, and how could it (and enhancement) be measured?... I choose a simple way out. I am going to assert that what I'm looking for is an increase in the variety of possibilities within which we can operate as (creative) people. I assume, as a matter of faith (I cannot prove it and I make no attempt to but it is the accepted "Wisdom of the Enlightenment"), that an increase in the variety of possibilities will lead to an increase in creativity.[49] (3_144)

As mentioned above, Ranulph explored Ashby's ideas surrounding variety often, advancing variety by enfolding an increasing array of processes and perspectives drawn from differing disciplines.

Beauty

Can beauty be allowed to play a part in representing and describing aspects of the world?

> Sometimes I am asked what is the value of describing the world we construct and thus find ourselves in. My answer is often taken to be odd. Most expect some comment about its usefulness, or perhaps, its truth. But for me what matters, what talks to and moves me, is its staggering beauty. The fact that we have become unused to talking of the products of our thinking in other than utilitarian and truth-based terms is, to me, a tragedy. That is why I believe it is one of the triumphs of Second Order Cybernetics that it permits and even encourages the sort of comment and valuation (i.e. beauty), and that I can say this, honestly, without fear or ridicule. (3_133)

Similar to what Ranulph has said above, one of the things I say when I give a talk about my art, is that beauty is allowed. I have a very broad concept of beauty. It can be figural, relating to the human body, found in landscape, design, fashion, conceptual art, music, playful exploration of poetic text, and it is at times entirely abstract in nature. There can also be humorous beauty in the perfect pun. In my youth, I had a geometry teacher who spent some time doing an equation on the board. He stood back from the board, chalk in hand, and said "do you see the beauty in that"? And I did!

In terms of composing composing, I might add, that beauty can also be aloud.

---

49. Glanville (1998).

Cybernetics Is My Art

Are there certain stories that Ranulph would return to? Ranulph often revisited the following text. It is also a fragment from the lines used in John Cage's work entitled *Indeterminacy*.

> It was the saying of D. T. Suzuki, doyen of teachers of Zen in the west, quoted by the composer John Cage:
>
>> Before studying Zen, men are men and mountains are mountains.
>> While studying Zen all is confused.
>> After studying Zen, men are men and mountains are mountains.
>> The difference is, that in the first case the feet were a little bit off the ground.[50]
>> (3_130)

I have heard Ranulph speak this Koan many times in different contexts…

> The second was that what he was saying and the way he was saying it mapped onto each other as I believed (and still do) the form and the content of works of art do: the form indicated the content, the content required the form. His argument went round, repeating from the same start/end point. Coming from a background in architecture and music, this meant a lot to me (and still does: cybernetics is my art). (3_130/131)

Of course, this circle being composed over time with an emergent end point, is the accretive circle which enfolds the interactant/composer as being built in to a circular causal feedback system.[51]

Different Observers/Composers

What was the central question related to difference that Ranulph sought to answer? It is clear that he was interested in language and meaning production, and in particular in the way we understand the world based on differing sets of lived experience. He was interested in emphasizing each person's individuality. He also was interested in asking the toughest of questions.

> I believe in the individual, and the distinction between each of us: I am convinced that each of us is different, and this difference is important. We cannot demonstrate we are or

---

50. This last line in the Zen text, when researched, has a few different possible ways of being understood [in differing quoted places as found on the internet]. The one presented here suggests the feet are on the ground "in the first place," as Ranulph wrote it… A second way to understand this is through the ambiguity of when the feet were off the ground (see the ending of The Long Conversation) for this version… and the third way to understand it has the feet "off the ground" at the end, which perhaps makes the most logical sense in terms of circular causal logic. For me this ambiguity makes the story even richer… but perhaps it was a typo above! [laughter]
51. Cybernetics is concerned with "circular causal feedback mechanisms" (see the Macy Conferences). Wiener's (1948) formulation in his eponymous book, writes of communication and control. (3_396)

think the same: attempting this requires assuming we are different, in order to be able to find the sameness. I have noticed we tend, in finding similarities, to forget the differences: the majority of discussions of our experience are focused toward a knowledge based on what we hold in common, ignoring difference...My question is how can we account for a world each of us sees differently, and which, as a result, we cannot be sure is the same world? (3_209)

It is this question that is often embodied in the research Ranulph undertakes, as well as in the systems he designs and/or authors. Perhaps such systems enable us to develop the ability to articulately register this difference. This navigates us back to where the Long Conversation started—to the creation of a system that embodies and celebrates difference.

Computation

How can I explore the computer not as a tool but as a partner? Can the computer become problematic? Can it become a nightmare as computational systems gain self-knowledge as well as the knowledge of their differences to that of the human? How can we avoid human/computer conflict in the future—the fruit of many dystopian film narratives! Perhaps we are close to huge cultural changes brought about by the self-awareness and autonomy of computers.[52] Ranulph suggests:

> It happens when the tool takes over and we are not aware. It happens because of that lack of awareness, but most of all because we treat the computer as a tool—a slave... We need rather to treat it as what it is, a medium in and through which our imaginations and imaginings are extended, collaboratively, instead of perverting it so it perverts us. (2_133)
>
> I would like to maintain that we can talk of the computer as:
> — An extension of ourselves: doing what we can do, but more so:
> — A partner for us: joining in and offering its own insights
> — A location or milieu in which we can act.
>
> My contention is that the more interesting interpretation is to collaborate with and/or in the computer, as an active partner. (2_181)
>
> The interface can admit circular activity and the creation of the new. Seen in this light, it is indeed both the site and the source of interaction. (2_32)
>
> What is meant by computable? To the cybernetician it means intentionally establishing a (productive) relation between things. (2_92)

In my teaching related to computation, I try to have students open up their minds to the potentials of what the computer can do and/or become. As

---

52. See Kurzweil (2005).

mentioned above, the computer is an open system with vast potential. I suggest that we author these systems (at this time) and they are only limited by our imaginations. Perhaps in the future new potentials will open up via the exploration of a new variety of computational imagination, creativity and intelligence, not to mention machinic self-coding. To my mind this will include what might be called machinic self-awareness. As in the manner Ranulph suggests, I often collaborate with the computer, seeking to author new systems that enable emergent creative potentials to arise via their use.

Computational Intelligence

How would you articulate computational intelligence?

I examine the relationship between human and computer through a consideration of intelligence, suggesting intelligence as we experience it is neither a (genetic) property of an object, nor an attributed quality (a la Turing, 1950) but comes about in the interaction between me and it. Thus, intelligence is shared, and sharing implies an essential equality. This is remarkably similar to Bateson's view of where the creativity is when a human is using a computer: it is in their sharing, within a particular context. (2_88)

This seems to follow from Ranulph's distinction making remarks above related to reciprocity. To paraphrase: If I am intelligent I can also potentially discern intelligence through interaction with the composer across from me.

Creative Computation

Can we imagine that computation could potentially feed the variety surrounding creativity through forms of unmanageability?

...I have argued extensively and continue to maintain that it is through admitting ideas other than our own into our worlds that we can increase the potential for us to be creative, and I will not repeat the argument here, except to say computers are, in this view, much more interesting when they do not fit our models too conveniently or too compliantly! (2_184)

Bill Seaman and John Supko, musician, composer and programmer, have collaborated with AIs in both the Audio work entitled S_traits and in their generative opera entitled *The Oper&*. In each case the computer provides its own variety of creative layering, arising via code authored by Supko, as well as through elaborate database design for ongoing variable substitutions by both Supko and myself.[53] I also play a role in code design by articulating particular functionalities we might bring to fruition via code

---

53. See Seaman (2007).

authorship, by defining the potential logic of pseudo-code structures. Supko and I also co-direct the Emergence Lab at Duke University, where we explore emergence in image, music and text production, both through computational and analogue means. I have also written a number of papers and presented invited talks on computational creativity.[54] Ranulph was also quite interested in computationally creative processes.

The Computer As Environment for Making

What are some of the processes and/or approaches surrounding the exploration of the computer as an environment for making?

> The computer as making. By this I mean the use of the computer to generate (novelty) to, from, and with us. For instance, I mean unlikely combinations or points of view, strange associations, unfamiliar descriptions or renditions, peculiar combinations and compositions, shocking fragmentations, automated, or created by, and between users working together or alone; serendipitous or intentional; through proper use and improper abuse of the computational environment; with exactness or imprecision of recording and processing. (2_124)

This brings to mind the notion of making strange, otherwise known as *defamiliarization*. *The World Encyclopaedia* provides this definition:

> The term was first coined in 1917 by Victor Shklovsky, one of the leading figures of the movement in literary criticism known as Russian Formalism. Formalism focused on the artistic strategies of the author and made the literary text itself, and not the historical, social or political aspects of the work of art, the focus of its study. The result was an appreciation for the creative act itself.[55]

I feel very close to Ranulph's thoughts above, given my interest in recombinant poetics,[56] the creative exploration of computationally driven combinatorics functioning in the service of creative production and emergence. I have explored combinatoric methodologies in my work, both analogue and digital, since the early 1980s.

As mentioned above, for me, the notion is that computers are open systems. We can author their input, output and functionality with a focus on generating output that can potentially be emergent in nature especially in terms of creating combinatoric fields of meaning, and exploring database potentials. This seems to point in a related manner to Ranulph's design of thinking through computational composing. Users become enfolded in a cybernetic loop while interacting with the computer, exploring circular

---

54. Seaman, (2013).
55. http://www.newworldencyclopedia.org/entry/Defamiliarization (retrieved July 4, 2018)
56. Seaman (1999, 2010).

causal interaction with such authored systems. When the system is working it functions as a whole—a second-order cybernetic continuum. I have become very interested in the authorship of generative computer systems that function in the service of exploring new forms of computationally sparked creative production. This kind of work takes an open mind—open to new forms of creation and new forms of creative reception. The output of these systems is always something to be reckoned with—somewhat unmanageable to use Ranulph's term—yet always defined with an eye open to the notion of creating this kind of expanded arena for novel forms of composition. Here emergence is central to learning and creative production.

Can I define my own re-definition for the computable (as Ranulph did) and explore the future of computing as it approaches thought or composing, by authoring a biologically inspired new form of bio-algorithm? What can we learn from historical research at the Biological Computer Laboratory related to this topic?

Conversation

Circling back, what is the importance of conversation to all of this?

How we might communicate is important, for unless we answer that, we cannot account for the type of conversation we are trying to explore. This is where conversation comes in. The mechanism which allows us to communicate without having to encode (and all that goes with it), which not only allows but requires that we construct and are separate, is the conversation. Conversation is an embodiment of feedback. Feedback is not strapped on, it is integral. Conversation is circular. Conversation, existing without coding, requires that we make our own meanings in our own understandings. Conversation leads necessity to (re-) interpretation, to construction [read composing, emphasis Seaman]: and conversation is the result of (re)interpretation and construction. … Conversation is at the heart of cybernetics, for conversation is a (perhaps the) form for communication when communication is considered in its own right rather than as a subsidiary requirement for control to occur, when communication is seen as communication per se. (3_76/77)

Thus, the role of both informal communication as embodied in our long conversation, and this more formal conversation with Ranulph's artifacts of thought, points to how a multiplicity of forms can become enfolded in the composing of composing.

Pask's Conversation Theory[57, 58]

What is Pask's conversation theory and how is it relevant to Ranulph's research and practice? A definition of conversation theory is presented on the InstructionalDesign.org site:

> The Conversation Theory developed by G. Pask originated from a cybernetics framework and attempts to explain learning in both living organisms and machines. The fundamental idea of the theory was that learning occurs through conversations about a subject matter which serve to make knowledge explicit. Conversations can be conducted at a number of different levels: natural language (general discussion), object languages (for discussing the subject matter), and metalanguages (for talking about learning/language).
>
> In order to facilitate learning, Pask argued that subject matter should be represented in the form of entailment structures which show what is to be learned. Entailment structures exist in a variety of different levels depending upon the extent of relationships displayed (e.g., super/subordinate concepts, analogies).
>
> The critical method of learning according to conversation theory is "teachback" in which one person teaches another what they have learned. Pask identified two different types of learning strategies: serialists who progress through an entailment structure in a sequential fashion and holists who look for higher order relation.[59]

Ranulph was highly enamored of Pask's conversation theory and abstracted its salient features in relation to his own interests, for example, the holding of conversational conferences as the President of the American Society for Cybernetics, as well as the use of conversation as an ongoing part of human mentoring, teaching, learning, design as well as music and art production:

> The body of what is collectively referred to as Conversation Theory is an astonishing achievement of imagination, nerve, rigor, ingenuity, and adumbration (to use one of the prof.'s (Pask's) favorite words. It is in many parts and particulars very hard to understand, because of a tendency to present it all, in its full complexity: which adds to

---

57. Pask's conversation theory is covered in three main books by him: Gordon Pask, *Conversation, Cognition and Learning*. New York: Elsevier, 1975; Gordon Pask, *The Cybernetics of Human Learning and Performance*, Hutchinson. 1975; and Gordon Pask, *Conversation Theory, Applications in Education and Epistemology*, Elsevier, 1976. Bernard Scott and Paul Pangaro, both students and later colleagues of Pask, have been able to provide less dense definitions of conversation theory which I find helpful, as did Glanville and Müller. Bernard Scott's paper, "Gordon Pask's Conversation Theory: A Domain Independent Constructivist Model of Human Knowing," *Foundations of Science*, special issue on "The Impact of Radical Constructivism on Science," edited by A. Riegler, 2001, vol. 6, no.4: 343–360, is one text which helps elucidate the subject. See also Ranulph Glanville and Karl H. Muller (Eds.), *Gordon Pask, Philosopher Mechanic—An Introduction to the Cybernetician's Cybernetician*, edition echoraum, 2007; Paul Pangaro, "Questions for Conversation Theory or Conversation Theory in One Hour," https://www.pangaro.com/published/Pangaro–Questions-for-Conversation_Theory_In_One_Hour-Kybernetes_2017.pdf (retrieved 29 November, 2019).
58. For a quick overview of conversation theory see Pangaro (1996).
59. Retrieved July 5, 2018 from http://www.instructionaldesign.org/theories/conversation-theory/

the inherent difficulties of subtle and unfamiliar concepts (although they appear much less strange nowadays). (3_84)

Design Conversation—A Conversation With the Self

What is this conversation with the self that in turn contributes to thought and accretive creative processes? Ranulph expanded on, and re-interpreted Pask's conversation theory, precisely in terms of conversations with the self as related to design processes and in exploring creativity in the arts. He was also interested in sharing these self-oriented conversations in critiques and gatherings, always exploring this oscillation related to conversations between the self and others.

> I maintain the circular act of conversing with oneself (normally through a medium of paper and pencil) with the concomitant switch between personae (often achieved so fast that both effectively co-exist), is the central activity in designing. I have argued it is fundamental to how we behave, and may be seen as the origin of cognitive activity. (3_430)

This kind of fast switching is also relevant to Ranulph's notion of being both the subject and object of Objects.

Conversation With the Self—Accretive Processes—"Try Again, Fail Again, Fail Better"

Can you imagine that Ranulph had some meaningful interactions with Samuel Beckett? (see the "Long Conversation")

> The Title "Try again, Fail again, Fail better" is taken from Beckett's 1984 novel *"Worstword Ho"* published by the GrovePress. In my view, it captures the conversational act at the heart of designing… (2_253)

It is interesting to think that Ranulph's employment of being temporarily out of control in terms of manageability could have emerged from this line of thought, which flows out of Beckett's writing.

Samuel Beckett. Copyleft

Different Modes of Conversation in Pask's Work

Pask Portrait (still from video). © Paul Pangaro, used with permission.

If we ask what are each of the modalities of conversation that were embodied in Ranulph's oeuvre, are there strong ties to Pask's differing uses of language? In my paper entitled "Language and Computational Creativity: Beyond the Expanded Conversation—Cybernetic Pattern Flows" mentioned above, I talk about differing kinds of conversation inherent to Pask's work across disciplines. In my book *Recombinant Poetics*, I discuss a parallel notion, that we are on the cusp of a non-logocentric linguistics that arises in computationally-oriented consensual domains.[60] An example of an analogical material conversation from Pask's explorations into electrochemical computing, was discussed in part in Pask's text "Physical Analogues to the Growth of a Concept":

> I can say what a solution means. This will be the case if, instead of talking about solutions and dynamic equilibria, I interact with the assemblage, regard it as similar in a functional manner, and employ it as an extension of my own thinking processes.[61]

We can continuously flow between these differing forms of conversation. In "A Comment, a Case History and a Plan," Pask discusses aesthetically potent environments:

> It may, ...respond to a man [or woman emphasis Seaman], engage him [her] in conversation and adapt its characteristics to the prevailing mode of discourse.[62]

Here, the notion of a prevailing mode of discourse is openly interpreted, where differing forms of conversation function as discourse vehicles for scientific research, for learning systems, and for artistic creation. Pask was already on to this notion of the expanded conversation in the service of interactive art production in the 1950s, exemplified in his work *Musicolour* (1956; 1st version, 1953). And later in his *Colloquy of Mobiles* (1968), where the computer functions as a vehicle of creative, if not sensual poetic human/machine conversation.

Ranulph explores multiple modes of conversation—via natural language, via analogy, and via metaphor, through artistic and/or design production, alternating between these differing conversational modalities as needed. So, in extending conversation theory, I believe these multiple forms of conversation, can also be vehicles of conversation with the self, especially in terms of art and design processes, exploring creative production of any variety imaginable!

Switching Roles

How can switching roles empower a conversation with the self?

---

60. Seaman (2010, p. 11).
61. Pask (1959, p. 915).
62. Pask (1971, p. 76).

> Now let us consider what happens in the case of the conversation with the self (as both speaker and listener, drawer and viewer). Is there a difference? Scarcely! The moment we consider (as I have argued above in the case of design) the paper and pencil (or whatever else) we use—together with our ability to switch roles—as behaving like our conversational partner, we are holding a conversation with ourselves. (2_227)

Here, the *whatever else* Ranulph mentions seems to suggest the openness of applying this idea, in particular in terms of exploring the potentials of computation in the contemporary creative context.

Object Theory

One asks, what role does Object theory take on in terms of unpacking meaning production as it is played out in terms of conversation with the self and others? Ranulph's short explanation of Object theory is quite fascinating. Object theory is also covered in depth elsewhere in this volume by Scott, and also in our "Long Conversation."

> I am less interested in what we have in common than what we have that supports us and keeps us different. I hold this difference to be self-evident. My question is, how can we account for a world which each of us sees differently, and which, as a result we cannot be sure is the same world? (3_209)

> As my response to this I built a framework (theory) in which I designed a structure to support this based in observation (a general term meaning more than visual looking!). The theory is built around what I call "Objects", which are taken to be self-observing. I know this sounds odd: if it worries you see the footnote. Objects are taken to have two roles: self-observing and self-observed. But each Object is just one Object, so these roles are seen as switching, which they do by generating time (they are oscillators). When the Object is self-observed, a slot is left open for observing that other Objects may look into (providing they are in the self-observed role, so are free to observe): an Object can observe another Object by occupying and "observing" slot while it is empty, which is when the (self) Object is not in the observing role, but in the observed role. Each Object generates its own time, which means their times appear different to each other, and that one Object might observe several different, other Objects simultaneously, allowing our observations of different Objects to relate several observed Objects together through our observations of such objects, which is the motivation for the work: I wanted to set up a structure that supported each of us seeing differently yet believing we see the same. I know of no one else who has put it this way: usually differences are to be explained away, not celebrated! (3_209/210)

One could say that Object theory fully supports the variety intrinsic to individual interpretation.

Varieties of Variety

> How many varieties of variety are there? (3_123)

I have often mentioned the idea of using a multi-perspective approach to knowledge production. Each vantage point provides a different spoke of variety holding up the wheel of form and content. I am just now realizing that the fact that these different perspectives are unmanageable, in that they do not fit into one another in any easy manner, generates new varieties as they are explored by being put in conversation with one another. This approach becomes a kind of generator of new relationality—increasing variety, and opening out new plains of understanding…

Variety Constantly Increasing…

How can infinite variety be entertained?

> I propose that the boundaries of universes should be seen as having metaphorical hands on them, constantly grabbing that which was outside but should be inside, placing it within the confines of the said Universe. Thus, the universe is in a state of constant expansion: that which had been being outside absorbed—leaving the outside bare, awaiting another external view (then to be absorbed in turn): variety constantly increasing. (3_122)

Understanding is a lifelong process, which is constantly being enacted. To my mind, everything, as we come to compose it has infinite potential depth.

Art and Design Variety

How does Ashby's law of requisite variety contribute to this conversation?

> The ability of any one system to control any other is determined by Ashby's Law of Requisite Variety, requiring that the variety of the states of the system that will control must equal or exceed that in the system to be controlled. The variety of a system made up of several people interacting vastly exceeds that in one person. Similarly, the variety in one person is vastly less than that in the world (using realist notions in shorthand). Thus, we live in an inherently unmanageable life. This may not be bad: it means there is always the potential of states as yet unimagined, if we cease restricting the world to what we already know. Openness can lead to the new, perhaps the main thing artists and designers believe they do. (3_390)

In this emergent world, we again must find our own path as we are constantly navigating, as well as promoting the wandering of others.

Variety and Novelty

Can Ranulph's use of the idea of the reciprocal help us re-understand aspects of experience through a form of inversion related to control?

> One *approach* [emphasis Seaman] is to close down (which is what dictators do): remove the possibilities and the freedoms which generate complexity. The other is to accept unmanageability as wonderful: which gives us endless novelty, a richness beyond our wildest imaginings, magic: and demands from us trust, open-mindedness and generosity, the sorts of qualities we would like to claim for ourselves. I love an understanding that promises me novelty and I try to make sure I set up situations so this is encouraged. (3_215)

Here, the acceptance of unmanageability as being wonderful, inverts the historical definition—the initial definition provided by Norbert Wiener at the birth of cybernetics, and perhaps feels quite different to our own personal definitions of control. Yet, it makes us understand more clearly the second-order cybernetic experience and the enfolding of circular causal interaction over time as a driver of both variety and novelty—finding a balance between being in control and being out of control in the service of creativity.

Error and Variety

How can we define new distinctions related to this line of the employment of cybernetic reciprocals? Ranulph posits error as sometimes being a positive attribute, and at times we certainly seek to alleviate error—the reciprocal.

> Error can help. It is important as a source of variety and novelty. Error tells us that the descriptions of the world we find ourselves in aren't in concurrence with that world, at the "moment". This allows/requires us to change these descriptions, either by reworking them or by finding something new. Both are strategies to accommodate the variety of the excluded. Error provides a mechanism for learning, part of which is in the inclusion of more variety. (3_224)

The acceptance of this re-purposing of the concept of error over the course of our conversations and creative processes is quite unique in the history of cybernetic thought. When viewed as a stage of a cybernetic cycle it certainly makes sense and points to how design and art processes often function as an accretive process. Yet, it is quite different to the original cybernetic steersman concept—the notion of error and direct error correction. In Ranulph's current definition, there might be a string of errors each suggestive of new forms, arriving at a final form through wondering and wandering. Through error we potentially move in a new direction that generates novelty and can feed creativity. To my mind this acceptance of error is both contrarian in nature, yet it makes sense in terms of this context of learning, and perhaps somehow is related to the Zen of understanding understanding,[63] and composing composing.

Variety and Unmanageability

How does Ranulph himself define the role of variety as it pertains to uncontrollability and in turn creativity?

> Variety and Creativity are, in this interpretation, connected. The more variety there is available to us, the more creative we are likely to be. So long as we let it remain unmanageable, and don't restrict the variety to what we have. Maybe this inversion is the sort of inversion that powers second order cybernetics, and which we can use to generate new understandings of precisely the type this column is addressing. (3_147)

Ranulph was always enfolding new understandings into his general understanding of the world as perhaps we all do. The difference is, he was deeply self-reflective about this mode of thinking and engaging with the operations which lead to it.

Control

How did Ranulph employ his conception of reciprocality in relation to control? Examining the controller ↔ system interrelationality he points out that the controller controls the system and the system controls the controller.

> Wiener's intended use *of the concept of control* [emphasis Seaman (expressed more clearly in his second cybernetics book)], is enabling control. This sort of control talks of the benefits of controlled movement in achieving aims: the purpose of enabling control is not to constrict, but to guide toward better performance.
>
> Being in control: so that a skier is in control as (s)he speeds down a mountain responding to the arbitrary surprises in the slope without falling. (2_265)

The only way to learn how to ski is to learn how to stop, as well as to learn how to fall without hurting yourself. The beginning of learning control is being slightly (and sometimes greatly) out of control. I remember learning how to ride a bike, which is in some ways similar. It seems to me that there

---

63. See Von Foerster (2003).

is a variety to differing kinds of control, especially control as it interplays with unmanageably in the pursuit of creativity. In terms of creativity, one might go out of the bounds of control during idea development, yet some technical, performative forms of control may slip in, once the variety and the breadth of concomitant creativity are explored.

When I am an author of a computer program (or work with a programmer to articulate the logic of one of my systems computationally) I might have a goal of empowering emergence through interaction—so here there is both a level of control related to loading specific elements into the database, and defining rules for their being called into a world, and a measure of out-of-controlness given the choices of the participant/interactant. This user-driven control becomes enacted through a range of interactant-oriented (read composer-oriented) choices.

Creativity

In my interactive works (as I mentioned earlier) I often explore the database aesthetic. I author a set of pointed variables that the interactant can be creative with (to a degree) through their navigation and interaction with that system. The database system increases variety and in fact potentially leads to an emergent outcome in each case. We can call this a form of generative computational composing. Can these differing kinds of interaction increase both variety and creativity in the light of Ranulph's definition? I will repeat here a fragment of a quote presented earlier. Ranulph states:

> I am going to assert that what I am looking for is an increase in the variety of possibilities within which we can operate as (creative) people. I assume, as a matter of faith (I cannot prove it and I make no attempt to, but it is the accepted "Wisdom of the Enlightenment"), that an increase in the variety of possibilities will lead to an increase of creativity. (3_145)

In my own interactive works, I often include chance processes within particular ranges of probability (canned chance in the spirit of Duchamp)[64] enabling one to explore emergent interaction within a given, authored system. In many cases I employ a database with an authored set of variables for substitution (and/or dynamic spatial interaction) by the interactant. This is a kind of middle ground where I am "loading" the dice a bit, heightening the probability of a certain variety of process-based emergent outcomes. Yet, it must be clear that I often use a very unmanageable set of processes early in the process of the creation of the variables, that might later potentially be housed in the interactive work. Starting the process with unmanageability potentially leads to the unique authorship of each variable that I finally load into the system of user potentials.

In terms of Ranulph's thought, we must remember this unmanageability is part of an accretive learning process, and is eventually assumed back into the creative trajectory or should I say goal of the overall system, and/or what the goal is, shifts in an ongoing emergent manner. The nascent goal of this kind of process, working in a Glanvillian manner, is often to find the goal though this form of practice.

Soane Museum Gallery. Copyleft.

Emergence

How does emergence impact the cybernetic steersman? Ranulph was interested in von Foerster's eigenforms[65, 66] as well as trivial and non-trivial

---

64. See *3 Standard Stoppages*, Marcel Duchamp (American, born France. 1887–1968) 1914. Wood box 11 1/8 x 50 7/8 x 9" (28.2 x 129.2 x 22.7 cm), with three threads 39 3/8" (100 cm), glued to three painted canvas strips 5 1/4 x 47 1/4" (13.3 x 120 cm), each mounted on a glass panel 7 1/4 x 49 3/8 x 1/4" (18.4 x 125.4 x 0.6 cm), three wood slats 2 1/2 x 43 x 1/8" (6.2 x 109.2 x 0.2 cm), shaped along one edge to match the curves of the threads. To make *3 Standard Stoppages*, Marcel Duchamp dropped three one-meter-long threads from the height of one meter onto three canvas strips. The threads were then adhered to the canvases, preserving the random curves they had assumed upon landing. Cut along the profiles of each fallen thread, the canvases served as templates for three draftsman's straightedges—wood tools that retain the length of the meter but paradoxically "standardize" the accidental curve.

    Duchamp's deliberately useless toolkit subverts standardized units of measure, while simultaneously poking fun at the scientific method. Though he glibly referred to *3 Standard Stoppages* as "a joke about the meter," his description of its outcome reads like a mathematical theorem: "If a straight horizontal thread one meter long falls from a height of one meter onto a horizontal plane twisting as it pleases [it] creates a new image of the unit of length." https://www.moma.org/learn/moma_learning/marcel-duchamp-3-standard-stoppages-1913-14/ (retrieved 30 November, 2019). Originally exhibited in Paris in 1913-1914.

machines. In the book *Complexification*, John Casti[67] writes of emergence as the science of surprise. What relevance did Ranulph find in the non-trivial machine to aspects of emergence?

> Because unpredictability means that we can anticipate surprise: all will not be as we had thought it would be. In other words, we had a model for the world we inhabit where what we observe may change in ways we cannot imagine. And that means we are never truly in control, that we can and must keep learning—maintaining our involvement. The world of the Non-Trivial Machine, as if by magic, creates surprises and cannot be tamed by us. (3_281)

Thus the metaphorical wheel of circular causation outlines a cycle/circle of organization and disorganization. Is this perhaps pointing to the perpetual motion of patterning, organizing and re-organizing, and for each of us continuing to learn over a lifetime? Von Foerster defines the difference between the trivial and non-trivial machine:

> A trivial machine is a machine whose operations are not influenced by previous operations. It is analytically determinable, independent from previous operations, and thus predictable. For non-trivial machines, however, this is no longer true as the problem of identification, i.e., deducing the structure of the machine from its behavior, becomes unsolvable.[68]

Ranulph provides the following observation about non-trivial machines:

> If the Eigen Form generates constancy (pattern) out of our disorder and allows us to credit objects in our experience, the Non-trivial machine breaks these patterns, allowing us to discover the new, anew: it disorganizes our ordering. (3_281)

This is where the steersman metaphor lurking historically behind the definition of cybernetics comes in, especially in terms of self-conscious growth in second-order cybernetics. The disordering again becomes ordered, as it is gathered and re-understood—the disordering creates a new patterning that is of itself, enabling the process to bring forth an emergent outcome.

---

65. Von Foerster (1981). Objects: tokens for (eigen-) behaviors. In *Observing Systems* (pp. 274–285). The Systems Inquiry Series. Seaside, CA: Intersystems Publications.
66. Kauffman (2003). Eigenforms—Objects as Tokens for Eigenbehaviors. *Cybernetics & Human Knowing*, Vol. 10, nos. 3-4, pp. 73–89. Retrieved from https://cepa.info/fulltexts/1817.pdf on November 30, 2019.
67. Casti (1994, p. ix).
68. The von Forester Page, https://www.univie.ac.at/constructivism/HvF.htm (retrieved 3 October, 2019)

Recursion

I seek to become introspective about the exploration of emergent processes, to empower them. As mentioned earlier, I co-direct, with John Supko, The Emergence Lab at Duke University, exploring all aspects of emergence in a mindfully aware manner, drawing students from many different disciplines including mathematics, computer science, music, the arts, and literature, and so forth. Of course, this class often focuses on being introspective about process. One asks: How can emergence arise through a variety of recursive analogue and digital processes? Ranulph provides this reflection related to thinking:

> Building on, composing and criticising are all varieties of thinking. So, put simply, academics think about thinking and have thought about (their) thoughts. They reflect, and they reflect on their reflections. This sort of thing—when we do something and then do it again on the outcome of that doing, potentially ad infinitum—is called recursive, and academics are involved in an essentially recursive activity. Almost everything we research, which involves us doing and thinking about what we do and think, is, at heart, recursive. (2_193)

Additionally, I find that when I circle back to ideas and processes that I have explored early on in my career, the knowledge of my lived experience, which includes knowledge of errors, mistakes and the employment of chance procedures, play into my current creative authorship. For me, the strongest works are those that let me return to them over and over again, each time adding a new layer of meaning to my field of understandings, be they authored by myself or others. I try to author works with this kind of goal in mind, finding the resonant edge of poetic legibility, an edge that leads to new readings/understandings on each return.[69]

My question to myself is: How can I devise an intelligent computational system that furthers my ability to make exciting generative, emergent works of art? Just how is intelligence defined in terms of this context?

Intelligence and the Turing Test

Might we say that the mind/brain/body/environment relationality is a not yet fully entailed computational environment of a biological nature? We were once called computers... Von Foerster described cognition as computations of computations (a recursive paradigm):

---

69. See Dworkin (2003).

Cognition: ⟶ computations of ⟲

How does Ranulph frame intelligence? I circle back to Ranulph's method in the following:

> Turing proposed changing from a test based on definitions of intelligence, to one based in recognizing intelligence in operation. (3_397)

> How do we come to consider intelligence? Traditionally we have thought of intelligence as a property of the individual in whom we have recognized it. However, our experience of intelligence is, I believe, through interaction. We assume the other may be intelligent, and confirm it by interacting with the other's behavior, recognizing the intelligence of the interaction. (3_401)

> So, intelligence is recognized as arising between us (as shared in this interaction, not in one participant or the other, no matter that we express it by a statement such as "You are clearly intelligent." We give it because we recognize (and can recognize) it in our shared behavior, in our interaction, and it is through this that we can come to consider ourselves intelligent. Thus, the attribution is a sort of mirroring activity. (3_401)

I would argue that it might be more of observing an abstraction of intelligence in machinic systems than a mirroring at this time. I can recognize a variety of intelligent processes at operation in machines, yet I think few machines at this time (if any) truly recognize their own intelligence, and/or truly recognize our intelligence as it falls in relation to them. The book *Neosentience | The Benevolence Engine* seeks to point at what might need to have happen for a deeper understanding of the intelligence in ourselves, others and in the machinic self, in terms of intelligence arising. Of course, this is where Pask's conversation theory as intelligent learning system becomes enfolded, being a piece of the puzzle.

Looping Back to Conversation Theory

Pask was extremely terse in his own texts concerning conversation theory. I believe it is important to re-see his concepts in terms of contemporary computational environments. I continue to ask, what importance can we attribute to Pask's conversation theory as it functions as a learning system?

> Pask evolved his Conversation Theory in the context of learning. Pask may be considered to be the first to develop machines that learnt, and which took part in a shared learning environment with learners. His conversations were originally intended to permit learners to study the ordered topics of a subject in a manner, and developing understandings, that suited each learner. (2_271)

I seek a specific variety of learner in neosentience research, which enfolds a kind of meta-learning—the ability to be recursive about learning processes and the forming of deep notions of contexts, and in so doing, developing the ability to fathom fast contextual shifts between meta-levels when approaching new contexts, be they linguistic or environmental. Perhaps the most difficult concept to apply to new forms of algorithm and/or new forms of computer science exploring biologically inspired bio-algorithms, is, among other things, in defining a computational sense of delight. Of course this points to the concept of engineering emotional computers! We must remember that in true intelligence emotion can play a part…

Delight

Can one empower the student (human or machinic) by learning along with them, and help them to define for themselves a personal sense of delight—to continuously rearticulate an individual definition of delight? I do not think machinic delight was what Ranulph was interested in this quote, but I do think it is interesting to contemplate what is at operation in how delight arises?

> As a researcher into and teacher of design, my interest is in the difficult stuff: how those who do design can understand their doing in a way that empowers, coupled with an insistence on the value of delight. I hold design research that fails to consider this is inadequate: a form of research in which what designers do is seen as material for other approaches to exploit, rather than as a source of a type of research and generation of knowing that comes from and is sensitive to the subject itself.[70]

Can computers become sensitive to the subject of delight itself? Ranulph had a particular love of delight.

> In my view, delight is for all. That includes the client, users, constructors, and designers—delight both in the object or process produced, and in the designing of it. (In English, the word design takes the form of both noun and verb. I am primarily interested in design as verb.) (3_426)

---

70. Glanville (2009a, p. 5)

It is perchance interesting here to present a quote from a fascinating article: "The Neural Mechanism Underlying Cognitive and Emotional Processes in Creativity." Given that I am quite interested in biomimetics, and its application to future computational systems, I present the following observation about creativity:

> Creativity is related to both cognition and emotion, which are the two major mental processes, interacting with each other to form psychological processes. Emotion is the major driving force of almost all creativities, sometimes in an unconscious way. Even though there are many studies concerning the relationship between creativity and cognition, there are few studies about the neural mechanisms of the emotional effects on creativity. Here, we introduce a novel model to explain the relationship between emotions and creativities: Three Primary Color model, which proposes that there are four major basic emotions; these basic emotions are subsided by three monoamines, just like the three primary colors: dopamine-joy, norepinephrine-stress (fear and anger), and serotonin-punishment. Interestingly, these three neuromodulators play similar roles in creativity, whose core features are value and novelty (surprise), like the characteristics of the core features of basic emotions (hedonic value and arousal value).[71]

In the same way that Ranulph posits the attribution of intelligence, we can perhaps consider thinking about a similar situation for acknowledging autonomous creativity in computational systems. This would include an emotion like delight.

It is clear that over a lifetime of experiences that Ranulph developed a unique, open sense of delight, which crossed many disciplinary domains, and he also wandered on occasion into the transdisciplinary realm. He continued by wandering in this set of worlds with a deep sense of wonder. Of course, this was also at operation in the mentors that most inspired him. How does wonderment also reveal an emotional realm in the self? Can we become mindfully aware of wonderment in operation?

Wonder

In speaking about Heinz von Foerster, which he did often, Ranulph also discussed wonder. Can one sum up what was at the center of von Foerster's sense of wonder?[72]

> When I look for one word to describe Heinz von Foerster, it is not magic, rather, it is wonder. This simple English word has so many meanings and applications, and all of them seem to fit. He made us wonder: he wrote and spoke in such a way and of such understandings that the only response seemed to be to wonder. What does this mean,

---

71. Gu et al. (2018, p.1).
72. See Von Foerster (2013).

> how can this be, I wonder what…? That sort of wonder: the wonder of being uncertain and trying to decide how to proceed. (3_253)

> I would hazard a guess that one way you could assess the creative potential of people is by looking at their ability to find wonder. (2_24)

It is interesting for me to posit wonderment in terms of computational creativity. And for me, this wandering in wondering (in the pluralistic sense of the word he describes above) is at the center of the wheel of Ranulph's method of composing composing.

Summary

Computation for Ranulph was just one of his many interests. Ranulph was deeply concerned with conversation, delight, ongoing processes—articulations continuously seeking clarity. His interest was to empower the drive for new knowledge production and learning, as well as owning up to the limits of this. To transcend the nature of our ignorance by pointing at it, and re-understanding it as something that is just there—again, pointing at the limits and simultaneously the incredible breadth of human knowing. Ranulph was living in cybernetic circles, in conversation with creativity, generating variety, exploring black boxes, rearticulating design and second-order cybernetics in relation to each other, and pointing at the complexities of, design, life and language in an ongoing second-order cybernetic manner.

He was driven to taking drives and talking, sometimes circling back to favorite destinations and sometimes just wandering in the landscape of possible destinations—be they literal or metaphorical. Ranulph spent much of his adult life articulating an ever growing sense of a second-order cybernetic set of understandings as they related to design, and design as it related to second-order cybernetics. This played itself out through the ongoing exploration of his expansive architecture of ideas and the broad re-definition of design that constantly bridged theory and practice—where "the form reflected the content, the content the form" (3_164). His was a wonderful life of composing composing.

I hope that I have brought Glanville's ideas and ways of thinking to a broader and long-lasting audience through the accretive composing of this book.

For the sake of circular causality, I am looping back to my list presented in the "Long Conversation" and adding *Architecture of conversation, Architecture of composing, Architechture of delight, architecture of inquiry*…and *Architecture of ideas*.

Architecture:

- of thought
- of knowing
- of physical properties—physics
- of proportion—art and music
- of sharing concepts—approach to second-order cybernetics
- of music
- of space
- of non-standard approaches to knowing
- of psychology/behavior
- of language
- of creativity
- of buildings
- of conversation
- of delight
- of inquiry
- of curiosity
- of form and content
- of emergence
- of composing
- of ideas

References for Composing Composing

Ashby, R. (1954). *Design for a brain*. New York: Wiley.
Bateson, G. (1972). *Steps to an ecology of mind*. New York: Ballantine Books.
Bateson, G. (1979). *Mind and nature*. New York: E. P. Dutton.
Brün, H. (1972). *For anticommunication*. Retrieved January 21, 2022 from https://sites.evergreen.edu/arunchandra/wp-content/uploads/sites/395/2018/05/anticom.pdf
Cage, J. (1967). *Silence*. Middletown, OH: Wesleyan University Press.
Casti, J. (1994). *Complexification: Explaining an illogical world through the science of surprise*. New York: Harper Collins.
Deleuze, G., & Guattari, F. (1987.) A thousand plateaus: Capitalism and schizophrenia (vol. 2; B. Massumi, Trans.). Minneapolis, MN: University of Minnesota Press.
Duchamp, M. (1912-1913) 3 Standard Stoppages. Collection at MoMA.
Dworkin, C. (2003). *Reading the illegible*. Evanston, IL: Northwestern University Press.
Ertl, M., Korn, W., & Müller, A. (2016). *Ranulph Glanville, Architecture | Art | Cybernetics | Design*. Vienna: echoraum press.
François, C. (Ed.). (2004). International encyclopaedia of systems and cybernetics (rev. ed.). Munich: Saur. (Originally published 1997)
Gaugusch, A. & Seaman, B. (2004). (Re)sensing the observer offering open order cybernetics. *Technoetic Arts: A Journal of Speculative Research, 2*(1), 17–31.
Glanville, R. (1966). Klee and Messiaen. *Arena, The AA Journal, 81*(898).
Glanville, R. (1994). The value of being unmanageable: Variety and creativity in CyberSpace. *Systems Research, 11*(3), page 10. Retrieved November 29, 2019 from https://www.univie.ac.at/constructivism/papers/glanville/glanville97-unmanageable.pdf

Glanville, R. (1998). A (cybernetic) musing: Variety and creativity. *Cybernetics & Human Knowing, 5*(3), 56–62.

Glanville, R. (2003). Review of von Foerster's and Poerksen's *Understanding Systems*. *Cybernetics & Human Knowing, 10*(3-4), pp. 183–187.

Glanville, R. (2007). Try again. Fail again. Fail better: The cybernetics in design and the design in cybernetics. *Kybernetes, 36*(9/10), 1173–1206.

Glanville, R. (2012, 2014, 2009). The Black B∞x, Volumes 1, 2, 3 (Limited edition/ numbered set). Vienna: edition echoraum.

Glanville, R. (2009a). A (cybernetic) musing: Design and cybernetics. *Cybernetics & Human Knowing, 16*(3-4), 175–186. Retrieved October 6, 2019 from http://asc-cybernetics.org/systems_papers/Glanville%20final.pdf p.5

Glanville, R. (2009). *Black B∞x, Vol. 3: 39 steps*. Vienna: edition echoraum.

Glanville, R. (2012). *Black B∞x, Vol. 1: Cybernetic circles*. Vienna: edition echoraum.

Glanville, R. (2014). *Black B∞x, Vol. 2: Living in cybernetic circles*. Vienna: edition echoraum.

Glanville, R. (2015). A (Cybernetic) Musing: Wholes and Parts, Chapter 1. *Cybernetics & Human Knowing, 22*(1), 81–92.

Glanville, R., & Müller, K. (2007). *Gordon Pask, Philosopher Mechanic: An introduction to the cybernetician's cybernetician*. Vienna: edition echoraum.

Gu, S., Gao, M., Yan, Y., Wang, F., Tang, Y., & Huang, J. (2018, October 28). The Neural Mechanism Underlying Cognitive and Emotional Processes in Creativity, *Front. Psychol.* (elec. pub.) Retrieved October 3, 2019 from https://www.frontiersin.org/articles/10.3389/fpsyg.2018.01924/full, p.1

Gunther, G. (2004). Cybernetic ontology and transjunctional operations. University of Illinois Engineering Experiment Station., Technical Report no. 4. Urbana, IL: Electrical Engineering Research Laboratory, University of Illinois. Retrieved March 13, 2018 from http://www.vordenker.de/ggphilosophy /gg_cyb_ontology.pdf (First published in M. C. Yovits & S. Cameron [Eds.]. [1962]. Self-Organizing Systems [pp. 313–392]. Washington, D.C.: Spartan Books.)

Heims, S. J. (1991). The Cybernetics Group: Constructing a social science for post war America. Cambridge, MA: The MIT press.

Kauffman, L. H. (2003). Eigenforms—Objects as tokens for eigenbehaviors, *Cybernetics & Human Knowing, 10*(3-4), 73–89. https://cepa.info/fulltexts/1817.pdf (retreived 30 November, 2019).

Koestler, A. (1964). *The act of creation*. New York: Macmillan Co.

Kostelanetz, R. (2003). *Conversing with Cage* (2nd ed.). New York: Routledge.

Kurzweil, R. (2005). *The singularity is near.* New York: Viking

Müller, A., & Müller, K. (2007). An Unfinished Revolution? Heinz von Foerster and the Biological Computer Laboratory: BCL 1958–1976. Vienna: edition echoraum.

Pangaro, P. (1996). *Cybernetics and conversation*. Retrieved January 27, 2017 from http://www.pangaro.com/published/cyb-and-con.html

Pangaro, P. (2017). Questions for conversation theory or conversation theory in one hour. Retrieved Nov. 29, 2019 from https://www.pangaro.com/published/Pangaro–Questions-for-Conversation_Theory_In_One_Hour-Kybernetes_2017.pdf

Pask, G. (1959). Physical analogues to the growth of a concept. In A. Uttley (Ed.), Mechanisation of Thought Processes, Proceedings of a Symposium held at the National Physical Laboratory on 24–27th November, 1958. National Physical Laboratory Symposium No. 10, Volume II, pp. 877-922. London: Her Majesty's Stationery Office.

Pask, G. (1961). *An approach to cybernetics*. Cambridge, MA: The MIT Press.

Pask, G. (1971). A comment, a case history and a plan. In J. Reichardt (Ed.), *Cybernetics, art and ideas* (pp. 76–99). Greenwich, CT: New York Graphic Society. Retrieved October 31, 2019 from https://pangaro.com/pask/Pask%20Cybernetic%20Serendipity%20Musicolour%20and%20Colloquy%20of%20Mobiles.pdf

Pask, G. (1975b). *The cybernetics of human learning and performance*. London: Hutchinson & Co Publishers.

Pask, G. (1975a). Conversation, cognition, and learning. New York: Elsevier.

Perriquet, O. & Seaman. W. (2011). Art ↔ Science Relationalities. The 17th *International Symposium on Electronic Art Proceedings, 2011* (pp. 1890-1895). Conference held September 14-21 in Istanbul. Retrieved January 16, 2021 from https://isea-archives.siggraph.org/wp-content/uploads/2019/03/ISEA2011_Full_Proceedings.pdf

Pias, C. (2016). *Cybernetics / The Macy Conferences 1946-1953: The Complete Transactions.* Chicago, IL: University of Chicago Press.

Scott, B. (2001). Gordon Pask's conversation theory: A domain independent constructivist model of human knowing. *Foundations of Science, 6*(4), 343–360.

Seaman, B. (1999). *Recombinant Poetics—Emergent Meaning As Examined & Explored Within a Specific Generative Virtual Environment.* Doctoral dissertation at The Center for Advanced Inquiry in Interactive Art, University of Wales. Retrieved July 1, 2018 from: http://billseaman.trinity.duke.edu /textsRecomb. (Later formally published with VDM Press, 2010; PDF of original)

Seaman, B. (2005). Pattern Flows | Hybrid Accretive Processes Informing Identity Construction *Convergence, 11*(4), 12–31.

Seaman, B. (2007). Recombinant poetics and related database aesthetics. In V. Vesna (Ed.), Database aesthetics (pp. 121–141). Minneapolis, MN: University of Minnesota Press.

Seaman, B. (2010). (Re)Thinking—the body, generative tools and computational articulation. *Technoetic Arts: A Journal of Speculative Research, 7*(3), 209–230.

Seaman, B. (2013). Neosentience and the abstraction of abstraction. *Systems: Connecting Matter, Life Culture and Technology, 1*(3), 50–67.

Seaman, B. (2017). Language and computational creativity: Beyond the expanded conversation—Cybernetic pattern flows. Retrieved July 7, 2018 from: https://noemalab.eu/ideas/language-and-computational-creativity/

Seaman, B. (2018). Towards a dynamic heterarchical ecology of conversations. *Cybernetics and Human Knowing, 25*(1), pp. 91–108.

Seaman, B., & Rössler, O. (2012). Neosentience | The Benevolence Engine. Chicago: Intellect Press.

Seaman, B., & Rössler, O. E. (2011). Neosentience | The Benevolence Engine. London: Intellect Press.

Umpleby, S. (2000). *Definitions of cybernetics.* Retrieved January 20, 2017 from http://asccybernetics.org/ definitions/ (revised version of 1982 document)

Varela, J., Thompson, E., & Rosch, E. (1993). *The embodied mind: Cognitive science and human experience.* Cambridge, MA: The MIT Press.

Von Foerster, H. (1981). Objects: tokens for (eigen-) behaviors. In *Observing systems* (pp. 274–285). The Systems Inquiry Series. Seaside, CA: Intersystems Publications.

Von Foerster, H. (2013). *The beginning of heaven and earth has no name. Seven days with second order cybernetics* (A. Müller & K. Müller, Eds.). New York: Fordham University Press.

Von Foerster, H. (Ed.). (1974). *Cybernetics of cybernetics. The control of control and the communication of communication.* Urbana: University of Illinois.

Von Foerster, H. (Ed.). (1995). Cybernetics of cybernetics. The control of control and the communication of communication. Minneapolis, MN: Future Systems. (Reprint of 1974 edition).

Von Glasersfeld, E., (n.d.). An exposition of constructivism: Why some like it radical. Retrieved January 19, 2017 from: http://www.oikos.org/constructivism.htm

Wiener, N. (1961). *Cybernetics: Control and communication in the animal and the machine* (2nd ed.). Cambridge, MA: The MIT Press.

Wiener, N. (1950). *The human use of human beings: Cybernetics and society.* Boston: Houghton Mifflin.

Wittgenstein, L. (1958). *Philosophical investigations* (3rd ed.; G.E.M. Anscomb, Trans.). Upper Saddle River, NJ: Prentice Hall.

Ranulph Photographing Finnish Architecture.
Image Courtesy of Aartje Hulstein
and the Glanville Archive, University of Vienna.

Ranulph Photographing Islamic Architechtural Features.
Image Courtesy of Aartje Hulstein
and the Glanville Archive, University of Vienna.

Appendix 1:
Curriculum Vitae

Ranulph Glanville
as of 24 August 2014

Personal Details

Born: 13 June 1946, London, UK
Marital Status: Married Aartje Hulstein on March 22nd, 1997
Children: One son, Severi, born 6 June 1974

Education

1959–1964	Bryanston School, Blandford, Dorset, UK
1964–1967	Architectural Association School, London, UK
1969–1971	Architectural Association School, London, UK
1971–1975	Brunel University, Uxbridge, UK: Department of Cybernetics
1976–1988	Brunel University, Uxbridge, UK: Centre for the Study of Human Learning

Qualifications/Degrees

1967	RIBA part I (Architectural Association, London)
1971	AA Diploma (Architectural Association, London)
1971	RIBA part II (Architectural Association, London)
1975	PhD in Cybernetics (Brunel University, Uxbridge)
1988	PhD in Human Learning (Brunel University, Uxbridge)
2006	DSc in Cybernetics and Design (Brunel University, Uxbridge)

Honours/Service

The Cybernetics Society, Fellow, Trustee 1979–; Systeemgroep Nederlands, Foreign Affairs Officer 1981–2005; Architectural Association, Life Member1990–; Elected to the Board of the International Institute for Advanced Study in Systems Research and Cybernetics, 1994–; Board member, International Federation for Systems Research 1996–; Portsmouth Arts in the Environment Committee, 1994–1999; Board of ACCOLADE and ADVOCAAD EU funded research projects1995–2001; Life Fellow of the Royal Society for the Arts, London, 1999–; Invited speaker at retirement celebrations for Prof. Gerard de Zeeuw, 2001; Consultant professor, Computer Aided Design, University of Buenos Aires, 2001; Invited to join board of Heinz von Foerster Archive, Vienna, 2001; Valedictorian speaker, prize giving and graduation, School of Architecture, University of Western Australia, 2003; Elected External Relations Officer, the Cybernetics Society, London (2004–); Board of Gordon Pask Archive, Vienna, 2007–. Elected Vice President, American Society for Cybernetics, Washington DC (2005–2008) and President, 2009–2011, second term 2012–2014; Facilitator, Workshop on Designing Solutions to Wicked Problems, Design Research Institute, Melbourne. Served on the Interdisciplinary Panel of the Research Council of Flanders (Belgium) 2009–2013. External Examiner Masters Degrees, Mackintosh School of Architecture, Glasgow School of Design 2010–2014. Advisory Board, DART+ foundations for early career researchers in design, 2013–. External Examiner, Masters Degree, Estonian Academy of Arts, 2013–2014.

Public Lectures

Recent major public lectures: "Freedom and the Machine", Royal College of Art (2014); "Slow", Estonian Academy of Arts (2013); "Cybernetics and Design", Ontario College of Art and Design University (2011); "Cybernetics, Systems and Design", University of Sichuan, China (2010); Public Lecture Series, "Freedom in Psychotherapy" (2009), "Five Cybernetic Principles" (2006) Rijeka Town Hall, Croatia; Deakin Lecture series in Melbourne, Australia (2005) "Innovation: Creativity, Intelligence and Cybernetics;" Stanislaw Lem Festival, Hebbel-am-Ufer Theater Berlin (2005) "Solaris and the Cybernetics of Saying Goodbye to Samuel Beckett;" Public Lecture, University of Hamburg (2005) "Observing the Observer. Science and Knowledge in the Age of Cybernetics;" (Deakin Lecture, Melbourne) and the Heinz von Foerster Lecture, Vienna, Austria (2004) "Grounding Difference: Panoramas and Perspectives from the Impossible Silence."

I gave a public professorial inaugural lecture ("Freedom and the Machine") at University College London in 2010.

Keynote Lectures

Recent Conference Keynotes: "Living in Cybernetics" American Society for Cybernetics (2014); "Designing and Researching" (2014) Conference on Research and Design; "To be Decided", Heinz von Foerster Centenary; Conference, "Cybernetics in Design" United Kingdom Systems Society (2013); "Blind" Making visible the invisible: Art, design and science in data visualisation; "Reflecting between Observing and Acting" (2012), "Error is not always bad" (2011), "Darkening the Black Box" (2010), "Cybernetics and Design" (2009), "Cybernetics and Ethics" (2008) World Multi-Conferences on Cybernetics, Systemics and Informatics, Orlando; "Designing Learning—Constructing your own Construct" (2009), "Pask at the Centre" (2007), "Design and Construct: Living as Designing" (2005), Conference of the Heinz von Foerster Society, Vienna; "Giving Up Control", UK Systems Society, Oxford (2006); "The Circle in Four Diagrams", American Society for Cybernetics, Washington DC (2005); "Notes towards a Group Design", sixth meeting of the European Academy of Design, Bremen (2005); "Control, Imagination and Addiction", American Society for Cybernetics, Toronto (2004); "Personal Wonder", The Biological Computer Lab Conference, Vienna (2003), "Practising Reflection for Reflective Practice", Reflective Research in Practice, Melbourne (2003).

Awards

Science Research Council Studentship (to study for PhD in cybernetics); Best paper, European Meeting on Cybernetics and Systems Research 1984, 1990; Best paper and best presentation awards, Orwellian Symposium in Karlovy Vary and International Institute for Advanced Study in Systems Research and Cybernetics conference in Baden-Baden, 1994; Distinguished Service Award and Special Scholarship Award, International Institute for Advanced Study in Systems Research and Cybernetics, 1994; Sundry research and travel grants and awards; Recent British Council travel and sponsored research exchange awards to Bialystok (three times), Vienna (twice), Helsinki and Tampere (four visits), and to Karlovy Vary and Baden-Baden, British Academy and Department of Trade and Industry awards for travel (Sao Paolo); Adjunct Professor, Royal Melbourne Institute of Technology University (1997–2005), Senior Professor of Research Design, St Lucas Architectuur, University of Leuven, 2005–; Visiting Professor of Research, Innovation Design Engineering, Royal College of Art and Imperial College, London (2008–2014); Founding Member and Vice

President, International Academy of Cybernetic and Systems Sciences (2010–2013, re-elected 2013–); Awards for outstanding keynote presentations (WMSCI, 2008, 2009, 2010, 2011); Award for lifetime achievement (WMSCI, 2011). Special Award of the American Society for Cybernetics, 2014. Permanent plenary on cybernetics at ISSS conferences named "Ranulph Glanville Plenary on Cybernetics," 2014.

Directorships

BKS+ and CybernEthics Research (2 distinct but related companies), Southsea; System Research, London

Memberships

Architectural Association (Life Member); American Society for Cybernetics (Vice President, currently President); Cybernetic Coalition (founder, chair); The Cybernetics Society (Trustee); Design Research Society; The Folly Fellowship; International Academy for Systems and Cybernetic Sciences (Vice President); International Association for Cybernetics; International Institute for Advanced Study in Systems Research and Cybernetics; International Federation for Systems Research; International Society for Systems Science; Royal Society for the Arts (Life Fellow); Systeemgroep Nederlands (Committee); MELA Foundation, New York.

Main Positions Held

1967–1968	*Assistant Art Master*, Bryanston School, Blandford, Dorset, UK
1968–1969	*Architectural Assistant,* Edgington Spink and Hyne, Windsor, UK
1969	*Architect*, JKMP Arkkitehdit, Helsinki, Finland
1971–1972	*Tutor*, School of Architecture, Cambridge University, UK
1972–1978	*Tutor, Unit Master*, Architectural Association School, London, UK
1974–1978	*Technical Tutor*, Architectural Association School, London, UK
1978–1982	*Lecturer*, School of Architecture, University of Portsmouth, UK
1982–1997	*Senior Lecturer*, School of Architecture, University of Portsmouth, UK
1994–1997	*Senior Research Fellow*, New Media Research Centre, Design Department, University of Portsmouth (secondment)
1995–	*Self Employed Consultant and Teacher*, Faculty of the Constructed Environment, and Design Research Institute, Royal Melbourne Institute of Technology University, Melbourne, Australia (continuing as a visitor since)
1996–2005	*Senior Research Fellow & Adjunct Professor*, Faculty of the Constructed Environment, and Design Research Institute, Royal Melbourne Institute of Technology University, Melbourne, Australia (continuing as a visitor since).
2005–2012	*Professor of Architecture and Cybernetics*, The Bartlett, University College London, London, UK
2006–	*Senior Professor of Research Design Visiting Professor of Research*, Sint Lucas Architectuur, Catholic University of Leuven, Ghent and Brussels, Belgium.
2008–2014	*Visiting Professor of Research*, Innovation Design Engineering, Royal College of Art, London, UK.
2010–	*Professor*, School of Architecture and Design, University of Newcastle, Newcastle, NSW, Australia.

2012–	*Adjunct Professor*, Digital Futures Initiative, Ontario College of Art and Design, Toronto, Canada.
2014–	*Research Senior Tutor and Professor*, Innovation Design Engineering, Royal College of Art, London.

Commercial

Canberra Biennial;
Department of Education and Training, New South Wales (Australia);
Issey-Media (the Netherlands);
Lab 3000—Digital Design Biennale (Australia);
Marchwood Priory Hospital (UK);
SwedBank (Sweden);
US Air Force;
inform ab (Sweden);
Foundation 2020 (Croatia)

Academic

Aalto University, MediaLab, Helsinki, Finland
Aarhus University, Faculty of Media and Technology, Denmark
Curtin University, Faculty of Built Environment and Design, Perth, Australia
Estonian Academy of Arts, Tallinn, Estonia
Gordon Pask Archive, Institute for Contemporary History, Vienna, Austria
Heinz von Foerster Archive, Institute for Contemporary History, Vienna, Austria
Hong Kong Polytechnic University, School of Design, Hong Kong
Hong Kong Design Institute, Hong Kong;
Monash University, Faculty of Art and Design, Melbourne, Australia
Ontario College of Art and Design University, Digital Futures Initiative, Toronto, Canada
Rensselaer Polytechnic Institute, Faculty of Architecture, Troy NY, USA
Royal Melbourne Institute of Technology University, Faculty of the Constructed Environment, Melbourne, Australia
Sint Lucas Hogeschool, Faculty of Architecture, Brussels, Belgium
University College London, The Bartlett, London, UK
University of Buenos Aires, Faculty of Architecture, Argentina
University of Canberra, School of Design and Architecture, Canberra, Australia;
University of Dundee, Duncan of Jordanstone College, Dundee, UK
University of New South Wales, Faculty of Built Environment and Design, Australia;
University of Newcastle, Faculty of Architecture and Design, Newcastle, Australia;
University of Western Australia, Faculty of Architecture and Fine Art, Perth, Australia;
University of Western Sydney, Department of Design, Sydney, Australia;

I am actively involved in attempts to set up a web-co-ordinated collective of design schools, generated by Chuck Pelly, Owner of Design Works and Chief Designer for BMW; and an international working group concerned with ethics and the web.
I founded the Cybernetic Coalition, an affiliation of various small cybernetics societies.

I have built academic bridges between the American Society for Cybernetics and the International Federation for Systems Science.

Other Jobs and Positions (part time, occasional, consultancy, etc.)

1972–1976	*Research Assistant*, System Research, Richmond, UK
1976–1984	*System Designer/Architect*, Peter Jackson Associates, London, UK
1978–1982	*Consultant*, US Army Research Institute, Washington DC, USA
1978–1984	*Special Project Organiser*, Architectural Association School, London, UK
1979–1985	*Visiting Professor*, SubFaculty of Andragology, University of Amsterdam, The Netherlands
1980	*Advisor on Education*, British Council, Caracas, Venezuela
1980–1982	*Consultant*, Government of Republic of Mexico
1983	*Consultant*, Libyan Government (University of Garyounis)
1984–1992	*Co-Director, Computing*, Architectural Association School, London, UK
1985–1993	*Advisor and Vice-Director, Senior Research Fellow*, Research Programme "Support, Survival and Culture" (funded by Dutch Government), University of Amsterdam, The Netherlands
1990–	*Consultant*, Marchwood Priory Hospital, Marchwood, UK
1991–	*Associate,* Centre for Innovation and Co-operative Technology, University of Amsterdam, The Netherlands
1992–1997	*Tutor*, Foundation Course (BA), Portsmouth College of Art, Design and Further Education (since August 1994, part of the University of Portsmouth's School of Design and Media)
1993–1995	*Advisor*, Dutch Institute for Design, Amsterdam, The Netherlands
1993–2000	*Diploma Technical Tutor*, Bartlett School of Architecture, University College London
1994–1999	*Supervisor*, Supervising PhD's at Liverpool John Moore's University and University of Wales College, Newport, UK: consultant and supervisor aiding underqualified departments develop their experience and expertise.
1995, 1996, 1997	*Visiting Scholar*, Shaw College, Chinese University of Hong Kong, Hong Kong (advisor on design of CD ROM's project for teaching Putonghua to Cantonese speakers)
1996	*Advisor,* Faculty of the Constructed Environment, Royal Melbourne Institute of Technology, Melbourne, Australia (Development of Three Peaks model)
1995–2001	*Consultant*, Clinton River Project/Watershed Project, Oakland University/ Rochester Community, Michigan USA
1996–1997	*Visiting Scholar*, Departmen of Design, Polytechnic University of Hong Kong, Hong Kong (advisor on the development of research)
1996–1997	*Visitor,* School of the Built Environment, De Montfort University, Leicester, UK (developing research and cross departmental collaboration, co-ordinating TMR research grant application to the EU for a network of Universities)
1996	*Consultant,* Fachhochschule NiederRhein, Mönchengladbach, Germany (developing strategy for designing and authoring CD ROMs, leading to successful CD publication)
1997	*Assessment Panel,* Department of Design, Polytechnic University of Hong Kong, Hong Kong
1997–2000	*Associate*, Centre for Advanced inquiry in the interactive Arts (CAiiA), UCWN, Newport, Wales

1998	*Visiting Tutor*, Faculty of Architecture, Sint Lucas Hoogeschool, Brussels, Belgium
1998–	*Board Member*, Global Village, Vienna, Austria.
1999–	*Industrial Partner*, ACCOLADE Project, funded by European Union
1999–2000	*Advisor,* Melbourne Festival, Australia.
1999-2001	*Consultant and Workshop Organiser*, TILT, DET, Government of New South Wales, Australia
2000-2001	*Supervisor*, Department of Psychology, University of Stockholm, Sweden
2000–	Thesis tutor, Bartlett School of Architecture, University College London
2001–	*Supervisor*, Graduate School, The Bartlett, University College London, UK
2002	Advisor and Editor, School of Design, Hong Kong Polytechnic University, Hong Kong
2002–	*Associate and Member*, Spatial Information Architecture Lab, Melbourne, Australia
2002–	*Advisor*, Planetary Collegium. Plymouth, UK
2002–	*Advisor*, Faculty of Architecture, Sint Lucas Hoogeschool, Brussels, Belgium: doctoral education (Senior Professor since 2006)
2003–2005	*Advisor*, Faculty of Architecture, Landscape and Visual Arts, University of Western Australia, Working on course development templates and research development
2003–2009	*Visiting Scholar*, School of Design and Architecture, University of Canberra, developing research and the Canberra Architecture Biennial
2003–2010	*Visiting Scholar*, School of Architecture, Rensselaer Polytechnic University, Troy, NY, USA
2004–2012	*Founder and Chair*, Symposium on Cybernetics, Interaction and Conversation, European Meeting on Cybernetics and Systems Theory, Vienna, Austria
2004	*International Advisor,* Digital Design Biennale, Lab 3000, Melbourne, Australia: exhibition advisor and juror.
2004–2007	*Advisor*, Australian National Training Authority, Australian Flexible Learning Framework, Research Programme, Vocational Education in the Information Age
2004–	*Professor*, University of Newcastle (Australia), Faculty of Architecture and Design
2004–	*Consultant Researcher*, WISDOM Institute (City Institute for Statistics and Information), Vienna.
2005–	*Advisor*, Design Research Institute, RMIT University, Melbourne, Australia
2005–2009	*Visiting Research Fellow*, Curtin University, Faculty of Built Environment and Design
2006	*Visiting Professor*, University of New South Wales, Faculty of Building Environment and Design
2006–2009	*Consultant*, Learning Communities Research Programme, Canberra, ACT, Australia
2006—	*Senior Professor*, Sint Lucas Architectuur, University of Leuven, Brussels and Ghent, Belgium
2010	*Advisor*, International Cybernetics PsychoTherapy Society
2010–2013	*Panel Member*, Interdisciplinary Panel, Research Foundation of Flanders
2010–2013	*Visiting Fellow*, Hong Kong Poytechnic University, Hong Kong, China
2011–2012	*Advisor*, School of Architecture, University of Antwerp

2011–2012	*Visiting Fellow*, Hong Kong Design Institute, Hong Kong, China
2011–2013	*Visiting Professor*, University of Newcastle (Australia), School of Design and Computing
2011–2014	*External Examiner*, Mackintosh School of Architecture, Glasgow School of Art, Glasgow, UK
2012–	*Advisor*, Estonian Academy of Arts, Tallinn, Estonia
2012–	*Adjunct Professor & Advisor*, Ontario College of Art and Design University, Digital Futures Initiative

Various/Assorted

Visiting Lecturer, Professor, Critic, Course Assessor, Examiner and Reviewer (see lecture appendix, below) Reviewer for several Academic Journals. Occasional Freelance Journalism.

Editorial

Considerable experience compiling and editing a number of Journals, Conference Proceedings etc. (including design, layout and publishing—further details in next section)

On the Editorial Advisory Board of the Journals *Cybernetics & Human Knowing*; *Kybernetes*; and *Systemica*; the *Journal of Creative Writing* and *Nomads*. And Executive Editorial Boards: *Systems Research and Behavioural Science* and the electronic journals *Constructivist Foundations* and *The Radical Designist*.

Editor of various special issues of journals, especially festschrifts and memorial volumes. Editorial Board *International Encyclopaedia of Systems and Cybernetics* and of its electronic equivalent.

Supervisor

Extensive supervision work listed at end of CV. Note, however, that because of particular conditions connected with work conditions and university regulations, and the peculiar status of architecture as a university course, it is impossible to give a full account.

PRODUCTION SUMMARY

Publications:

Approximately 350 papers, books, writings and similar, in the general areas of Architecture, Cybernetics, Design Theory and Philosophy. [See Appendix 2]

Conferences: Attendance

Conferences regularly attended since 1976. An average of 3 or 4 per year. [See Appendix 5 for summary of main conferences]

Conferences: Committees

Committee member/chair, European Meetings on Cybernetics and Systems Research; Problems of .... Conferences; Architecture Schools Computing Association; The Cybernetics Society; The International Institute for Advanced Study in Systems Research and Cybernetics; ACCOLADE; American Society for Cybernetics; International Academy of Cybernetic and Systems Sciences.

Lectures, etc.

In the UK, most European Countries, the United States of America, Canada, Central and South America, Libya and South Africa, Australasia and the Far East. [See Appendix 7 for a summarising list]

Performances

Live Electronic Music and Events, with the Pierrot Players, the Fires of London, the Scratch Orchestra, Cornelius Cardew, Overcoat and HalfLanding, at many venues (e.g., Queen Elizabeth Hall, Electric Garden, Mercury Theatre London) and on BBC and SudWest Deutsch Rundfunk; recently, "48–52" (for academic choir), Baden-Baden, Amsterdam, Brussels and Bremen. Music for dance, Sao Paolo, 1999. Performances, Amsterdam 1997, 1999, Sao Paolo 1999, Newport 1997, 1998, 2000): "Generator" for electronically projected sound and vision (with Severi Glanville) was show at BEAP Perth, Australia, 2002, and has since been projected in several venues. "Beckett's Farewell to Stanislaw Lem" at the Hebbel am Ufer Theatre, Berlin (2005)—part of a festival celebrating the work of Stanislaw Lem. Recent projections of electronic pieces at various venues and festivals.

Interviews

Several radio and television interviews, including a series of Radio Interviews, Karlovy Vary and a video with the Fachhochschule, Mönchengladbach, Germany and, most recently with the editor of Patterns (see publications).
Internet interviews for Anomalie, France and the BASIC Paradox, Germany.
Three interviews for the Estonian Academy of Arts.
Three series of video film projects in several parts, made by the Royal College of Arts; the American Society for Cybernetics; and the video artist Bill Seaman.

Exhibition/Performance

"Light Sphere" at Heals Gallery, London;
Installation "A Wall for Sir Geoffrey Vickers" in the Alps above St. Moritz, Switzerland and (in modified form) at the Portsmouth Polytechnic.
"Slalom"—Installation plus Collages at Marchwood Priory, Southampton.
"Record Becoming Apparent," Installation for the Inauguration of the University of Portsmouth.
"Earth, Air, Fire, Water," Installation in Fort Purbrook, Portsmouth and, later, (modified) at the Portsmouth City Museum.
"Amsterdam Discovered," Installation at Conference on "Problems of Values and (In)variants," Amsterdam and, later, (modified) at Third Biennial of Art Schools, Maastricht.
"eMail Art" (graphic and text versions), Bratislava.
"Aldous Huxley, the Deception of Poverty," Helsinki.
"Installing the Context", Vienna, "Arty-Fax" (with Trash Treasure Artists), Aachen.
"Möbius Boundary Distinction" (with Barney Townsend) in the Museum of Cybernetic Objects, Washington DC.

During the 1960s my music was regularly performed, and I was active in the Avant Garde Music Scene in London. Returning to occasional sound works, "Generator," (see above, with my son Severi Glanville), for sound with images, was performed at the

Biennial of Electronic Arts, Perth, Western Australia, August 2002. Blind has been performed in many venues. I am working on other pieces.

Editor

Extensive experience and service.

Judge

Competition judge, International Awards for Best Papers on "Collective Support Systems and their Users" and "Interface Design and Shared Working."
International Advisor for the First Digital Design Biennale, Melbourne, 2004–5.
Judge and Chair of Judges, Heinz von Foerster Award, 2006—. Competition designer and Chair of judges, International Open Competition "The Cybernetics of Cybernetics" 2010.

Films

I am currently involved in 3 separate film projects based around my way of acting.

Appendix 2:
Papers, Publications and Writings

Ranulph Glanville
as of 24 August 2014

Before 1970

1966.01	Klee and Messiaen. Arena, The AA Journal,[1] 81(898), 1966 [#1]
1966.02	"The Godot Game" ACC Symbols, London, no 2, 1966 [#2]
1967.01	"Kirbymoorside" (with C. Hambury & G. Woolson), in J Donat (Ed.), World Architecture 4, Studio Vista, London, 1967 [#3]
1969.01	"Architectural Education is a Paper Tiger ... Expose Yourselves ... It's Legal" presented to Suomen Arkkitehtiliito Finlands Arkitektsförbundet Education Conference, Helsinki, 1969 [#4]
1969.02	"Tapiola is a Paper Tiger ... Expose Yourselves ... It's Legal" report on the proposed new centre development for Tapiola, Finland and planning document JKMP Arkkitehdit, Helsinki, 1969 [#5]

1970

1970.01	"Hello Mother" (written in Finnish) Jousimies, Helsinki, 1970 [#6]
1970.02	"Some Relationships between the Formal Structure of Finnish Language and the Formal Structure of Finnish Architecture" (AA History Thesis, unpublished) 1970 [#7]

1972

1972.01	"Learning, Styles and Performance under Stress: and Experimental Set-up" Confidential Report to the USAF through System Research Ltd., Richmond, 1972 [#8]

1974

1974.01	"Alvar Aalto & the Failure of the Modern Movement" Open University Course Book & Lecture in The Modern Movement Series OU/BBC, Milton Keynes, 1974 [#9]

1975

1975.01	"A Cybernetic Development of Theories of Epistemology and Observation, with reference to Space and Time, as seen in Architecture" (PhD Thesis, unpublished) Brunel University, 1975. Also known as "The Object of Objects, the Point of Points,—or Something about Things" [#10]

---

1. The publication numbers of some items have been changed from earlier versions of this CV. Former publication numbers are to be found in square brackets [ ] at the end of each reference, preceded by a hash (#) sign. A (not completely up-to-date) CV in html, with links to a number of the papers mentioned in this section, may be found at: http://www.univie.ac.at/constructivism/ people/ glanville/cv.html. Selected papers may also be found at http://homepage.mac.com/ranulph/ FileSharing1.html, in the folders "papers" and "MCA". A proper web site is under construction, due to come on line at the end of 2014.

1976

1976.01	"Alvar Aalto, the Architect's Architect" Design Magazine, 1976 [#13]
1976.02	"Is Architecture just a Hollow Space, or is it the Empty Set?" Architectural Association Quarterly (AAQ), London, 8–4, 1976 [#14]
1976.03	"Meaning in Western Architecture" (a review article) Architectural Association Quarterly (AAQ), London, 8–3, 1976 [#12]
1976.04	"What is Memory, that it can remember what it is ?" in Trappl, R et al. (Eds.) "Recent Progress in Cybernetics & Systems Research, Volume 7" (Proceedings 3 European Meeting on Cybernetics and Systems Research) Hemisphere Press, Washington DC, 1976 [#11]

1977

1977.01	"Acoustic Report on Southwark Cathedral" (with P. Gillieron), unpublished report to the Cathedral Dean and Chapter, 1977 [#15]
1977.02	"Amazing Space: for the Architectural Stimulus-response Rat?" Architectural Association Quarterly (AAQ), London 9—2/3, 1977 [#17]
1977.03	"Experiential Projects: Learning to Perceive, and Perceiving to Learn" presented to 2 ICIUE, Newcastle, 1977 (Centre for the Study of Human Learning, Brunel University) [#18]
1977.04	"Finnish Vernacular Architecture" Architectural Association Quarterly (AAQ), London, 9—1, 1977 [#16]
1977.05	"Learning more by Error than by Trial" presented to 2 International Conference on Personal Construct Theory, Oxford, 1977 (Centre for the Study of Human Learning, Brunel University) [#19]
1977.06	"The Logic of Description: or why Physics won't work" presented to International Conference on Applied General Systems: Recent Developments and Trends, Binghamton, NY, 1977 [#21]
1977.07	"The Nature of Fundamentals, applied to the Fundamentals of Nature" in Klir, G (Ed.) Proceedings 1 International Conference on Applied General Systems: Recent Developments & Trends, Plenum, New York, 1977 [#20]

1978

1978.01	"800 Years of Finnish Architecture" (review article), Architectural Association Quarterly (AAQ), London, 10—3, 1978 [#33]
1978.02	"Alvar Aalto—Totality and Detail—a Style of Completeness" Architectural Association Quarterly (AAQ), London, 10—3, 1978 [#35]
1978.03	"Leaving Space for Design" presented to North London Polytechnic Design Research Group, 1978 [#28]
1978.04	"The (W)holes in the Grid: Spatial Training for Architects" paper presented to Conference on Applications of the Repertory Grid, Uxbridge, 1978 [#32]
1978.05	"The Concept of an Object and the Object of a Concept" presented at 4 European Meeting on Cybernetics and Systems Research, Linz, 1978 [#25]

1978.06	"Theory of Model Dimensions applied to Relational Data Bases" (with Jackson, P). Proceedings 4 World Organisation for General Systems and Cybernetics, 1978 [#27]

1979

1979.01	"Beyond the Boundaries", in Ericson, R (Ed.), "Proceedings Society for General Systems Research Silver Jubilee Conference, London" Springer Verlag, London, 1979 [#40]
1979.02	"Constriction by Diction and Construing by Doing" presented at 3 International Congress on Personal Construct Psychology, Breukelen, 1979 [#38]
1979.03	"Construct Games" AANE, London, 1979 [#39]
1979.04	"MOTIF 8, The History of a Self-conscious Design Course" exhibition, London, 1979 [#47]
1979.05	"Ruing & Construing" (with A. Pedretti & P. Jackson), in Ericson, R. (Ed.)— "Proceedings Society for General Systems Research Silver Jubilee Conference, London" Springer Verlag, London, 1979 [#41]
1979.06	"The Form of Cybernetics: Whitening the Black Box" in Miller, J (Ed.), "Proceedings of 24 Society for General Systems Research/American Association for the Advancement of Science Meeting, Houston, 1979" Society for General Systems Research, Louisville, 1979 [#36]

1980

1980.01	"A Chorus of Universes and a Couple of Small Points about the Universe" in Trappl, R., et al. (Eds.) "Progress in Cybernetics & Systems Research" vol. 8, Hemisphere Press, Washington DC, 1980 [#44]
1980.02	"All Thoughts of Things" in Trappl, R., et al. (Eds.) "Progress in Cybernetics & System Research" vol. 8, Hemisphere Press, Washington DC, 1980 [#43]
1980.03	"Consciousness, and so on" Journal of Cybernetics, vol. 10, pp 301–12, 1980 [#24]
1980.04	"Construct Heterarchies" International Journal of Man Machine Studies, vol. 13, 1980 [#49]. See 1980.05 and 1980.14
1980.05	"Construct Heterarchies" (in revised form, slightly expanded with technical note) in Shaw, M (Ed.), "Recent Advances in Personal Construct Technology" Academic Press, London, 1981 [#49a]
1980.06	"Description, Invention, Reality" guest Editor, special issue of Architectural Association Quarterly (AAQ), London, 12—4, 1980. Contributors: Frances Yates, Ernst Gombrich, Ranulph Glanville, Michael Evans, J. Christopher Jones [#67]
1980.07	"Entertaining the Environment" in de Zeeuw, G, and van der Eeden, P, (Eds.) "Proceedings Conference on Problems of Context" Vrije Universiteit Druck, Amsterdam, 1980 [#37]
1980.08	"Mapping Realities " Architectural Association Quarterly (AAQ), London, 12—4, 1980 [#55], [#67a]
1980.09	"Measure Meant" m/s , since incorporated in modified form in "The Mechanics of Togetherness: Measure Meant" in Glanville, R and de Zeeuw, G (Eds.) "Problems of Support, Survival and Culture" (number 118) [#30]. See 1993.14

1980.10	"The Architecture of the Computable" Design Studies, vol. 1, No. 4, 1980 [#34]
1980.11	"The Domain of Language" (with A. Pedretti), in Trappl, R et al. (Eds.), "Progress in Cybernetics & Systems Research" vol. 11, Hemisphere Press, Washington DC, 1980 [#29]
1980.12	"The Model's Dimensions: A Form for Argument" Proceedings 4 World Organisation for General Systems and Cybernetics, 1978, and International Journal of Man Machine Studies, vol. 13, 1980 [#26]
1980.13	"The Same is Different" in Zeleny M (Ed.) "Autopoiesis" Elsevier, New York, 1980 [#22]
1980.14	Translation of "Construct Heterarchies" into Polish for publication (full details of reference unavailable) [#49b]
1980.15	"Whither Design Science" internal memorandum, Portsmouth Polytechnic, 1980 [#74]

1981

1981.01	"A Precise Methodology for an Imprecise Method" presented at 5 World Organisation for General Systems and Cybernetics, Mexico City, 1981 [#61]
1981.02	"A Programme for Programmed Learning" internal memorandum, Portsmouth Polytechnic, 1981 [#65]
1981.03	"Alvar Aalto", "Le Corbusier" & "Ross Ashby" in Wintle, J (Ed.), "The Makers of Modern Culture" Routledge & Kegan Paul, London, 1981 [#45]
1981.04	"Calculator Saturnalia" (with Pask G. & Robinson M.), Wildwood House, London, 1980 and Random House, New York, 1981 [#23]
1981.05	"Levels & Boundaries of Problems" presented at conference on Problems of Levels and Boundaries, Amsterdam, 1981 [#*59]. An extended version is published as 2000.10.
1981.06	"Occam's Adventures in the Black Box" in Lasker, G (Ed.), "Applied Systems & Cybernetics" vol. II, Pergamon, Oxford, 1981 [#51]
1981.07	"Social Systems & Socialised Science" presented at 5 World Organisation for General Systems and Cybernetics, Mexico City, 1981 [#60]
1981.08	"Some Initial Comments on Word Proceedings" internal memorandum and discussion document, Portsmouth Polytechnic, 1981 [#71]
1981.09	"The Office Project" the OFFICE, North Harrow, 1979 (see Appendix 4: Software) [#54]
1981.10	"What a Waste" including "Cybernetics: the Art of the State", & "Why not the New Cybernetics?" presented at American Society for Cybernetics Conference, Washington DC, 1981 [#64]
1981.11	"Why Design Research?" presented at Design Research Society, Portsmouth, 1980 in Jacques, R and Powell, J "Design/Method/ Science" Westbury House, Guildford, 1981 [#53]
1981.12	"Your Inside is out and your Outside is in" (with F Varela), in Lasker, G, (Ed.), "Applied Systems & Cybernetics" vol. II, Pergamon, Oxford, 1981 [#52]. See 1981.13
1981.13	"Your Inside is out and your Outside is in" privately circulated one page summary and technical note [#52a]

1982

1982.01	"A Cybernetic View of Education and a View of Cybernetic Education" keynote address to 17 American Society for Cybernetics Meeting, Columbus, Ohio, 1982 [#79]
1982.02	"A Note on Numbers & Marking" open memo, Portsmouth Polytechnic, 1982 [#81]
1982.03	"A Stool to Meditate on" (with Powell, J, and de Zeeuw, G), video tape presented at 27 Society for General Systems Research/American Association for the Advancement of Science, Washington DC, 1982, also video-script [#69]
1982.04	"Against Fuzzy Thinking" invited contribution to "Systemica" vol. 2 no 5 (in Dutch translation), 1982 [#57]
1982.05	"Architecture and its Education" briefing document for projected special issue of Arkkitehti (Finnish Architectural Review) on Architectural Education & Creativity, 1982 [#73]
1982.06	"Charles Barry", "Friedrich Froebel", "George Gilbert Scott" & "Alfred Jarrry", in Wintle, J, (Ed.), "The Making of Nineteenth Culture" Routledge & Kegan Paul, London, 1982 [#62]
1982.07	"In Deference to Self-reference" Design Research Society Notes, Jan/Feb., 1982 [#56]
1982.08	"Inside Every White Box there are two Black Boxes trying to get out" presented at Conference of the Cybernetics Society, London, 1979, in Behavioural Science vol. 12, no 1, 1982 [#42]
1982.09	"Is it Real, Daddy?" book project, now incorporated in "Wittgenstein's Monument: Is it Real, Daddy?" BKS+, Southsea, 1994 (number 151) [#66]. See 1994.07
1982.10	"Jumping" working paper for University of Amsterdam, 1986 [#58]
1982.11	"Little Logicals" (notes on some logical & illogical topics), unpublished [#*63]
1982.12	"Method in Methodology" in Troncale, L (Ed.) "General Systems Methodology" (Proceedings 27 Society for General Systems Research/ American Association for the Advancement of Science), Society for General Systems Research, Louisville, 1982 [#68]
1982.13	"Motivation" (report on a reflexive design project), unpublished, 1982 [#72]
1982.14	"Pask's Bones—or a Poor Man's Guide to Gordon Pask" invited contribution to book on Gordon Pask edited by Doreen Steg, abandoned [#80]
1982.15	"Reasons to be Cheerful, Pts.. 1, 2 and 3" presented at 6 European Meeting on Cybernetics and Systems Research, Vienna, 1982, (includes "Abbreviation", "It's not that I ain't never said no nothing worth saying ...", "Paper" & "Lessness", papers presented as limited edition art works) [#70]

1983

1983.01	"The Theory of Model Dimensions, applied to the Computer Solution of a Syllogism" (with Jackson, P) International Journal of Man Machine Studies, vol. 19, 1983 (a reworking of "Theory of Model Dimensions applied to Relational Data Bases" number 27) [#84]

1983.02 "Very Rarely Stable" in van Gigch, J & Ramon, J (Eds.) "Meta-models and Meta-systems" Academic Press, London 1983 [#75]

1984

1984.01 "Cedric Price, Precisely" in works II, The Architectural Association, London, 1984 [#89]

1984.02 "Distinguished and Exact Lies" in Trappl, R, (Ed.) "Cybernetics & Systems Research 2." Elsevier North Holland, New York, 1984 [#87]

1984.03 "The work of Gordon Pask" Editor, special Edition of Cybernetic. (Manuscript lost by publishers in production. Project re-surfaced with new material as "Gordon Pask: a Festschrift" (collector and editor), Systems Research vol. 10, no 3, 1993, number 135 [#86]. See 1993.04

1984.04 "Emptiness" in Trappl, R, (Ed.) "Cybernetics & Systems Research 2" Elsevier North Holland, New York, 1984 [#88]. This paper exists in a form attached to 1984.92.

1984.05 "Get your Teeth into a Small Slice—the Cake of Liberty" invited paper for the 1984 George Orwell Conference, Hasselby, Sweden. Presented at American Society for Cybernetics, Philadelphia, 1984. (Hasselby conference cancelled: a reworked version appears as [#82].) See 1994.06

1984.06 "The One Armed Bandit" invited paper for the Conference on "Building Utilisation and Assessment" Portsmouth, 1983, in Powell, J, Cooper, I, & Lera, S. "Designing for Building Utilisation" Spon, London, 1984 [#83]

1985

1985.01 "I am Sitting in the Buffet of a Train" (incorporating "Designing Modern Life"). Contributions to Jones, C. "The Design of Modern Life" in Jones, J. C. "Essays in Design" Wiley, Chichester, 1985 [#90]

1985.02 "The Oulu School" m/s for Arkkitehti, 1985 [#91]

1986

1986.01 "As Cybernetics is" presented at International Association for Cybernetics Conference, Namur, 1986 [#97]

1986.02 "Cybernetics & Design" video presentation at 21 American Society for Cybernetics Conference, Virginia Beach, USA, 1986 [#94]

1986.03 "Designing Cybernetics" presented at 8 European Meeting on Cybernetics and Systems Research, Vienna, 1986, Journal of the Polish Design Research Society, publication details untraceable [#95]

1986.04 "Mechanical Trees" in de Zeeuw, G (Ed.), "Knowledge (Dis)Appearance" Systemica, Delft University Press, 1986 [#92]

1986.05 "The Cybernetics of Ethics & the Ethics of Cybernetics" tutorial at 21 American Society for Cybernetics Conference, Virginia Beach, USA, 1986, video publication [#93]

1987

1987.01 "Alvar Aalto: the Formative Years" and "Alvar Aalto: the Decisive Years", review article, AA files no 16, 1987 [#100]

1987.02	"Cybernetics and Utility" in Continuing the Conversation 5, Gravel Hill, Kentucky, 1987 [#101]
1987.03	"The Question of Cybernetics" presented at 8 European Meeting on Cybernetics and Systems Research, Vienna, 1986, published in Cybernetics, an International Journal, vol.18, 1987 [#96]. See 1988.07

1988

1988.01	"Architecture and Space for Thought" (Ph D Thesis, unpublished) Brunel University, Uxbridge, 1988 [#99]
1988.02	"Einführung: die Leere der Black Box—und Andere Mechanical Trees" ("The Emptiness of the Black Box—and Other Mechanical Trees"), introductory article in "Objekte" (translated by D Baecker), Merve Verlag, Berlin, 1988 (see number 104): only published in English, 1988 [#103]
1988.03	"Impossible Worlds and other Mythical Beasts" in "Problems of (Im)possible Worlds" (Edited by Glanville, R and de Zeeuw, G), Thesis Publishers, Amsterdam, 1988 [#104]
1988.04	"Impossible Worlds!" (with de Zeeuw, G) introduction to "Problems of (Im)possible Worlds" (Edited by Glanville, R and de Zeeuw, G), Thesis Publishers, Amsterdam, 1988 [#105]
1988.05	"Objekte" (selected papers translated by D Baecker), Merve Verlag, Berlin, 1988 [#102]
1988.06	"Problems of (Im)possible Worlds" (Edited, with G de Zeeuw), (conference Proceedings), Thesis Publishers, Amsterdam, 1988 [#106]
1988.07	"The Question of Cybernetics" republished in the General Systems Yearbook, Society for General Systems Research, Louisville, 1988 [#96a]

1989

1989.01	"Generosity: an Ethos: an Endpiece" in Glanville, R and de Zeeuw, G (Eds.) "Mutual Uses of Cybernetics and Science" Amsterdam, 1989, Thesis Publishers, Amsterdam, 1991 (2 Vols.) [#108]

1990

1990.01	"Sed Quis Custodient Ipsos Custodes" in Heylighen, F, Rosseel, E and Demeyere, F (Eds.) "Self-Steering and Cognition in Complex Systems" Gordon and Breach, London, 1990 [#107]
1990.02	"The Self and the Other: the Purpose of Distinction" in Trappl, R "Cybernetics and Systems '90" the Proceedings of the European Meeting on Cybernetics and Systems Research, World Scientific, Singapore, 1990 [#111]. See 1994.04 for German translation.

1991

1991.01	"Collective Support Systems and their Users" (Edited, with de Zeeuw, G), Thesis Publishers, Amsterdam, 1991 [#114]
1991.02	"Design" in de Jong, J, (Ed.) "Addenda and Errata" Thesis Publishers, Amsterdam, 1991 [#112]
1991.03	"Generosity" in de Jong, J, (Ed.) "Addenda and Errata" Thesis Publishers, Amsterdam, 1991 [#113]

1991.03	"Mechanical Trees" introductory review article in Systems Research, vol. 8 no 3 (1991) [#121]
1991.04	"Mutual Uses of Cybernetics and Science" (Edited, with de Zeeuw, G) Proceedings of the conference on Mutual Uses of Cybernetics and Science, Amsterdam, 1989, Thesis Publishers, Amsterdam, 1991 (2 Vols.) [#110]
1991.05	"Mutual Uses of Cybernetics and Science—a Preface" in Glanville, R and de Zeeuw, G (Eds.) "Mutual Uses of Cybernetics and Science" Amsterdam, 1989, Thesis Publishers, Amsterdam, 1991 (2 Vols.) [#109]
1991.06	"Interaction of Actors Theory" Pask, G and Zeeuw, G de, compiled, composed and edited by Glanville R, OOC Program, University of Amsterdam, 1991

1992

1992.01	"CAD Abusing Computing" in Proceedings eCAADe 1992, Polytechnic University of Catalonia, Barcelona, 1992 [#122]
1992.02	"The Second Disorder: the Amsterdam School" in Trappl, R "Cybernetics and Systems Research '92" the Proceedings of the European Meeting on Cybernetics and Systems Research, World Scientific, Singapore, 1992 [#120]

1993

1993.01	"An Introduction to Natural Magic" (by Heinz von Foerster), edited, partially transcribed and presented, and with introductory comments, Systems Research, vol. 10 no 1, 1993 [#129]
1993.02	"Exploring and Illustrating" in Proceedings eCAADe 1993, Eindhoven University of Technology, Eindhoven, 1993 [#128]
1993.03	"Gordon Pask, a Festschrift" (commissioner, contributor and editor), Systems Research vol. 10, no 3, 1993 [#135]
1993.04	"Gordon Pask, a Sketch for an Unofficial Biography" in Glanville, R (Ed.), "Gordon Pask, a Festschrift", Systems Research vol. 10, no 3, 1993 [#131]
1993.05	"Gordon Pask: Publications and Projects" in Glanville, R (Ed.), "Gordon Pask, a Festschrift", Systems Research vol. 10, no 3, 1993 [#134]
1993.06	"Interactive Interfaces and Human Networks" (edited with de Zeeuw, G), Thesis Publishers, Amsterdam, 1993. (This publication was originally and briefly published, erroneously, under the title "Interactive Interfaces Between Collective Support Systems and their Users".) [#115]
1993.07	"Introduction to Gordon Pask, a Festschrift" in Glanville, R (Ed.), "Gordon Pask, a Festschrift", Systems Research, vol. 10 no 3, 1993 [#130]
1993.08	"Introduction to Problems of Support, Survival and Culture" (with de Zeeuw, G), in "Problems of Support, Survival and Culture" (Edited with G de Zeeuw), Thesis Publishers, Amsterdam, 1993 [#118]
1993.09	"Introduction: Behind the Screen" in "Interactive Interfaces and Human Networks" (Edited with de Zeeuw, G), Thesis Publishers, Amsterdam, 1993 [#116]

1993.10	"Looking into Endoscopy: the Limitations of Evaluation in Architectural Design" in Aura, S, Alavalkama, I and Palmqvist, H (Eds.) "Endoscopy as a Tool in Architecture", TUT • A •(Tampere University of Technology), Tampere, 1993 [#126]
1993.11	"On becoming Gordon's Student" in Glanville, R (Ed.), "Gordon Pask, a Festschrift", Systems Research vol. 10, no 3, 1993 [#132]
1993.12	"Pask: a Slight Primer" in Glanville, R (Ed.), "Gordon Pask, a Festschrift", Systems Research vol. 10, no 3, 1993 [#133]
1993.13	"Problems of Support, Survival and Culture" (Edited with de Zeeuw G), (conference Proceedings), Thesis Publishers, Amsterdam, 1993 [#119]
1993.14	"The Mechanics of Togetherness: Measure Meant" in Glanville, R and de Zeeuw, G (Eds.), "Problems of Support, Survival and Culture" Proceedings (Edited with de Zeeuw G), Thesis Publishers, Amsterdam, 1993 [#117]
1993.15	"The Tampere Meeting of the European Architectural Endoscopy Association" Report for Dutch Institute of Design, internally circulated, 1993 [#127]

1994

1994.01	"A (Cybernetic) Musing", Cybernetics and Human Knowing vol. 2, no 4, 1994. Also to be found at http://www.ingenta.com/ under "Cybernetics & Human Knowing" [#151]
1994.02	"as if (Radical Objectivism)" in Trappl, R "Cybernetics and Systems Research '94" the Proceedings of the European Meeting on Cybernetics and Systems Research, World Scientific, Singapore, 1994 [#138]. See 1994.03
1994.03	"as if (Radical Objectivism)" privately circulated one page summary and technical note [#138a]
1994.04	"Das Selbst und das Andere: Der Zweck der Unterscheidung" (translated of 1990.02 by D. Baecker) in Baecker, D (Ed.) "Kalkül der Form" Suhrkamp, Frankfurt am Main, 1994 [#111a]
1994.05	"eCAADe '94, Strathclyde" Building Design October 14, 1994. Token article marking first of a series of occasional journalistic articles. None of the further articles are mentioned in this list of publications. [#152]
1994.06	"(Get Your Teeth into a Small Slice—) the Cake of Liberty" in Lasker, G (Ed.) "Procs. 2nd Orwellian Symposium—Advances in Studies of Societal Control Karlovy Vary 1994" IIASSRC, Windsor, Ontario, 1995 (see no 82: this is a reworked version) [#143]
1994.07	"In the World: Wittgenstein's Monument 1" BKS+, Southsea, 1994 [#153]
1994.08	"Prezentacje Jasne, Uczciwe i Prawdziwe", in Kadysz A (Ed.) "Materialy Konferency Jne, Projektowanie Wspomagane Komputerowo—Bariery i Inspiracje Twórcze" Polytechnic University of Bialystok, 1994 [#137a]. (Translation into Polish of number #141, see 1994.12)
1994.09	"Prezentacje Jasne, Uczciwe i Prawdziwe" special modified version in Polish Computer World (full details unavailable) [#137b]. See 1994.12
1994.10	"Reima Pietilä: an Appreciation" in Architectural Design, vol. 64 no 3/4, April/May 1994 [#146]

1994.11	"Remoteness and the Value of Sharing" in Maver, T and Petric, J (Eds.) "The Virtual Studio", Proceedings of eCAADe 1994, University of Strathclyde, Glasgow, 1994 [#149]
1994.12	"Representations Fair, Honest and Truthful" in Kadysz A (Ed.) "Materialy Konferency Jne, Projektowanie Wspomagane Komputerowo—Bariery i Inspiracje Twórcze" Polytechnic University of Bialystok, 1994 [#137]
1994.13	"Some Walls are Thick", since revised, appearing as "Living in Lines" [#147]. See 2000.08
1994.14	"Variety in Design" Systems Research, vol. 11, no 3, 1994 [#148]

1995

1995.01	"A (Cybernetic) Musing: Communication 1: Coding" In Cybernetics & Human Knowing vol. 3 nos 3, 1995. Also to be found at http://www.ingenta.com/ under "Cybernetics & Human Knowing" [#163]
1995.02	"A (Cybernetic) Musing: Control 1" In Cybernetics & Human Knowing vol. 3 no 1, 1995. Also to be found at http://www.ingenta.com/ under "Cybernetics & Human Knowing" [#161]
1995.03	"A (Cybernetic) Musing: Control 2" In Cybernetics & Human Knowing vol. 3 no 2, 1995. Also to be found at http://www.ingenta.com/ under "Cybernetics & Human Knowing" [#162]
1995.04	"Architecture and Computing: a Medium Approach" in Procs. 15th Meeting of Association for Computing in Architectural Design in America, University of Washington, Seattle, 1995 [#157]
1995.05	"Chasing the Blame" in Lasker, G (Ed.) "Research on Progress—Advances in Interdisciplinary Studies on Systems Research and Cybernetics" Vol. 11, IIASSRC, Windsor, Ontario, 1995 [#144]
1995.06	"Comparing CAD and Endoscopy in Exploring, Generating and Illustrating Architectural Designs" (with Petri Siitonen) in Martens, B (Ed.) "Proceedings of 2 EAEA", TU Wien, Vienna, 1995 [#155]
1995.07	"Creativity and Media in Architecture" in Martens, B (Ed.) "Proceedings of 2 EAEA", TU Wien, Vienna, 1995 [#154]
1995.08	"Installing the Context" report on a joint project between Technical University of Vienna and University of Portsmouth, 1995 [#156]
1995.09	"Introduction to Problems of Values and (In)variants" in Glanville, R and de Zeeuw, G (Eds.) "Problems of Values and Invariants" Thesis Publishers, Amsterdam, 1995 [#124]
1995.10	"Learning and MultiMedia" (with Robin McKinnon-Wood) in Procs. 13th Meeting of European Computer Aided Architectural Design in Education, University of Palermo, Palermo, 1995 [#158]
1995.11	"Problems of Values and (In)variants" Conference Proceedings (Edited with de Zeeuw, G), Thesis Publishers, Amsterdam, 1995 [#125]
1995.12	"The Cybernetics of Value and the Value of Cybernetics: the Art of Invariance and the Invariance of Art" in Glanville, R and de Zeeuw, G (Eds.) "Problems of Values and (In)variants" (conference Proceedings) Thesis Publishers, Amsterdam, 1995 [#123]

1996

1996.01	"A (Cybernetic) Musing: Robin McKinnon-Wood and Gordon Pask: a Lifelong Conversation" In Cybernetics and Human Knowing vol. 3 nos

	4, 1996. Also to be found at http://www.ingenta.com/ under "Cybernetics & Human Knowing" and at http:// www.venus.co.uk/gordonpask/candhk.htm [#166]
1996.02	"Communication without Coding: Cybernetics, Meaning and Language (How Language, becoming a System, Betrays itself)", invited paper in Modern Language Notes, Vol. 111, no 3 (ed. Wellbery, D), 1996 [#159]
1996.03	"Computing, Education and Architecture for the Lost Profession" in af Klerker, J (Ed.) Procs. 14th Meeting of European Computer Aided Architectural Design in Education, Technical University of Lund, Lund, 1996 [#172]
1996.04	"Creativity and HyperMedia, MulitMedia, the InterNET and Virtuality" In Asanowicz A and Sawicki R (Eds.) "CAD Creativeness, Proceedings of the IV Conference on Computer in Architectural Design", Technical University of Bialystok, Bialystok, 1996 [#167]
1996.05	"Gordon Pask 1928–1996, an Obituary" Bulletin of the International Federation for Systems Research, available at web site http://www.venus.co.uk/gordonpask/ifsr.htm
1996.06	"Guest Editorial" in Glanville, R (Ed.) "Heinz von Foerster, a Festschrift", Systems Research Vol. 13 no 3, 1996 [#168]
1996.07	"Heinz von Foerster, a Festschrift", (commissioner, contributor and editor), Systems Research Vol. 13 no 3, 1996 [#171]
1996.08	"Heinz von Foerster: a Biographical Outline" in Glanville, R (Ed.) "Heinz von Foerster, a Festschrift", Systems Research Vol. 13 no 3, 1996 [#169]
1996.09	"Heinz von Foerster: the Form and the Content" in Glanville, R (Ed.) "Heinz von Foerster, a Festschrift", Systems Research Vol. 13 no 3, 1996 [#170]
1996.10	"Justice Love and Wisdom—Graham Barnes" (review article) in Systems Research vol. 13 no 2, 1996 [#139]
1996.11	"NOAH: the Ark of Knowing in a Learning Environment" (with Robin McKinnon-Wood) in Trappl, R (Ed.) Procs. 13 EMCSR, Vienna, University of Vienna and Austrian Society for Cybernetic Studies, 1996 [#160]

1997

1997.01	"A (Cybernetic) Musing: Communication—Conversation 1" Cybernetics & Human Knowing vol. 4 no 1, 1997. Also to be found at http://www.ingenta.com/ under "Cybernetics & Human Knowing"
1997.02	"A (Cybernetic) Musing: Communication—Conversation 2" Cybernetics & Human Knowing vol. 4 no 2-3, 1997. Also to b e found at http://www.ingenta.com/ under "Cybernetics and Human Knowing" [#174]
1997.03	"A (Cybernetic) Musing: the Animal and the Machine " Cybernetics & Human Knowing vol. 4 no 4, 1997. Also to be found at http://www.ingenta.com/ under "Cybernetics & Human Knowing"
1997.04	"A Ship without a Rudder" in Glanville, R and de Zeeuw, G (Eds.) "Problems of Excavating Cybernetics and Systems", BKS+, Southsea, 1997 [#140]
1997.05	"Behind the Curtain" in Ascott, R (Ed.) Procs. First Conference on Consciousness Reframed, UCWN, Newport, 1997

Appendices

1997.06	"Bernard Scott—Kybernetik im Verborgenen" in Bardmann, T (Ed.) "Zirkuläre Positionen" Westdeutscher Verlag, Opladen, 1997
1997.07	"Gordon Pask: a Summary of His Work", InterNet publication commissioned by the School of Business Administration, St. Gallen University, placed on the ISSS web site under http:// www.iss.org. luminaries, 1997 [#180]
1997.08	"Introduction" in Glanville, R and de Zeeuw, G (Eds.) "Problems of Excavating Cybernetics and Systems", BKS+, Southsea, 1997 [#141]
1997.08	"Nicht wir führen die Konversation, die Konversation führt uns" interview in Bardmann, T (Ed.) "Zirkuläre Positionen" Westdeutscher Verlag, Opladen, 1997
1997.09	"Pictures at an Exhibition: a HyperNavigator for CyberSpace" (with C Ferris), (see Appendix: Software), 1997 [#145]
1997.10	"Problems of Excavating Cybernetics and Systems", Edited with G de Zeeuw, BKS+, Southsea, 1997 [#142]
1997.11	"The Value when Cybernetics is Added to CAAD" in Nys, K, Provoost, T, Verbeke, J and Verleye, J (Eds.) "The Added Value of Computer Aided Architectural Design" Brussels, Hogeschool voor Wetenschap en Kunst Sint-Lucas, 1997 [#179]

1998

1998.01	"A (Cybernetic) Musing: Language and Science and the Language of Science" Cybernetics & Human Knowing vol. 5 no 4, 1998 (with Sima Sengupta and Gail Forey). Also to be found at http://www.ingenta.com/ under "Cybernetics and Human Knowing"
1998.02	"A (Cybernetic) Musing: The Gestation of Second Order Cybernetics 1968–1975: a Personal Account" Cybernetics and Human Knowing vol. 5 no 2, 1998. Also to be found at http://www.ingenta.com/ under "Cybernetics & Human Knowing"
1998.03	"A (Cybernetic) Musing: Variety and Creativity" Cybernetics & Human Knowing vol. 5 no 3, 1998. Also to be found at http:// www.ingenta.com/ under "Cybernetics & Human Knowing"
1998.04	"A (Cybernetic) Musing: Varieties of Variety" Cybernetics & Human Knowing vol. 5 no 1, 1998. Also to be found at http:// www.ingenta.com/ under "Cybernetics & Human Knowing"
1998.05	"John Hamilton Frazer— von der Arkitektur zur Autotektur" in Bardmann, T "Zirkuläre Positionen 2" Westdeutscher Verlag, Opladen, 1998
1998.06	"Keeping Faith with the Design in Design Research" de Montfort DRS, 1998 published as part of proceedings on conference web site at http://www.dmu.ac.uk/ln/4dd/drs9.html
1998.07	"Triads" on web site http://iasl.uni-muenchen.de/discuss/lisforen/lisforen.htm#bewuflt click on bewusstsein, kommunikation, zeichen)

1999

1999.01	"A (Cybernetic) Musing: Encyclopaedias and the Form of Knowing. A Celebration of Charles Francois' 'International Encyclopaedia of Systems and Cybernetics: a Sort of Self-Referential Work of Reference'" Cybernetics & Human Knowing vol. 6 no 1, 1999. Also to be found at http://www.ingenta.com/ under "Cybernetics & Human Knowing"

1999.02 "A (Cybernetic) Musing: the Millennium Bug" Cybernetics & Human Knowing vol. 6 no 3, 1999 (with Bob Barbour, Michael Schreiber and Stuart Umpleby). Also to be found at http://www.ingenta.com/ under "Cybernetics & Human Knowing"

1999.03 "A (Cybernetic) Musing: Thinking the New Millennium" Cybernetics & Human Knowing vol. 6 no 4, 1999. Also to be found at http://www.ingenta.com/ under "Cybernetics & Human Knowing"

1999.04 "Acts Between and Between Acts" in Ascott, R (Ed.) "Reframing Consciousness", Intellect, Exeter, 1999. See 1999.09 for German translation

1999.05 "As Is and As If" Procs. Conference Invençao Thinking the New Millennium Sao Paolo, Brazil, web site http://www.itaucultural.org.br/invencao/papers/Glanville.htm, 1999

1999.06 "Constructive Realism" (edited with van Dijkum, C and de Zeeuw, G), BKS+, Southsea, 1999

1999.07 "Researching Design and Designing Research" Design Issues vol. 13 no 2, 1999

1999.08 "Scenes" Commemoration for Niklas Luhmann in Bardmann, T and Baecker, D (Eds.) "Gibt es Eigentlich den Berliner Zoo noch?", UVK Universitätsverlag Konstanz, Konstanz, 1999

1999.09 "Zwischenhandlungen und Zwischen Handlungen" trans. Pfister, Michael, in Heitz, AV and Pfister, M (Eds.) "Da Zwischen" Museum für Gestaltung, Zürich, 1999 (German translation of 1999.04)

2000

2000.01 "A (Cybernetic) Musing: The State of Cybernetics" Cybernetics & Human Knowing vol. 7 no 2–3, 2000. Also to be found at http://www.ingenta.com/ under "Cybernetics & Human Knowing"

2000.02 "Afterword" in Glanville, R and de Zeeuw, G (Eds.) "Problems of Action and Observation", Procs. 1997 conference, BKS+, Southsea, 2000

2000.03 "Architecture, Computers and Information: a View from the High Plains of Cybernetics" invited seed article and discussion, ISEA Chatterbox internet discussion "Defining Architecture, Defining Information", 2000

2000.04 "(Die Relativität des Wisswens)—Ebenen und Grenzen von Problemen" trans. Ort, Nina from number 1981.05 in Jahraus, O and Ort, N (Eds.) "Beobachtungen des Unbeobachtbaren" Velbrück Wissenschaft, Göttingen, 2000

2000.05 "First Sight" in Glanville, R and de Zeeuw, G (Eds.) "Problems of Action and Observation", BKS+, Southsea, 2000 [#176]

2000.06 "Inbetweenies" in Ascott, R (Ed.) Procs. Third Conference on Consciousness Reframed, UCWN, Newport, 2000

2000.07 "Introduction" in Glanville, R and de Zeeuw, G (Eds.) "Problems of Action and Observation", BKS+, Southsea, 2000 [#*177]

2000.08 "Living in Lines" in McLeod, R (Ed.) "Interior Cities", RMIT Press, Melbourne, 2000

2000.09 "Problems of Action and Observation", edited by Glanville, R and de Zeeuw, G, BKS+, Southsea, 2000 [#*178]

2000.10	"(The Relativity of Knowing)—Levels and Boundaries of Problems" on web site htp:// iasl.uni-muenchen.de/discuss/lisforen/lisforen.htm#bewuflt (click on bewusstsein, kommunikation, zeichen) 2000 (developed from an original presented at the Conference on Problems of Levels and Boundaries, Amsterdam, 1981)
2000.11	"The Value of being Unmanageable: Variety and Creativity in CyberSpace" in Eichmann, H, Hochgerner, J and Nahrada, F (Eds.) "Netzwerke", Falter Verlag, Vienna, 2000 (Procs. Conference "Global Village '97", Vienna, 1997) [#*175]

2001

2001.01	"A Cybernetic Musing: Constructing My Cybernetic World" Cybernetics & Human Knowing vol. 8 no 1–2, 2001
2001.02	"About Gordon Pask" (with Bernard Scott), Kybernetes vol. 30 issue 5/6
2001.03	"Afterthought and Future Scenario" in Stellingwerff, M and Verbeke, J (Eds.) "Accolade: European Workshop" Delft University Press—Science, Delft
2001.04	"Afterword" (with Scott, B) in Kybernetes vol. 30 nos 7/8
2001.05	"An Intelligent Architecture" invited paper, Convergence vol. 7, no 2
2001.06	"And He was Magic" special issue: Gordon Pask remembered and celebrated (edited with Bernard Scott), Kybernetes vol. 30 issue 5/6
2001.07	"Between Now and Then: the Auto-Interview of a Lapsed Musician" Leonardo Music Journal, vol. 11
2001.08	"Gordon Pask remembered and celebrated. Part 1" special issues of Kybernetes, vol. 30 issues 5/6, edited (with Bernard Scott)
2001.09	"Gordon Pask remembered and celebrated. Part 2" special issues of Kybernetes, vol. 30 issues 7/8, edited (with Bernard Scott)
2001.10	"I Could…but I Won't: a tribute to Heinz von Foerster" web site 2001, url: http://www.univie.ac.at/constructivism/HvF/festschrift
2001.11	"Introduction" (with Bernard Scott), Kybernetes vol. 30 issue 5/6
2001.12	"Listen!" in de Zeeuw, G, Vahl, M and Mennuti, E (Eds.) "Problems of Participation and Connection", Lincoln Research Centre, Lincoln, 2001
2001.13	"Nona Meyeah Teay" electronic music composition (1967), Leonardo Music Journal, vol.11, CD
2001.14	"Not Aping the Past: Mirror Men" in Stellingwerff, M and Verbeke, J (Eds.) "Accolade: European Workshop" Delft University Press—Science, Delft, 2001
2001.15	"Triads" in Jahraus, O and Ort, N (Eds.) "Bewusstsein—Kommunikation—Zeichen", Max Niemeyer Verlag, Tuebingen, 2001
2001.16	"An Observing Science" invited paper, Foundations of Science, vol. 6 nos 1–3 (a reworking and extension of 1999.05)
2001.17	"The Man in the Train: Complexity, UnManageability, Conversation and Trust" in Würthrich, H, Winter, W and Philipp, A (Eds.) Grenzen Ökonomischen Denkens, Wiesbaden, Gabler, 2001

2002

2002.01	"A (Cybernetic) Musing: Cybernetics and Human Knowing" Cybernetics & Human Knowing, vol 9 no 1
2002.02	"Biographical Notes" in Glanville, R (guest editor) "Gerard de Zeeuw—a Festschrift" Special Issue of Systems Research and Behavioural Science vol19 no 2
2002.03	"Doing the Right Thing: the Problems of… Gerard de Zeeuw, Academic Guerilla" in Glanville, R (guest editor) "Gerard de Zeeuw—a Festschrift" Special Issue of Systems Research and Behavioural Science vol19 no 2
2002.04	"Festschrift for Gerard de Zeeuw" special issue of System Research and Behavioural Science vol 19 no 2 (guest editor)
2002.05	"Francisco Varela: a Working Memory" Cybernetics & Human Knowing, vol 9 no 2
2002.06	"Generator" (with Severi Glanville) audio visual piece played at BEAP, Perth, and since.
2002.07	"Heinz von Foerster," Soziale Systeme vol 8 no 2
2002.08	"Heinz von Foerster—a Personal Farewell" Cybernetics & Human Knowing vol 9 no 3–4
2002.09	"Introduction" in Glanville, R (guest editor) "Gerard de Zeeuw—a Festschrift" Special Issue of Systems Research and Behavioural Science vol 19 no 2
2002.10	"Second Order Cybernetics" commissioned article in "Encyclopaedia of Life Support Systems," EoLSS Publishers, Oxford: net publication at http://www.eolss.net
2002.11	"The Incorporated Observer" in Ascott, R (Ed.) Proc Concsiousness Reframed 2002, UCWN Newport.

2003

2003.01	"In Prise of Buffers", Cybernetics and Human Knowing vol 10 nos 3–4
2003.02	" A (Cybernetic) Musing: Some Examples of Cybernetically Informed Educational Practice" Cybernetics and Human Knowing, vol 9 no 3
2003.03	"A Note on Knowing," published as "Ein Wort über Wissen—A Note on Knowing" (trans Ort, N), in Jahraus, O and Orrt, N (Eds.) "Theorie—Prozess—Selbstreferenz," Konstanz, UVK Verlagsgesellschaft
2003.04	"An Irregular Dodecahedron and a Lemon Yellow Citroën" in van Schaik, L "Practice of Practice, Melbourne RMIT Press
2003.05	"Architecture and the Embodiment of Knowledge" in Grün, E and del Cano, E (Eds.) "Ensayos Sobre Sistemica y Cibernetica." Buenos Aires, Editorial Dunken
2003.06	"Behaving Well" in Lasker, G et al, Procs 17th IIASSS Conference, IIASSS, Windsor, Ontario
2003.07	"Cybernetics & Human Knowing, Memorial to Heinz von Forster" editor, with Brier, S, Cybernetics and Human Knowing vol 10 no 3–4
2003.08	"Designing Reflections: Reflections on Design" (with Leon van Schaik), in Durling, D and Sugiyama, K (2003) Proceedings of the third conference, Doctoral Education in Design, Chiba, Chiba University.

2003.09	"Foreword: the Ouroboros and the Glass Bead Game" (with Brier, S) Cybernetics and Human Knowing vol 10 no 3–4
2003.10	"Inter View: Designing Interfaces and Inter-facing Design" (with Madeleine AkTypi), Anomalie Digital Arts no 3
2003.11	"Lost" invited chapter in Lo, A (Ed.) "Navigating Design: a Voyage of Discovery" HongKong SAR, School of Design, Hong Kong Polytechnic University
2003.12	"Machines of Wonder and Elephants that Float through Air" in Brier, S and Glanville, R (Eds.) "Cybernetics and Human Knowing, Memorial to Heinz von Forster", Cybernetics & Human Knowing vol 10 no 3–4
2003.13	"Obituary—Heinz von Foerster" Systems Research vol 20 no 1 (one of five obituaries written for Heinz von Foerster)
2003.14	"Understanding Systems: a Review Article" Cybernetics & Human Knowing vol 10 no 3–4

2004

2004.01	"A (Cybernetic) Musing: Control, Variety and Addiction" Cybernetics & Human Knowing vol 11 no 4
2004.02	"A Conscience for Cybernetics" Procs 17th EMCSR, Vienna, Austrian Society for Cybernetic Studies
2004.03	"A Cyber_Reader: Neil Spiller + Understanding Systems: Heinz von Foerster with Bernhard Poerksen—a review article" Architectural Design vol 74 no 4, July/August
2004.04	"Control, Imagination and Addiction" (keynote address to American Society for Cybernetics Conference, Toronto, August 2004): http://www.asc-cybernetics.org/2004/GlanvillePaper.pdf.
2004.05	"Desirable Ethics" Cybernetics and Human Knowing vol 11 no 2
2004.06	"Implications of Glanville's perspective for the professional development of knowledge workers" transcript of presentation, in Working and Learning in Vocational Education and Training in the Knowledge Era. Final Report of the Professional Development for the Future Project, Australian Flexible Learning Framework.
2004.07	"International Encyclopaedia of Systems and Cybernetics, second edition" various entries for Francois, C (compiler and editor), Munich, KG Saur
2004.08	"Interview with Ranulph Glanville: the other way round—science as design" (with Wolfgang Jonas and Jan Meyer-Veden) in Jonas, W and Meyer-Veden (Eds.) "Mind the Gap" Bremen, HG Hauschild
2004.09	"Mai von Foerster and Herbert Brün: together and apart. An addendum to Heinz von Foerster's Obituary" Systems Research and Behavioural Science vol 21 issue 1 pp. 101–102
2004.10	"The Purpose of Cybernetics" Kybernetes vol 33 no 6

2005

2005.01	"Appropriate Theory" Proceedings of FutureGround Conference of the Design Research Society, Melbourne, Monash University (on CD)
2005.02	"As If" hip hop setting of paper As If by Scientific Fly on CD, dataworld, number dr–940413–22–28
2005.03	"Certain Propositions concerning Prepositions" Cybernetics & Human Knowing vol 12 no 3

2005.04	"Cybernetics" in Mitcham, C (Ed.) Encyclopedia of Science, Technology, and Ethics, Woodbridge Connecticut, Macmillan Reference USA
2005.05	"Dark Hero of the Information Age: In Search of Norbert Wiener, Father of Cybernetics" review article, Cybernetics & Human Knowing vol 13 no 2
2005.06	"International Encyclopaedia of Systems and Cybernetics, second edition: a Review Article" Cybernetics and Human Knowing vol 12 no 1–2
2005.07	"Knowledge Creation and Research in Design and Architecture," with Verbeke, J, in Ameziane, F (Ed.) Procs EURAU'04, European Symposium on Research in Architecture and Urban Design, Marseilles, Université de Marseilles
2005.08	"Lerner ist Interaktion: Gordon Pask's 'An Approach to Cybernetics'" in Baecker, D (Ed.) "Schlüsselwerke der Systemtheorie" Wiesbaden, Verlag für Sozialwissenschaften

2006

2006.01	"Construction and Design" Constructivist Foundations vol 1 no 3 (web published journal at www.univie.ac.at/constructivism/journal/1.3)
2006.02	"Design and Mentation: Piaget's Constant Objects" The (Radical) Designist, 07/2006 zero issue (web publication at iade.pt/designist)
2006.03	"Invisibility and Silence" Cybernetics & Human Knowing vol 13 no 1
2006.04	"Life is a verb: review of Poerksen, B and Maturana, H 'From Being to Doing'" Cybernetics & Human Knowing vol 13 nos 3–4
2006.05	"Ranulph Glanville, a Conversation" (with Barbara Vogl, editor) Patterns July–September 2006
2006.06	"Ranulph Glanville, a Conversation" (with Vogl, B) in Vogl, B (Ed.) (2006) "Patterns" News Journal of the American Society for Cybernetics spring/summer 2006 pp 1–8
2006.07	"The IFSR, Diagrammes and Inclusive Logic" Cybernetics & Human Knowing vol 13, no 3–4
2006.08	"Visual Logic" in Trappl, R et al (Eds.) Cybernetics and Systems 2006, Vienna, Austrian Society for Cybernetic Studies
2006.09	"What Makes the Difference—Reflecting on Reflecting (on Reflecting on Reflecting)" in Verbeke, J and Janssons (Eds.) "Reflections + 3" Brussels, Sint Lucas Architectuur.
2006.10	"Designing Professional Development for the Knowledge Era" (Working Group Report(2006)): http://www.icvet.tafensw.edu.au/resources/life_based_learning.htm

2007

2007.01	"A Cybernetic Serendipity" Exhibition Review, http://maverickmachines.com/WordPress/?page_id=88
2007.02	"A t' Tribute: Qualities: Properties and Attributes" in Zeeuw, G de, Vahl, M and Mennuti (Eds.) (2007) Problems of Individual Emergence, Systemica vol 14 nos 1–6, pp 123–137
2007.03	"An Approach to Cybernetics" in Glanville, R and Müller, KH (Eds.) "Gordon Pask, Philosopher Mechanic: an introduction" Vienna, edition echoraum, pp 13–27

2007.04	"And He was Magic" in Glanville, R and Müller, KH (Eds.) "Gordon Pask, Philosopher Mechanic: an introduction" Vienna, edition echoraum, pp 119–141
2007.05	"Ashby and the Black Box" Cybernetics and Human Knowing, vol 14 nos 2–3, pp 189–96
2007.06	"Comparison" (with Müller KH) e-Wisdom 4 Construction Comparisons/ Comparing Constructions, pp 18–31
2007.07	"Cybernetics and Design" (edited) special double issue of Kybernetes, vol 36 no 9–10
2007.08	"Cybernetics and Design" compiled and edited special double issue of Kybernetes, vol 36 nos 9–10, 443 pp
2007.09	"Design Prepositions" in Belderbos, M, and Verbeke, J (Eds.) (2007) "The Unthinkable Doctorate" Brussels, Sint Lucas, pp 115–126
2007.10	"Design, the User, and Klaus Krippendorff's 'The Semantic Turn'" Cybernetics & Human Knowing vol 14 no 4, pp 107–12
2007.11	"Designing Complexity" Performance Improvement Quarterly, vol 20 no 2, pp 75–96
2007.12	"Editorial: Ninety Years of Constructing" (with Riegler, A) in Glanville, R and Riegler A (Eds.) "Ernst von Glasersfeld, a Festschrift" Constructivist Foundations vol 2 nos 2-3 http://www.univie.ac.at/constructivism/journal/2.2/
2007.13	"Editorial: Ninety Years of Constructing" (with Riegler, A) in Glanville, R and Riegler A (Eds.) The Importance of being Ernst" Vienna, edition echoraum, pp 11–20
2007.14	"Ernst von Glasersfeld, a Festschrift" compiled and edited with Riegler, A Constructivist Foundations vol 2 nos 2–3, http://www.univie.ac.at/constructivism/journal/2.2/, 146 pp
2007.15	"Executive Report, Die Planung / A Terv, no 247 06/07 2048" in Die Planung / A Terv Berlin/Budapest, pp 27–30 (published under a pseudonym)
2007.16	"Five Machines and One Pask" http://www.maverickmachines.com/WordPress/wp-content/uploads/2007/07/ranulp hglanville.pdf
2007.17	"Foreword" (with Müller, KH) in Glanville, R and Müller, KH (Eds.) Gordon Pask, Philosopher Mechanic: and introduction, Vienna, edition echoraum, pp 6–10
2007.18	"Foreword" in Jung, R (2007) Experience and Action—selected items in systems theory, Vienna, edition echoraum, pp 11–14
2007.19	"Gordon Pask (ISSS Luminary)" in Glanville, R and Müller, K (Eds.) "Gordon Pask, Philosopher Mechanic: an introduction" Vienna, edition echoraum, pp 53–63
2007.20	"Gordon Pask, Philosopher Mechanic: an Introduction" edited with Müller, KH, Vienna, edition echoraum, 237 pp
2007.21	"Grounding Difference" in Müller, A and Müller, KH, "An Unfinished Revolution? Heinz von Foerster and the Biological Computer Laboratory 1958-1976" Vienna, edition echoraum, pp 361–406
2007.22	"Guest Editorial" in Glanville, R (Ed.) (2007) Cybernetics and Design, special double issue of Kybernetes, vol 36 no 9–10, pp 1153–1157
2007.23	"List of Publications (Gordon Pask)" (with Scott, BCE) in Glanville, R and Müller, K (Eds.) "Gordon Pask, Philosopher Mechanic: and Introduction" Vienna, edition echoraum, pp 201–218

2007.24	"Nothing Compares" e-Wisdom 4 Construction Comparisons/Comparing Constructions, pp 25–32
2007.25	"Personal Wonder" in Müller, A and Müller, KH, "An Unfinished Revolution? Heinz von Foerster and the Biological Computer Laboratory 1958-1976" Vienna, edition echoraum, pp 131–141
2007.26	"Still" Audio-visual piece presented at BEAP, Perth Australia, September 11 to 16
2007.27	"The Edge of Stillness" in Worden, S (Ed.) "Proceedings of the Conference on Computers in Art and Design Education" Perth, Curtin University Press (published on CD) pp 68–74, and also on the web at http://cedar.humanities.curtin.edu.au/conferences/cade/pdf/CADE_STILLNESS.pdf
2007.28	"The IFSR, Diagrams and Inclusive Logic" Cybernetics & Human Knowing vol 13 nos 3–4, pp 135–143
2007.29	"The IFSR, Diagrams and Inclusive Logic" The Systemist vol 29 no 3, pp 168–176
2007.30	"The Importance of being Ernst" edited with Riegler, A, Vienna, edition echoraum, 507 pp
2007.31	"The Importance of Being Ernst" in Glanville, R and Riegler, A (Eds.) (2007) "Ernst von Glasersfeld, a Festschrift" Constructivist Foundations vol 2 nos 2–3, http://www.univie.ac.at/constructivism/journal/2.2/
2007.32	"The Importance of Being Ernst" in Glanville, R and Riegler, A (Eds.) (2007) "The Importance of being Ernst" Vienna, edition echoraum, pp 29–35
2007.33	"Try again. Fail again. Fail better. The cybernetics in design and the design in cybernetics" in Glanville, R (Ed.) (2007) "Cybernetics and Design" special double issue of Kybernetes, vol 36 nos 9–10, pp 1173–1206
2007.34	"What Makes the Difference—Reflecting on Reflecting (on Reflecting on Reflecting)" in Janssens, N and Verbeke, J (Eds.) (2007) "Reflections + 3 (Research Training Seminars 2006)" Brussels, Sint Lucas
2007. 35	"Alvar Aalto, Ross Ashby", "Timothy Berners-Lee", "Bill Gates", "Alfred Jarry", "Steven Jobs", "Le Corbusier", "Gordon Pask", "George Gilbert Scott", "Heinz von Foerster", in Wintle, J (Ed.) "New Makers of Modern Culture", Abingdon, Routledge

2008

2008.01	"A (Cybernetic) Musing: Five Friends" Cybernetics & Human Knowing, vol 15 nos 3–4
2008.02	"A Cybernetic Serendipity" in Glanville, R and Müller, A (Eds.) "Pask Present" Vienna, echoraum
2008.03	"All the 8's" Cybernetics and Human Knowing, vol 15 no 1, pp 86–94
2008.04	"A (Cybernetic) Musing: All the 8's" in Glanville, R and Müller, A (Eds.) "Pask Present" Vienna, echoraum
2008.05	"Blind" Electronic music, performed at the Melbourne Festival, October 2008
2008.06	"Conversation and Design" in Luppicini, R (Ed.) "Handbook of Conversation Design for Instructional Applications" Hershey, NY, Information Science Reference

Appendices 311

2008.07	"I have nothing to say", Mixed media piece the Cybernetic Coalition, presented echoraum, Vienna, November 2008
2008.08	"Introduction" in Glanville, R and Müller, A (Eds.) "Pask Present" Vienna, echoraum
2008.09	"Pask at the Centre" in Trappl, R (Ed.) (2008) "Cybernetics and Systems '08" Procs 19 EMCSR, ÖSGK. Vienna
2008.10	"Pask Present" (exhibition catalogue) edited with Müller, A, Vienna, edition echoraum, 196 pp
2008.11	"Slow" in Glanville, R and Müller, A (Eds.) "Pask Present" Vienna, echoraum
2008.12	"Slow" Video piece presented at Pask Present Exhibition, Vienna, March 24 to April 4

2009

2009.01	"39 Steps", volume 3 of "The Black Boox", selected papers, Vienna, edition echoraum
2009.02	"A Cybernetic Development of Theories of Epistemology and Observation, with reference to Space and Time, as seen in Architecture" (PhD Thesis) published on EThOS as http:// ethos.bl.uk/ OrderDetails.do?did=2&uin=uk.bl.ethos.456747
2009.03	"Architecture and Space for Thought" (PhD Thesis) published on EThOS as http://ethos.bl.uk/OrderDetails.do?did=1&uin=uk.bl.ethos.232936
2009.04	"A (Cybernetic) Musing: Black Boxes" Cybernetics & Human Knowing, vol 16, nos 1–2
2009.05	"A (Cybernetic) Musing: Design and Cybernetics" Cybernetics & Human Knowing vol 16 no 3–4
2009.06	"Problems wicked and undecideable" in Burry, M and Cutler, T, "Designing Solutions to Wicked Problems; A Manifesto for Transdisciplinary Research and Design", Melbourne, RMIT Press
2009.07	"Reflecting and Acting: reflecting on acting and acting on reflecting", in Janssons, N and Verbeke, J (Eds.), Reflections 79, Brussels, Sint Lucas

2010

2010.01	"A (Cybernetic) Musing: Architecture of Distinction and the Distinction of Architecture", Cybernetics and Human Knowing, vol 17 no 3
2010.02	"ASC 2009" (edited, conference proceedings with Guddemi, P), Cybernetics & Human Knowing, vol 17, nos 1–2
2010.03	"ASC 2009" (with Guddemi, P, Fischer, T and Kauffman, L), Cybernetics & Human Knowing, vol 17, nos 1–2
2010.04	"Constructing the Photograph" in Roxburgh, M "Light Relief (Part II)", Sydney, UTS Press
2010.05	"Creativity, Design, Education" in Williams, A, Ostwald, M and Askland, HH (Eds.) "From theory to practice—39 opinions", Sydney, ATLC
2010.06	"Design and Mentation—Piaget's Constant Objects", The Radical Designist Book, Corte-Real, E (Ed.),
2010.07	"Doing the Cybernetics of Cybernetics. Cybernetics: Art, Design, Mathematics—A MetaDisciplinary Conversation", Cybernetics & Human Knowing, vol 17 no 3

2010.08	"Error" in Trappl, R (Ed.) (2008) "Cybernetics and Systems '10" Procs 20 EMCSR, ÖSGK, Vienna
2010.09	"Introduction: ASC 2009" (with Guddemi, P), Cybernetics & Human Knowing, vol 17, nos 1–2
2010.10	"Quasi Entailment Mesh" (with Pak, B) in Trappl, R (Ed.) (2008) "Cybernetics and Systems '10" Procs 20 EMCSR, ÖSGK. Vienna
2010.11	"Reflecting", in Verbeke, J (Ed.) "Reflections 11", Brussels, Sint Lucas

2011

2011.01	"A (Cybernetic) Musing: The Boundaries of Distinction? The Distinction of Boundaries? ", Cybernetics and Human Knowing vol 18 nos 1–2
2011.02	"Cybernetic Circles—The Black Box volume 1", selected papers with 7 new essays, Vienna, echoraum|WISDOM
2011.03	"Introduction: a conference doing the cybernetics of cybernetics", Kybernetes vol 40 nos 7–8
2011.04	"Notes from the Reflecting Workshop, St Lucas, June 23–25, 2011" (with Hohl, M), in Verbeke, J (Ed.) "Reflections 16", Brussels, Sint Lucas
2011.05	"Obituary: Ernst von Glasersfeld 1917-2010 ", Cybernetics & Human Knowing vol 18 nos 1–2
2011.06	"To Be Decided", Cybernetics and Human Knowing vol 18 nos 3–4
2011.07	"Special Issue: Cybernetics: Art, Design, Mathematics—A Meta-Disciplinary Conversation" Kybernetes, vol 40 nos 7–8

2012

2012.01	"A (Cybernetic) Musing: Cybernetics and Circularities" Cybernetics & Human Knowing vol 19 no 4
2012.02	"A (Cybernetic) Musing: Wicked Problems", Cybernetics & Human Knowing vol 19 no 1
2012.03	"A Dialogue in Design" (with Kauffamn, L) in Glanville, R (Ed.) "Trojan Horses", Vienna, echoraum|WISDOM
2012.04	"Creativity and Learning in Architecture and Design", in Askland HH, Ostwald M, Williams AP, (Eds.) "Assessing Creativity: Supporting Learning in Architecture and Design", Sydney, Australian Government Office for Learning and Teaching
2012.05	"Cybernetics of Cybernetics" in Glanville, R (Ed.) "Trojan Horses", Vienna, echoraum|WISDOM
2012.06	"Klaus Krippendorff, a Directory", (compiled and edited), web publication by American Society for Cybernetics
2012.07	"Making invisible the visible" in Hohl, M (Ed.) "Making visible the invisible: Art, design and science in data visualisation", Huddersfield, University of Huddersfield Press
2012.08	"Prelude and Foreword" in Glanville, R (Ed.) "Trojan Horses", Vienna, echoraum| WISDOM
2012.09	"Radical Constructivism = Second-order Cybernetics", Radical Constructivism = Second-order Cybernetics", Cybernetics & Human Knowing vol 19 no 4
2012.10	"Trojan Horses" (compiled, edited and contributed), Vienna, echoraum|WISDOM

2013

2013.01	"A (Cybernetic) Musing: Anarchy, Alcoholics Anonymous and Cybernetics: Chapter One", Cybernetics and Human Knowing vol 20 nos 3–4
2013.02	"An Ecology of Ideas: Guest Editorial" (with Griffiths, D) Kybernetes vol 42 nos 9–10
2013.03	"Cybernetics", in Arnold, D (Ed.) "Traditions of Systems Theory. Major Figures and Contemporary Developments" New York and London, Routledge
2013.04	"Cybernetics of Cybernetics Competition: The Official Guidelines", Cybernetics and Human Knowing vol 20 nos 1–2
2013.05	"Introduction to the Unrefereed Papers" (with Griffiths, D) Kybernetes vol 42 nos 9–10
2013.06	"Listening: Proceedings of ASC Conference 2011", (with Delfina Fantini van Ditmar, Cybernetics and Human Knowing vol 20 nos 1–2
2013.07	"Special Issue: An Ecology of Ideas" (compiled and edited, with Griffiths D), Kybernetes vol 42 no s 9–10
2013.08	"Ülikool annab aega enda harimiseks" ("University gives you the time to educate yourself"), Interview with Vaikla, T-K, Maja 1–013
2013.09	"When Good Enough is Better than Best" paper presented to Dupont Circle Conference, Washington DC, December 2013. Web version at https://www.youtube.com/watch?v=m-8GmlNb6EY

2014

2014.01	"A (Cybernetic) Musing: Anarchy, Alcoholics Anonymous and Cybernetics: Chapter Two", Cybernetics and Human Knowing vol 21 nos 1–2
2014.02	"Interview—On the Individual" in Joekalda, J, Tali, J and Tiksam, S (2014) "Interspace, Essays on the Digital and the Public", Architecture Centre of Estonia and Lugemik, Tallinn (bilingually published in Estonian under the same imprint as "Vapa Ruum")
2014.03	"Introduction to Acting, Learning Understanding Conference Proceedings" Kybernetes vol 43 nos 9–10
2014.04	"Introduction", in Goldt, K. (in press),"The Bibliography of Gregory Bateson's Steps to an Ecology of Mind in Alphabetic Order of the Titles Part 1: A – O"
2014.05	"Living in Cybernetic Circles—the Black Boox volume 2", more selected papers including 7 new essays, Vienna, echoraum|WISDOM
2014.06	"Loovusele tuleb anda ruumi ja jõudu", ("We should give space and strength to foster creativity') Interview with Valk, V, Sirp, no 26, 4 July 2014 at http://www.sirp.ee/index.php?option=com_content&view=article&id=22311:2014-07-03-19-44-28&catid=20:arhitektuur&Itemid=25&issue=3496
2014.07	"Special Issue: Acting, Learning, Understanding", (compiled and edited, with Griffiths, D and Baron, P) Kybernetes vol 43 nos 9–10
2014.08	"Stuart Umpleby, a Directory", (compiled and edited with Karl H Müller and Pille Bunnell), web publication by American Society for Cybernetics

2014.09	"The Black Boox", 3 volume limited edition signed boxed set, Vienna, echoraum|WISDOM
2014.10	"The Sometimes Uncomfortable Marriages of Design and Research", in Rodgers, P and Yee, J (Eds.) (in!press) "Design0Research", London, Routledge

In Press and Pending

"Comparison II" (with Müller, K.) commissioned paper
"Comparison I" commissioned paper
"From (art) object to (art) observer" (2004) to be published in MetaMute
"Innovation, Creativity, Intelligence, Cybernetics" Deakin lecture transcript, Melbourne, delivered May 2005 (web version on http://www.deakinlectures.com/transcripts/2005.php)
"Notes towards a Group Design" keynote lecture at the sixth meeting of the European Academy of Design, Bremen, April 2005, to be published in the proceedings
"The Object of Objects, the Point of Points—or, something about Things", PhD Thesis (1975), and "Architecture and Space for Thought", PhD Thesis 1987, being scanned by the British Library for their collection of scanned PhDs.
"Why are our Aliens Anthropoids?" to be published in Intelligent Agent.
"A Gordon Pask Reader: texts from the cybernetician's cybernetician", (with Scott, BCE and Pangaro, P) edition echoraum, Vienna.

Other Publications

Eulogy for Robin McKinnon-Wood, Windsor, 1995
Conference Eulogy for Gordon Pask, Vienna, 1996
Obituary for Frank Tindall, *The Guardian*, London, 1998
Memorial address for Alfred Janes, Royal Society of Arts, London, 1999
Valete Gerard de Zeeuw, Amsterdam, 2001
Eulogy for Elizabeth Pask, London, 2001
Cecil Peters; an Unofficial Biography, Southsea, 2001
Obituaries for Heinz von Foerster (5 in total: various publications, some appearing as publications, above) Learning with Locker invited memorial tribute, 2005
Obituary for Ernst von Glasersfeld, *Cybernetics & Human Knowing*, 2011
Eulogy for John Allan, 2011
Celebration of 40th birthday of the journal, *Kybernetes*, 2011
Eulogy for Rodie Peters, September 2013

Assorted pieces and reports, *Architects' Journal*, *Building Design Journal*, London and elsewhere, and on Web Site homepage.mac.com/ranulph/Filesharing1.htm

Appendix 3:
Research Experience

Ranulph Glanville
as of 24 August 2014

1970	Technical Evaluation and Review of "Domino" Timber Building System for oy Alstrom ab, Varkaus, Finland
1971–1975	Research for Ph D (Cybernetics) under Prof. Gordon Pask at Brunel University
1972	Experimental Design "The Effect of Learning Style on Performance under Stress" for System Research Ltd., under contract to USAF
1972–1978	Development of Theory of Model Dimensions and its Realisation in a Computer Programme with PJA
1973–1991	Research into design education and spatial perception at Architectural Association and Portsmouth Polytechnic (now the University of Portsmouth)
1974–now	Research into fundamental cybernetic Objects at Brunel University, Architectural Association and Portsmouth Polytechnic
1976	Research into active analogue memories with Professor R Gregory, Brain & Perception Laboratory, Bristol University
1976–1988	Research for Ph D (psychology) under Dr. L Thomas at Centre for the Study of Human Learning, Brunel University
1978–1988	Research into theories of observation and representation, with special reference to Black Box Theory, Portsmouth Polytechnic
1979	Development of software design for Architect's Offices with The OFFICE
1980–now	Research into the relationship pertaining between cybernetics, design & science, Portsmouth Polytechnic
1981–1985	Initiation of, making and directing students in making research videos about research methodology in Architecture and the Social Sciences, with applications, University of Amsterdam and Portsmouth Polytechnic
1984–1988	Development of notions +, 0 and – space, Portsmouth Polytechnic
1985–1993	University of Amsterdam: member of development team for the Dutch Government funded research programme, OOC Program ("Support, Survival and Culture"), and project member consultant and advisor, Ph D selection and supervision panel
1985–1989	Research into "Impossible Worlds", University of Amsterdam, Portsmouth Polytechnic
1986	Research into children's perception of home, Portsmouth Polytechnic
1991	Senior Resident Research Fellow, OOC Program ("Support, Survival and Culture"), University of Amsterdam, The Netherlands
1991–now	Research into the Development of the Uses of Computing as a Medium: and the Comparison of different Architectural Media of Representation, University of Portsmouth (and various informal partners abroad)

1992–now	Investigation of Sharing, Design and Creativity Amplification through Computing, University of Portsmouth (and various informal partners abroad)
1993–1995	Implementations of Aspects of Pask's Learning Theory (ThoughtSticker/CASTE), University of Portsmouth
1993	Cross-platform Use of Computers in the Development of Architectural Design Ideas, Bialystok Polytechnic University, Poland
1994–1997	Survey of Use of Representational Media in Architecture, University of Portsmouth, carried out with Universities abroad
1994	Assessment of Full Scale Laboratory, at Vienna University of Technology, University of Portsmouth
1995	Comparison of Effectiveness of CAD and Endoscopic Models in Developing and Assessing Design Proposals, at Tampere University of Technology, University of Portsmouth
1995–1997	Origination, Development and Co-ordination of Research Application to European Union (Training and Mobility of Researchers Programme under the 4th Directive) on "The Use of Computers to Enhance Creativity", University of Portsmouth and de Montfort University, together with the following: Liverpool John Moore's University (all in the UK), Technical Universities of Tampere (Finland), Lund (Sweden), Vienna (Austria), Chalmers Technical University (Sweden), Sint Lucas (Belgium), Centre for Innovation and Co-operative Technology (Holland); IBM UK scientific research centre and KOZO (Europe) ab; and architectural practices. The application (not funded) also involved a number of workshops on, eg, methodology and research in design and the arts.
1996–1997	Advise, Direct and Implement Research programme on the Organisation of Accounts of Architectural Designs at de Montfort University, with the School of Architecture and Educational Technology Group.
1999–2001	Development of ACCOLADE programme (workshop conference and publication) as industrial partner, in collaboration with the Saint Lucas Architecture School, Brussels, with funding from the European Union.
2001–2003	Marriage of ACCOLADE to form a super consortium bidding for EU funds.
2004–2006	Advisor and theorist to the Working and Learning in Vocational Education and Training in the Knowledge Era. Final Report of the Professional Development for the Future Project, Australian Flexible Learning Framework.
2005	MIN(e)D: application to EU for a programme to develop Knowledge for (action/change), to complement the normal output of research, Knowledge of (what is).
2005	Member of team developing strategies for architectural innovation based in local sources, with Prof Leon van Schaik (RMIT) and Prof Geoffrey London (Government Architect, Western Australia).
2006–	Advisor on Research Training Sessions, and on a number of research programs, Sint Lucas Architectuur, University of Leuven, Brussels and Ghent.
2006–2013	Advisor to Design Research Institute, RMIT University, Melbourne, Australia. Personal advisor to Prof Mirk Burry, director and Ms Katherine Wilkinson, executive director.

2009–2013	Member of Interdisciplinary Panel, FWO (Research Organisation of Flanders)
2010–2012	Invited to design and advise on provision of research training, and to oversee research projects Hong Kong Design Institute
2010–2013	Invited member, SPIRES Architecture Research Group (British Government Funded)
2012–	Developing a research program on disabled mobility and home, with Treloars School

Several projects have been supported by the British Council. Others have been supported by in house competitive bids or discretionary grants. The OOC Program (University of Amsterdam) was funded by direct grant from the Dutch Government—at the same time, the largest grant ever awarded in the social sciences in the Netherlands.

Appendix 4:
Software Developed

Ranulph Glanville
as of 24 August 2014

- Development of software design for Architect's Offices with The OFFICE (1979)

- Pictures at an Exhibition (with Christopher Ferris)
  A HyperNavigator for CyberSpace (written in HyperTalk for HyperCard for Mac), indexing and connecting drawings developed from each other (1993)

- Slide Catalogue (with Robin McKinnon-Wood)
  A program (written in HyperTalk for HyperCard for the Mac) permitting the loose structuring of a slide catalogue, the construction of lecture notes around slide collections, and searching for slides by keywords (which may be added to the collection at any time) (1994)

- NOAH (with Robin McKinnon-Wood)
  A program (written in HyperTalk for HyperCard for the Mac) embodying and extending key concepts from Pask's "Conversation Theory", aiding authors and learners navigate within a learning environment through the productive relationship of topics to be studied. This program may form the basis of a multimedia authoring tool for CAL and Distance Learning (1995).

- Research Connections
  A proposal and sketch design for a computer programme that searches for similarities in publications and advises authors. Based on Apple computer system software

- Rich View (with Dimitrios Lenis)
  Software to allow multi-linked, multi-authored course work to be developed and studied (2010–)

Appendix 5:
Major Conferences Attended (& Major Sponsors)

Ranulph Glanville
as of 24 August 2014

1968

- Suomen Arkkitehtiliito Finlands Arkitektsförbundet (Finnish equivalent of the RIBA). Architecture & Education, Helsinki

1973

- Design, York, (RIBA sponsored)

1976

- 3rd European Meeting on Cybernetics and Systems Research, Vienna
- 2nd ICIUE., Newcastle, (Centre for the Study of Human Learning sponsored)
- 1st International Conference on Applied General Systems, Recent Developments and Trends, Binghamton, (NATO sponsored)

1977

- 2nd International Congress on Personal Construct Psychology, Oxford, (Centre for the Study of Human Learning sponsored)

1978

- 4th European Meeting on Cybernetics and Systems Research, Linz
- Applications of the Repertory Grid, Uxbridge
- 4th World Organisation for General Systems and Cybernetics., Amsterdam
- 5th Cybernetics Society, Uxbridge

1979

- 24th Society for General Systems Research/American Association for the Advancement of Science, Houston, (Portsmouth Polytechnic sponsored)
- Conference on "Problems of Context", Amsterdam, (Systeemgroep Nederlands sponsored)
- 3rd International Congress on Personal Construct Psychology, Breukelen, (University of Amsterdam sponsored)
- Workshop Conference on Personal Construct Theory, London
- Society for General Systems Research, Silver Jubilee Conference, London
- Cybernetics Society, London,
- 1st Workshop Conference on Self-Reference, Isle of Wight

1980

- European Meeting on Cybernetics and Systems Research, Vienna, (Austrian Government sponsored)
- 2nd Workshop Conference on Self-Reference, Schwarzau-im-Gebirge

- 10th Anniversary Meeting of Systeemgroep Nederlands, Volendam (Systeemgroep Nederlands sponsored)
- 3rd Workshop Conference on Self-Reference, Amsterdam
- 7th Cybernetics Society, London
- Conference of Design Research Society on Design/Science/Method, Portsmouth (video presence) (Portsmouth Polytechnic sponsored)
- 1st International Congress in Applied Cybernetics & Systems Research, Acapulco, (British Academy sponsored)
- 4th Workshop Conference on Self-Reference, Acapulco

1981

- 1st Problems of Levels & Boundaries, Amsterdam
- 5th World Organisation for General Systems and Cybernetics., Mexico City (Mexican Government sponsored)
- 18th American Society for Cybernetics Annual Meeting on "The New Cybernetics" Washington DC

1982

- 27th Society for General Systems Research/American Association for the Advancement of Science, Washington DC (University of Amsterdam sponsored)
- 6th European Meeting on Cybernetics and Systems Research, Vienna, (Portsmouth Polytechnic sponsored)
- 4th United Kingdom Systems Society, Milton Keynes
- 19th American Society for Cybernetics, Columbus, Ohio, USA

1983

- Conference on "Building Utilisation and Assessment", Portsmouth, (SSRC sponsored)
- 20 American Society for Cybernetics, Los Altos, California, (American Society for Cybernetics sponsored)

1984

- 7th European Meeting on Cybernetics and Systems Research, Vienna,
- International Conference on "Systems and Design, Budapest"
- Gordon Research conference on "Fundamentals of Cybernetics", New Hampton, New Hampshire, USA (Gordon Research Foundation sponsored)
- 21st American Society for Cybernetics, Philadelphia, USA (funded by teaching introductory tutorials)

1985

- 11th Cybernetics Society, London,
- Conference on "Problems of Disappearing Knowledge", Amsterdam,, (invited speaker, Systeemgroep Nederlands sponsored)

1986

- 22nd American Society for Cybernetics, Virginia Beach, USA (funded by video of lecture"Cybernetics & Design" and contribution to inductory tutorials)

- 8th European Meeting on Cybernetics and Systems Research, Vienna, (Portsmouth Polytechnic & Austrian Society for Cybernetic Studies sponsored)
- International Association for Cybernetics Namur, Belgium.

1987

- Conference on "Problems of (Im)possible Worlds", Amsterdam (sponsored by University of Amsterdam)
- International Meeting on "Self-Steering and Cognition in Complex Systems—the NewCybernetics", Brussels.

1988

- 9th European Meeting on Cybernetics and Systems Research, Vienna (sponsored by University of Amsterdam)

1989

- Conference on "Mutual Uses of Cybernetics and Science", Amsterdam (sponsored by University of Amsterdam)

1990

- 10th European Meeting on Cybernetics and Systems Research, Vienna (sponsored by Portsmouth Polytechnic)

1991

- Conference on "Problems of Support, Survival and Culture", Amsterdam (sponsored by University of Amsterdam)
- Conference on "Tractatus Cybernetico-Philosophicus Project", Brussels (sponsored by CICT)

1992

- 11th European Meeting on Cybernetics and Systems Research, Vienna (sponsored by University of Amsterdam)
- eCAADE '92, Barcelona (sponsored by British Council)

1993

- Conference on "Problems of Values and (In)variants", Amsterdam, (sponsored by CICT)
- 1st Meeting of the European Architectural Endoscopy Association, Tampere (sponsored by the Dutch Institute of Design)
- Workshop on Models for Human Action, 6th Congress of the International Institute for Advanced Studies in Systems Research and Cybernetics, Baden-Baden (sponsored by CICT)
- eCAADE '93, Eindhoven (sponsored by University of Portsmouth)

1994

- 12th European Meeting on Cybernetics and Systems Research, Vienna (sponsored by British Council and University of Portsmouth)

- Conference on "Projektowanie Wspomagane Komputerowo—Bariery i Inspiracje Twórcze", Bialystok (sponsored by British Council)
- 2nd International Orwellian Symposium, Karlovy Vary, (British Council sponsored)
- 7th Congress of the International Institute for Advanced Study in Systems Research and Cybernetics, Baden-Baden (sponsored by University of Portsmouth)
- Workshop on Emergence, 7th Congress of the International Institute for Advanced Studies in Systems Research and Cybernetics, Baden-Baden (sponsored by British Council)
- eCAADe '94, Glasgow

1995

- ASCA annual Conference, Oxford
- Conference on Problems of Excavating Cybernetics and Systems, Amsterdam (co-chair, sponsored by the Conference)
- Conference "Einstein Meets Magritte", the 25th anniversary conference of the Free University, Brussels, Brussels (invited speaker, scholarship from Free University, Brussels)
- 6th Congress of International Institute for Advanced Study in Systems Science, Baden-Baden (sponsored by the University of Portsmouth)
- 2nd Meeting of European Architectural Endoscopy Association, Vienna (sponsored by the University of Portsmouth)
- 15th Meeting of Association for Computing in Architectural Design in America (ACADIA), Seattle (sponsored by the University of Portsmouth)
- 13th Meeting of European Computer Aided Architectural Design in Education, Palermo

1996

- 13th European Meeting on Cybernetics and Systems Research, Vienna (sponsored by University of Portsmouth)
- Conference on "CAD Creativeness", Bialystok (sponsored by British Council)
- 14th Meeting of European Computer Aided Architectural Design in Education, Lund

1997

- The "Global Village '97" Conference, Vienna, 1997 (invited speaker, sponsored by the conference)
- Conference on Problems of Actions and Observations, Amsterdam, 1997 (co-chairman, sponsored by the conference)
- Conference on the Added Value of Computer Aided Architectural Design, Brussels, 1997 (keynote speaker, sponsored by the conference)
- Consciousness Reframed I, Newport, Wales, 1997

1998

- Designing Design Research 2, Meeting of the Design Research Society, Leicester, 1998 (invited speaker)
- Meeting on Design Intelligence, Melbourne, 1998 (chair, sponsored by RMIT University)
- Consciousness Reframed II, Newport, Wales, 1998 (sponsored by RMIT University)

1999

- Conference on Problems of Connection and Participation, Amsterdam, 1999 (co-chairman, sponsored by the conference)
- Invencao, Sao Paolo, 1999 (sponsored by the British Academy, the Department of Trade and Industry)
- ACADIA, Snowbird Utah, 1999

2000

- Chatterbox Conference, ISEA 2000 web (invited speaker)
- ACCOLADE European Workshop Conference, Brussels 2000 (invited speaker, sponsored by conference/European Union)
- Consciousness Reframed III, Newport, Wales 2000 (RMIT University sponsored) 2001

- Conference on Problems of Individual Emergence, Amsterdam, 2001 (co-chairman, sponsored by the conference)

2002

- Consciousness Reframed 2002, in association with the Biennial of Electronic Arts, Perth, Perth Western Australia 2002 (sponsored by RMIT)
- Heinz von Foerster Symposium Vienna 2002 (invited speaker, sponsored by the City of Vienna)
- Second-Order Cybernetics, Cybernetics Society Conference, London, 2002 (invited speaker)
- Cities Edge, Melbourne, Australia, 2002.

2003

- Research on Research—Practice-Based Research: Recognition, Relevance, Rigour, Melbourne, Australia, 2003 (keynote speaker, sponsored by RMIT)
- 15th Congress of the International Institute for Advanced Study in Systems Research and Cybernetics, Baden-Baden, invited speaker, 2003
- 3rd Doctoral Education in Design Conference, Tsukuba/Tokyo, 2003
- Conference of the Heinz von Foerster Society in Memoriam Heinz von Foerster, Vienna, keynote speaker, sponsored by the City of Vienna 2003

2004

- 17th European Meeting on Cybernetics and Systems Research, Vienna, keynote speaker, symposium founder and chair, 2004
- 5th Annual Conference of the Foundation 2020, Brijuni, Croatia, invited speaker and panel; member, sponsored by Foundation 2020, 2004
- American Society for Cybernetics, Toronto, keynote speaker, 2004
- FutureGround, International Conference of the Design Research Society, Melbourne, committee member, 2004

2005

- European Academy of Design 06, Bremen, Germany, keynote speaker, session chairman, 2004
- The Unthinkable Doctorate, Brussels, keynote speaker, session chairman, 2004

- American Society for Cybernetics, Washington DC, keynote speaker, session chairman, 2004
- Conference of the Heinz von Foerster Society on Constructivism, Vienna, keynote speaker, sponsored by the City of Vienna, 2005
- Finding Fluid Form, Brighton, conference of AHRB/ERC group, 2005
- More is More, London, conference of AHRB/ERC group, keynote speaker, sponsored by AHRB/ERC, 2005

2006

- 18th European Meeting on Cybernetics and Systems Research, Vienna, symposium founder and chair
- Annual Conference of the American Society for Cybernetics, Champaign Urbana, IL, USA, judge, workshop
- Annual Meeting of UK Systems Society, keynote speaker

2007

- Conference of the Heinz von Foerster Society on Conversation and Gordon Pask, Vienna, keynote speaker, sponsored by the Austrian Government and the City of Vienna
- 11th World Multi-Conference on Systemics, Cybernetics and Informatics, Orlando, keynote speaker

2008

- 12th World Multi-Conference on Systemics, Cybernetics and Informatics, Orlando, keynote speaker
- 19th European Meeting on Cybernetics and Systems Research, Vienna, symposium founder and chair
- Annual Conference of the American Society for Cybernetics, Champaign Urbana, IL, USA, judge, workshop

2009

- 13th World Multi-Conference on Systemics, Cybernetics and Informatics, Orlando, keynote speaker
- Annual Conference of the American Society for Cybernetics, Olympia, WA, USA, conference chair, best paper judge
- Annual Meeting of UK Systems Society, keynote speaker and respondent
- CADFutures, Montreal, Canada, keynote speaker
- Conference of the Heinz von Foerster Society on Learning, Creativity and the Interdisciplinary, Vienna, keynote speaker, sponsored by the Austrian Government and the City of Vienna
- Conference on Communication (by) Design, St Lucas Brussels, session chairman and wind up
- Workshop on Designing Solutions to Wicked Problems, Design Research Institute, Melbourne, facilitator

2010

- 1st conference of the International Academy for Cybernetic and Systems Sciences, Chengdu, China, keynote speaker
- 14th World Multi-Conference on Systemics, Cybernetics and Informatics, Orlando, keynote speaker
- 20th European Meeting on Cybernetics and Systems Research, Vienna, symposium founder and chair
- Annual Conference of the American Society for Cybernetics, Troy, NY, USA, "Cybernetics: Art, Design, Mathematics—A MetaDisciplinary Conversation" conference chair and organiser, best paper judge
- Annual Meeting of the International Society for Systems Science, Waterloo, Canada, keynote speaker
- Design for Social Business conference, Milan, invited attendee

2011

- 15th World Multi-Conference on Systemics, Cybernetics and Informatics, Orlando, keynote speaker
- Annual Conference of the American Society for Cybernetics, Richmond, Indiana, USA, conference chair and organiser, best paper judge
- Heinz von Foerster Centenary Conference, keynote speaker
- Design for Social Business conference, Barcelona, invited attendee

2012

- 16th World Multi-Conference on Systemics, Cybernetics and Informatics, Orlando, keynote speaker
- Annual Conference of the American Society for Cybernetics, Asilomar, California, USA, conference chair and co-organiser, best paper judge
- Design for Social Business conference, Milan, invited attendee
- Annual Conference of the United Kingdom Systems Society, keynote speaker

2013

- Annual Conference of the American Society for Cybernetics, Bolton, UK, conference chair and organiser, best paper judge
- Dupont Circle Meeting, Washington DC, invited speaker
- DFG Roundtable: Design Research in Germany and in International Comparison, Berlin, invited participant
- Conference on Research by Design, St Lucas Brussels, session chairman and wind up

2014

- Fuschl Meeting of International Federation for Systems Research, invited attendee
- Conference on Designing and Researching, keynote speaker
- American Society for Cybernetics, Washington DC, chair, organiser and speaker Best paper judge. Presented ASC President's farewell address.. Recipient of ASC Special Award

Appendix 6:
Main (Regular) Teaching Experience

Ranulph Glanville
as of 24 August 2014

- O and A level Art (including Architecture), Bryanston School, Blandford, Dorset
- First Year Architecture Tutor, Cambridge University
- First Year Architecture Tutor (Urban Studies), Intermediate School Unit Master, Technical Studies Tutor, Architectural Association School
- First Year Architecture Tutor (Urban Studies), Intermediate School Unit Master, Technical Studies Tutor, Architectural Association School
- Visiting Student Special Tutor, Architectural Association School
- Special Projects Organiser and Tutor, Architectural Association School, London
- Year One and Year Two Tutor, Year One Co-ordinator and Tutor, Year Five Option Co-ordinator and Tutor, Year Six Tutor (Thesis), University of Portsmouth
- Dissertation Tutor, University of Portsmouth
- Computer Introducer and Tutor, Architectural Association School
- Computer Introducer and Tutor, University of Portsmouth
- Computer Introducer and Tutor, University of Portsmouth
- Foundation Course Tutor, Portsmouth College of Art, Design and Further Education
- Studio teacher, Graphics Masters Degree, Central St. Martin's School of Art, London
- Technical Dissertation Tutor and Course Organiser, the Bartlett School of Architecture, University College London
- Masters and Doctoral Programme, School of Architecture + Design, Royal Melbourne Institute of Technology University
- Architectural Studies Dissertation Tutor and Course Organiser, the Bartlett School of Architecture, University College London.
- Special Research Study Tutor, Masters of Architecture Programme, the Bartlett School of Architecture, University College London.
- Research senior tutor and staff research mentor, Innovation Design Engineering, Royal College of Art, London.
- Working recently as a consultant, I have gained considerable experience in visiting Universities and designing, organising and directing workshops, and working with Industry and Government.

Appendix 7:
Lecture Venues

Ranulph Glanville
as of 24 August 2014

United Kingdom

- Architectural Association, London
- Bartlett School of Architecture, University College, London
- Bristol University, Brain and Perception Laboratory and School of Architecture
- British Broadcasting Corporation
- Brunel University, Uxbridge, Departments of Cybernetics and Psychology
- Cambridge University, Department of Architecture
- Cybernetics Society, London
- De Montfort University, School of Architecture, School of Design
- Independent Television
- Goldsmiths College, University of London, Department of Art and Design
- Liverpool John Moore's University, Division of Architecture and Planning
- Newcastle University, Department of Architecture
- North London Polytechnic, School of Environmental Studies
- Open University, Systems Group, and Arts Faculty
- Polytechnic of Central London, Architecture Department
- Portsmouth Polytechnic, Department of Biology
- Portsmouth Polytechnic, School of Architecture
- Portsmouth Polytechnic, School of Information Science
- Ravensbourne College, Interactive MultiMedia Programme
- Reading University, Department of Cybernetics
- Royal College of Art, Innovation Design Engineering
- Royal Chartered Society of Civil Engineers, London
- Sheffield University, Department of Architecture
- Ulster Polytechnic, Belfast, Department of Art History, Criticism and Research (ex Northern Ireland Polytechnic)
- University College London, London
- University of Dundee, Duncan of Jordanstone College School of Architecture
- University of Hull, Centre for Systems Science
- University of Humberside, School of Architecture
- University of Northumbria, School of Design
- University of Nottingham, School of Architecture
- University of Portsmouth, School of Architecture
- University of Wales College, Centre for Interactive Arts

Europe

- Aalto University Helsinki, Finland
- Aarhus University, Faculty of Media and Technology, Denmark
- Åbo University of Economics and Management, Finland
- Amsterdam University, Centre for Innovation & Co-operative Technology, The Netherlands

- Amsterdam University, Subfaculty of Andragology, The Netherlands
- Bialystock Polytechnic University, Faculty of Architecture, Poland
- Budapest College of Art, Hungary
- Delft University of Technology, Department of Computing, The Netherlands
- Ecole de Beaux Arts, Paris, France
- Estonian Academy of Arts
- Free University, Amsterdam, Psychology Department, The Netherlands
- Graz University of Technology, Department of Architecture and Planning, Austria
- Hamburg University, School of Journalism and Management
- Hebbel am Ufer Theatre, Berlin, Germany
- Helsinki Technical University, Architecture Department, Finland
- Humbolt University, Berlin, Faculty of Aesthetics, Germany
- inform ab/Swedbank, Stockholm, Sweden
- Joensuu Technical University, Department of Urban Geography, Finland
- Johannes Keppler University, Linz, Department of Systems Engineering, Austria
- Lisbon Institute of Technology, Department of Mechanical Engineering, Portugal
- Oulu University, Architecture Department, Finland
- Polish Academy of Sciences, Warsaw, Poland
- Rijeka Town Hall, Croatia, Public Lecture Series
- Royal Art Academy, Hilversum, The Netherlands
- Royal Danish Academy of the Arts, School of Architecture, Copenhagen, Denmark
- Royal Veterinary University, Copenhagen, Denmark
- School of Cybernetic Psychotherapy, Riejka, Croatia
- Sint Lucas Hoogeschool, Faculty of Architecture, Brussels, Belgium
- Systeemgroep Netherlands, Utrecht and Amsterdam, The Netherlands
- TAIK, University of the Arts, Helsinki, Finland
- Tampere University, Department of Architecture and Planning, Finland
- Technical University Berlin, Department of Multimedia, Germany
- Technical University Delft, Department of Architecture, The Netherlands
- University of Antwerp, School of Architecture
- University of the Aegean, Department of Design Systems, Syros, Greece
- University of Oslo, School of Architecture and Design, Oslo, Norway
- University of Vienna, Heinz von Foerster Lecture
- University of Vienna, Department of Contemporary History
- Vienna University of Technology , Department of Architecture and Planning, Austria
- Vienna University of Technology , Department of Theoretical Biophysics, Austria
- Warsaw University of Technology, Faculty of Architecture, Poland
- Witten/Hardecke University, Faculty of Fundamental Studies, Germany
- Wroclaw University of Technology, Department of Mechanical Engineering and Systems Cybernetics, Poland

The Americas
- Antioch University, Seattle, School of Architecture
- Bennington College, Bennington
- Columbia University, New York, School of Architecture
- Concordia University, Montreal, Institute of Educational Technology
- Cooper Union, New York, School of Architecture
- Duke University, Rayleigh Durham, Department of Art Theory and History
- Educational Technology Research Group

- George Washington University, Washington DC, School of Government and Business Management
- Georgia Institute of Technology, Atlanta, Department of Systems Studies
- Human Interface Technology Laboratory of the University of Washington, Seattle
- Indiana University West, Richmond Indiana, Department of the Executive Vice Chancellor
- Iona College, New Rochelle, Department of Business Management
- Massachusetts Institute of Technology, Architecture Machine Group
- Oakland University, Rochester, Michigan, Department of Business
- Old Dominion University, Norfolk, Department of Systems Engineering
- Ontario College of Art and Design, Digital Futures Initiative, Toronto, Canada
- Rensselaer Polytechnic Institute, School of Architecture
- San Jose State University, Cybernetics & Systems Program
- San Jose State University, Music Department
- State University of New York, Binghampton, Systems Science Department
- State University of New York, Geneseo, Philosophy Department
- Torcauto de Tella University, Buenos Aires, Argentina, Contemporary Architecture Centre
- University of Buenos Aires, Argentina, CAO Centre for Computer Aided Design
- University of Illinois, Urbana, Biological Computer Laboratory
- University of Massachusetts, Amherst, Department of Astronomy and Physics
- University of Mexico, Mexico City, Department of Education and
- University of Ohio, Department of Education
- University of Oregon, Eugene, Department of Architecture
- University of Pennsylvania, Philadelphia, Department of Architecture
- University of Texas at Austin, Austin, School of Architecture
- University of Zulia, Maracaibo, Venezuela, Department of Architecture

Africa

- Cape Town University, Cape Town, Republic of South Africa
- University of Garyounis, Benghazi, Libya
- University of Lesotho, Maseru, Lesotho
- University of Natal, Durban, Republic of South Africa
- University of the Orange Free State, Bloemfontein, Republic of South Africa
- University of the Witwatersrand, Johannesburg, Republic of South Africa

Australasia and the Far East

- Chinese University of Hong Kong, Hong Kong; School of Architecture and Shaw College
- Curtin University, Perth, Australia, Faculty of the Humanities
- Deakin University, Geelong, Australia, School of Architecture
- Department of Education and Training, Government of New South Wales, Australia
- Hong Kong Design Institute, Hong Kong
- Hong Kong University, Faculty of Architecture and Building
- Melbourne Town Hall, Deakin Lecture, Australia
- Monash University, Faculty of Art and Design
- Polytechnic University of Hong Kong, Hong Kong, Department of Design
- Public Services Agency, Sydney, Australia

- Royal Melbourne Institute of Technology University, Melbourne, Australia, School of Architecture + Design
- University of Canberra, Canberra, Department of Design and Architecture
- University of Newcastle, Newcastle, Australia, School of Architecture
- University of Newcastle, Newcastle, Australia, School of Design
- University of New South Wales, Faculty of the Built Environment and Design
- University of Technology, Sydney, Faculty of Architecture and Design
- University of Western Australia, Perth, Department of Architecture
- Victoria University of Wellington, New Zealand, School of Architecture
- Waikato University, Hamilton, New Zealand, Department of Educational Technology

Special Events Arranged at Architectural Association, London

1965–1967	AA concert series (contemporary music)
1966	Analysis & contemporary music
1971	Review of Finnish Architecture
1974	Knowledge of Knowledge
1977	Perception, Cognition and Learning
1979	Invention of Realities
1980	Description, Invention, Reality
1981	Concepts of Space and Time
1982	Space: Place in Architecture ("sPlace")
1983	What we don't know about …
1983–1986	(Micro) Computing for Architects

Appendix 8:
Visiting Critic/Professor
(Shorter Stays, Irregular)

Ranulph Glanville
as of 24 August 2014

1968	Barry Summer School, Electronic Music Workshop
1972	International Institute for Design, Summer School, London
1972–1973	Cambridge University, Department of Architecture
1972–1973	Bennington College, Bennington, Vermont, USA
1977	Bennington College, Bennington, Vermont, USA
1978	Special Projects & Events (see below) at Architectural Association
1979	Ulster Polytechnic (formerly Northern Ireland Polytechnic), Department of Art History, Criticism and Research
1979	Architectural Summer School, London
1979	Amsterdam University, Subfaculty of Andragology, The Netherlands
1980	Amsterdam University, Subfaculty of Andragology, The Netherlands
1981	Amsterdam University, Subfaculty of Andragology The Netherlands
1981	West of England Architecture Schools Liaison Design Weekend
1981	Pre 5 World Organisation for General Systems and Cybernetics Conference Summer School in Cybernetics, Mexico City
1983	Amsterdam University, Subfaculty of Andragology, The Netherlands
1983	Cape Town University and University of Natal the Orange Free State and the Witwatersrand, Republic of South Africa
1983	University of Garyounis, Benghazi, Libya
1984	Bennington College, Bennington, Vermont, USA
1985	Amsterdam University, Subfaculty of Andragology, The Netherlands
1985–1993	Programma OOC, Amsterdam University, The Netherlands
1986	Old Dominion University, Norfolk, Virginia, USA
1986	State College of New York, Department of Music, Geneseo, NY, USA
1993	Bialystok Polytechnic University, Poland
1994	Slovak Academy of Arts, Bratislava, Slovakia
1994	Institute for Spatial Studies, Vienna University of Technology, Austria
1994–1996	Centre for Architecture, Liverpool John Moore's University
1995	Architecture Department, Tampere University of Technology, Finland
1995	Interdisciplinary work based in Department of Architecture, Helsinki University of Technology, Finland (visiting researcher)
1995	Institute for Spatial Studies, Vienna University of Technology, Austria
1995	Visiting Scholar, School of Architecture and Shaw College, Chinese University of Hong Kong, Hong Kong
1995	Visiting Critic, Faculty of the Constructed Environment, Royal Melbourne Institute of Technology University, Melbourne, Australia
1996	Visiting Scholar, School of Architecture and Shaw College, Chinese University of Hong Kong, Hong Kong
1996	Visiting Professor, Department of Design, Polytechnic University of Hong Kong, Hong Kong

1996	Advisory Visiting Researcher, Faculty of the Built Environment, De Montfort University, UK
1996–now	Senior Research Fellow and Adjunct Professor, Faculty of the Constructed Environment, Royal Melbourne Institute of Technology University, Melbourne, Australia
1997	Visiting Scholar, School of Architecture and Shaw College, Chinese University of Hong Kong, Hong Kong
1998	Visiting Teacher, Sint Lucas Hogeschool, Brussels, Belgium
1998	Visiting Teacher, Faculty of Architecture, Landscape and Visual Arts, University of Western Australia, Perth, Australia
2001	Guest Critic, Victoria University, Wellington, New Zealand
2002	Guest Critic, Victoria University, Wellington, New Zealand
2002	Guest Editor and Scholar, School of Design, Hong Kong Polytechnic University, Hong Kong, SAR, China
2003	Advisor, Faculty of Architecture, Landscape and Visual Arts, University of Western Australia, Perth, Australia
2003	Visiting Teacher, Sint Lucas Hogeschool, Brussels, Belgium
2003	Visiting Scholar, University of Canberra, School of Design and Architecture
2004	Guest Critic, Victoria University, Wellington, New Zealand
2004	Advisor, Faculty of Architecture, Landscape and Visual Arts, University of Western Australia, Perth, Australia
2004	Visiting Scholar, University of Canberra, School of Design and Architecture
2005	Visiting Professor, Monash University, Faculty of Art and Design
2005	Visiting Research Fellow, Curtin University, Faculty of Architecture and Design
2005	Advisor, Faculty of Architecture, Landscape and Visual Arts, University of Western Australia, Perth, Australia
2005	Visiting Scholar, University of Canberra, School of Design and Architecture
2005	Visiting Professor, University of Newcastle (Australia), Faculty of Architecture and Building
2005	Visiting Professor, Rensselaer Polytechnic Institute, School of Architecture
2005	Advisor, University of Dundee, Duncan of Jordanston College, School of Architecture
2005	Visitor, George Washington University, School of Business and Mangement, Washington DC
2006	Visiting Research Fellow and Advisor, Curtin University, Faculty of Architecture and Design
2006	Visiting Scholar, University of Canberra, School of Design and Architecture
2006	Visiting Professor, University of Newcastle (Australia), Faculty of Architecture and Building
2006	Visiting Professor, Rensselaer Polytechnic Institute, School of Architecture
2006	Advisor and Master Planner, University of Dundee, Duncan of Jordanston College, School of Architecture
2006	Visiting Scholar, University of the Aegean, Syros, Greece
2007	Visiting Research Fellow and Advisor, Curtin University, Faculty of Architecture and Design

2007	Visiting Scholar, University of Canberra, School of Design and Architecture
2007	Advisor and Master Planner, University of Dundee, Duncan of Jordanston College, School of Architecture
2007	Visiting Professor, University of Newcastle (Australia), Faculty of Architecture and Building
2007	Advisor, WISDOM (Viennese Statistics Office)
2008	Visiting Research Fellow and Advisor, Curtin University, Faculty of Architecture and Design
2008	Visiting Scholar, University of Canberra, School of Design and Architecture
2008	Advisor and Master Planner, University of Dundee, Duncan of Jordanston College, School of Architecture
2008	Visiting Professor, University of Newcastle (Australia), Faculty of Architecture and Building
2008	Advisor, WISDOM (Viennese Statistics Office)
2009	Visiting Research Fellow and Advisor, Curtin University, Faculty of Architecture and Design
2009	Visiting Scholar, University of Canberra, School of Design and Architecture
2009	Visiting Professor, University of Newcastle (Australia), Faculty of Architecture and Building
2009	Advisor, WISDOM (Viennese Statistics Office)
2009	Visiting Professor, Hong Kong Design Institute
2010	Visiting Research Fellow and Advisor, Curtin University, Faculty of Architecture and Design
2010	Visiting Scholar, University of Canberra, School of Design and Architecture
2010	Visiting Professor, Hong Kong Polytechnic University, School of Design
2010	Visiting Professor, University of Newcastle (Australia), Faculty of Architecture and Building
2010	Visiting Professor, University of Sichuan, Faculty of Business and Administration
2010	Visiting Professor, Monash University, School of Architecture, Art and Design
2011	Visiting Professor, University of Newcastle (Australia), Faculty of Architecture and Building
2011	Visiting Professor, Monash University, School of Architecture, Art and Design
2011	Visiting Professor, Hong Kong Polytechnic University, School of Design
2011	Visiting Professor, Hong Kong Design Institute
2012	Visiting Professor, University of Newcastle (Australia), Faculty of Architecture and Building
2012	Visiting Professor, Hong Kong Polytechnic University, School of Design
2012	Visiting Professor, Hong Kong Design Institute
2013	Visiting Professor, University of Newcastle (Australia), Faculty of Architecture and Building
2013	Visiting Professor, Monash University, School of Architecture, Art and Design
2014	Aarhus University, Faculty of Media and Technology

I have continued my association with RMIT University (since 1995) and with The Bartlett, UCL (since 1994, and more formally since 2000. Since 2005 my connection with Sint Lucas has changed and is now permanent. These institutions do not appear in the above list, once my relationship with them becomes fixed).

Appendix 9:
Supervisory Experience

Ranulph Glanville
as of 24 August 2014

Extensive supervisory and examining experience at Masters and Doctoral level.

Supervision

Supervised and/or is supervising 60 PhD students at the following universities:

- Estonian Academy of Arts, Estonia—Architecture
- Hong Kong Polytechnic University, Hong Kong SAR—School of Design
- Liverpool John Moore's University, UK—School of Architecture and Building
- Portsmouth Polytechnic (now University of Portsmouth), UK—School of Architecture
- Royal College of Art, London, UK—Innovation Design Engineering
- Royal Melbourne Institute of Technology University, Australia—Faculty of the Constructed Environment
- Stockholm University, Sweden—Department of Psychology
- St Lucas, Catholic University of Leuven, Belgium—LUCA faculty of Architecture
- University College London, UK—The Bartlett
- University of Amsterdam, Netherlands—OOC Programma/Support, Survival and Culture Research Program
- University of Canberra—Department of Design and Architecture
- University of Wales College, Newport, UK—CAiiA: Centre for Advanced investigation in the interactive Arts

(Since Architecture is a 5 year course in the UK and the UK influenced educational world, I have supervised a large number of post graduate students completing the final part of their architectural studies. According to the Bologna Agreement, these students are considered as Masters. An estimate is 300, + 30 more conventional masters by research candidates.)

I have taken part in the examination of Masters and Doctoral Candidates (as an external examiner) at the following Universities:

- Brunel University, Uxbridge, UK
- Cape Peninsular University of Technology, Cape Town, South Africa
- De Montfort University, Leicester, UK
- Royal Melbourne Institute of Technology University, Australia
- University College London, UK

Goodby Ranulph. Much Love.